Defects and Geometry in Condensed Matter Physics

This book describes the key role played by thermally excited defects such as vortices, disclinations, dislocations, vacancies and interstitials in the physics of crystals, superfluids, superconductors, liquid crystals and polymer arrays.

Geometrical aspects of statistical mechanics become particularly important when thermal fluctuations entangle or crumple extended line-like or surface-like objects in three dimensions. In the case of entangled vortices above the first-order flux lattice melting transition in high-temperature superconductors, the lines themselves are defects. A variety of theories combined with renormalization-group ideas are used to describe the delicate interplay among defects, statistical mechanics and geometry characteristic of these problems in condensed matter physics.

This indispensible guide has its origins in Professor Nelson's contributions to summer schools, conference proceedings and workshops over the past twenty years. It provides a coherent and pedagogic graduate-level introduction to the field of defects and geometry.

DAVID NELSON is Mallinckrodt Professor of Physics and Professor of Applied Physics at Harvard University. He received his Ph.D. in 1975 from Cornell University. His research focuses on collective effects in the physics of condensed matter, particularly on the connections between thermal fluctuations, geometry and statistical physics. In collaboration with his Harvard colleague, Bertrand I. Halperin, he has proposed a theory of dislocation- and disclination-mediated melting in two dimensions. Professor Nelson's other interests include the statistical mechanics of metallic glasses, the physics of polymerized membranes, vortex phases in high-temperature superconductors and biophysics.

Professor Nelson is a member of the National Academy of Sciences, a member of the American Academy of Arts and Sciences and a Fellow of the American Physical Society; he has been an A. P. Sloan Fellow, a Guggenheim Fellow and a Junior and Senior Fellow in the Harvard Society of Fellows. He is the recipient of a five-year MacArthur Prize Fellowship, the National Academy of Sciences Prize for Initiatives in Research and the Harvard Ledlie Prize.

Defects and Geometry in Condensed Matter Physics

David R. Nelson
Lyman Laboratory of Physics,
Harvard University

CAMBRIDGE UNIVERSITY PRESS
Cambridge, New York, Melbourne, Madrid, Cape Town, Singapore,
São Paulo, Delhi, Dubai, Tokyo

Cambridge University Press
The Edinburgh Building, Cambridge CB2 8RU, UK

Published in the United States of America by Cambridge University Press, New York

www.cambridge.org
Information on this title: www.cambridge.org/9780521004008

First published 2002

A catalogue record for this publication is available from the British Library

Library of Congress Cataloguing in Publication data

Nelson, David R.
  Defects and geometry in condensed matter physics/David R. Nelson.
    p.   cm.
  ISBN 0 521 80159 1
  1. Condensed matter–Defects. I. Title.

QC173.458.D43 N45 2001
530.4′12–dc21   2001025621

ISBN 978-0-521-80159-1 Hardback
ISBN 978-0-521-00400-8 Paperback

Transferred to digital printing 2010

To Patricia, Meredith, Christopher and Peter

Contents

Preface to the book

Considerable relief accompanies completion of a project like this, especially one pursued (intermittently) for a period of eighteen years! There are many people to thank, not the least of which are the numerous graduate students, postdocs and other colleagues mentioned in Chapter 1 and in the acknowledgements of the remaining chapters. However, I owe a special debt to Michael Fisher, Leo Kadanoff and Bert Halperin, who provided inspiring examples of how to do theoretical physics early in my career. I am also grateful to Harvard University and the National Science Foundation of the USA, for creating the environment and freedom to do curiosity-driven research for the past twenty-five years.

I was fortunate to find a sympathetic publisher in Cambridge University Press. Simon Capelin and Rufus Neal provided expert editorial guidance and extraordinary patience in face of distractions caused by, among other things, my three years as Chair of the Harvard Physics Department. The original documents on which the last eight chapters are based illustrate the recent history of scientific publishing. At least one chapter evolved from an old-fashioned typewritten manuscript. Others originally appeared via a photographic offset printing process. The later chapters were created using LaTex and at least one is available (in an early form) on the World Wide Web (http://arXiv.org/abs/cond-mat/9502114). I appreciate the willingness of the original publishers to allow me to adapt my contributions to various proceedings, summer schools and workshops. I owe a special debt to Sally Thomas, Steven Holt, Jo Clegg and Jayne Aldhouse for the expert way in which they created a seamless high-quality book from these disparate media. Saul Teukolsky, Paul Horowitz and Renate D'Archangelo kindly provided advice and assistance with the index.

Farid Abraham generously assisted with the preparation of the illustration on the front cover. For more beautiful images created by Farid and his collaborators, see his gallery (http://www.almaden.ibm.com/vis/membrane/gallery.html). The pictures on the back cover are double-sided decorations of flux lines in high-temperature superconductors, due to Zhen Yao, Charles Lieber and their associates. I am grateful to Zhen and Charlie for permission to use their remarkable images, which led to the first experimental measurement of a bosonic "phonon–roton" spectrum for entangled vortex lines. For a more detailed discussion, see the article with George Crabtree in the April 1997 issue of *Physics Today* and references therein. While going over the final page proofs, I was struck again by the many striking and experimentally observable manifestations of geometry, defects and statistical mechanics in condensed-matter physics. I hope others will be able to capture some of this excitement while reading this book.

David R. Nelson
Rhinelander, Wisconsin
September, 2001

Acknowledgements

Renormalisation. Adapted with permission from *Encyclopedia of Physics*, edited by R. G. Lerner and G. L. Trigg (VCH Publishers, New York, 1991) pp. 1060–1063.

Defect-mediated phase transitions. Adapted with permission from *Phase Transitions and Critical Phenomena*, edited by C. Domb and J. L. Lebowitz, vol. 7 (Academic, New York, 1983) pp. 1–99.

Order, frustration and two-dimensional glass. Adapted with permission from *Topological Disorder in Condensed Matter*, edited by F. Yonezawa and T. Ninomiya (Springer, Berlin, 1983) pp. 164–180.

The structure and statistical mechanics of glass. Adapted with permission from *Applications of Field Theory to Statistical Mechanics*, edited by L. Garrido (Springer, Berlin, 1985) pp. 13–30.

The statistical mechanics of crumpled membranes. Adapted with permission from *Random Fluctuations and Pattern Growth: Experiments and Models*, edited by H. E. Stanley and N. Ostrowsky (Kluwer, Dordrecht, 1988) pp. 193–217.

Defects in superfluids, superconductors and membranes. Adapted with permission from *Fluctuating Geometries in Statistical Mechanics and Field Theory*, edited by F. David, P. Ginsparg and J. Zinn-Justin (North-Holland, Amsterdam, 1996) pp. 423–477.

Vortex line fluctuations in superconductors from elementary quantum mechanics. Adapted with permission from *Phase Transitions and Relaxation in Systems with Competing Energy Scales*, edited by T. Riste and D. Slurrington (Kluwer, Dordrecht, 1993) pp. 95–117.

Correlations and transport in vortex liquids. Adapted with permission from *Phenomenology and Applications of High Temperature Superconductors*, edited by K. S. Bedell *et al.* (Addison Wesley, New York, 1991) pp. 187–239.

The statistical mechanics of directed polymers. Adapted with permission from *Observation, Prediction and Simulation of Phase Transitions in Complex Fluids*, edited by M. Baus, L. F. Rull and J.-P. Rychaert (Kluwer, Dordrecht, 1995) pp. 293–335.

Chapter 1
Fluctuations, renormalization and universality

The idea of a book on defects and geometry in condensed matter physics slowly nucleated as I prepared contributions to various reviews, schools and workshops during the period 1982 to 1996. Although I have other interests in theoretical physics, I kept coming back at regular intervals to the statistical mechanics of defects and to related problems in the physics of flexible lines and surfaces. A consistent picture of these phenomena began to emerge and it transpired that work published in various forums during the early 1990s built on research I had described, e.g., at a summer school in the 1980s. Because considerable time and effort went into these reviews and all areas described are still active fields of investigation, it seemed reasonable to combine eight of them with this new introductory chapter in book form. The result, I hope, is a reasonably coherent account of the fascinating interplay among defects, geometry and statistical mechanics which has played such a central role in condensed matter physics during the past quarter century [1].

All chapters emphasize research in which I had a direct role. I have not attempted exhaustive reviews of these subjects and I apologize in advance to those whose work I have overlooked or neglected. These chapters are aimed at graduate students in physics, physical chemistry and chemical engineering as well as at more advanced researchers. Whenever possible, I tried to make the material intelligible to experimenters as well as theorists and to mention the many ingenious experiments which motivate the theories.

Condensed matter theorists owe a tremendous debt to our friends in the experimental community. They challenge us to *predict* and not merely *postdict* experimental phenomena. Our experimental colleagues will go to great lengths to design an experiment capable of testing an interesting new theoretical idea. Theorists must be careful what they say, because predictions in condensed matter physics can often be tested in a matter of months with relatively inexpensive table-top experiments. We owe our colleagues in the laboratory a great debt because they keep us honest and inspire us with their beautiful experiments.

I also owe a special personal debt to the numerous theoretical graduate students, postdoctoral fellows and scientific colleagues who have contributed to the ideas in this book. Without their enthusiasm, dedication and many crucial insights this work would not have been possible.

This book describes, among other things, the statistical mechanics of vortices, disclinations, dislocations, vacancies and interstitials. Excitation of these defects in crystals, superfluids, superconductors, liquid crystals and polymer arrays usually requires strong thermal fluctuations. Geometrical aspects of statistical mechanics, with or without defects, often become particularly interesting when these fluctuations entangle or crumple extended line-like or surface-like objects in three dimensions. Sometimes, as is the case for entangled vortices above the first-order flux lattice melting transition in type-II superconductors, the lines themselves are defects! Because modern ideas about the renormalization group and universality in the presence of fluctuations underpin most of the work on defects and the statistical mechanics of lines and surfaces in this book, the remainder of this chapter provides a brief introduction to this point of view with several illustrative examples. We conclude with a survey of subsequent chapters.

1.1 Fluctuations and universality in condensed matter physics

Condensed matter physics flourished in the second half of the twentieth century, due in part to the application of sophisticated tools for understanding thermal and quantum fluctuations to an astonishing variety of problems. The failure of uncontrolled "mean field" or decoupling approximations is particularly evident close to equilibrium critical points, where fluctuations occur over multiple length scales, from an atomic dimension of order ångström units to a (diverging) correlation length which can be micrometers or more. The renormalization group, which was exported from particle physics by Kenneth Wilson in the early 1970s, allows a systematic understanding of such nested length scales (see Appendix A for an elementary introduction to the renormalization group in the "hydrodynamic" context considered here). A particularly striking result is that most of the detailed physics of matter at microscopic length scales is *irrelevant* for critical-point phenomena. One can make precise quantitative predictions about certain "universal" critical exponents or scaling functions without getting the microscopic physics right in detail. What matters is symmetry, conservation laws, the range of interactions and the dimensionality of space. The physics of the diverging fluctuations at a critical point on large length scales is largely decoupled from the physics on atomic scales of order ångströms. The idea of new universal laws of physics governing fluctuations at a critical point was nicely summarized by A. Z. Patashinskii and V. L. Pokrovskii [2], two pioneers of scaling ideas at critical points in the USSR, who wrote that

> When fluctuations, these shapeless amoebas, overlap in large numbers to form a continuous, undescribable soup, new and sharply defined laws . . . come into play, cutting through the chaos.

It turns out that not just critical points but entire phases of matter are described by a "universal," coarse-grained, long-wavelength theory. This point was recognized by Wilson [3], who argued that Landau's hydrodynamic treatment of magnets *far* from critical points (carried out in the 1930s) was itself representative of a particularly simple renormalization group fixed point. One can make similar statements about the hydrodynamic laws derived for fluids in the nineteenth century. Upon systematically integrating out the high-frequency, short-wavelength modes associated with atoms and molecules, one should be able to arrive at, say, the Navier–Stokes equations. One does not have to be at a critical point to have universal physical laws insensitive to the microscopic details. We now have many concrete calculations well away from critical points that support this point of view. Ignorance about microscopic details is typically packaged into a few phenomenological parameters characterizing the "fixed point," such as the density and viscosity of an incompressible fluid like water in the case of the Navier–Stokes equations.

The modern theory of critical phenomena has interesting implications for our understanding of what constitutes "fundamental" physics. For many important problems, a fundamental understanding of the physics involved does *not* necessarily lie in the science of the smallest available time or length scale. The extreme insensitivity of the hydrodyamics of fluids to the precise physics at high frequencies and short distances is highlighted when we remember that the Navier–Stokes equations were derived in the early nineteenth century, at a time when even the discrete atomistic nature of matter was in doubt. The same equations would have resulted had matter been continuous at all length scales. The existence of atoms and molecules is irrelevant to the profound (some might even say "fundamental") problems of understanding, say, turbulence in the Navier–Stokes equations at high Reynolds numbers [4]. We would face almost identical problems in constructing a theory of turbulence if quantum mechanics did not exist, or if the discreteness of matter first became noticeable at length scales of order fermis instead of ångström units.

Many aspects of condensed matter physics, by which I mean the study of matter at everyday length and energy scales, do, of course, depend crucially on quantum mechanics and the particulate nature of matter. We cannot begin to understand phonons in solids, the specific heat of metals, localization in semiconductors, the quantum Hall effect and high-temperature superconductivity without knowing about the quantum mechanics of protons, neutrons, electrons and, occasionally, muons and positrons. There comes a point, however, when a more traditional reductionist approach burrows down to such short length scales and high energies that its conclusions become largely irrelevant to the physics of the world around us. This is why most condensed matter physicists are not aiming to discover the "fundamental" laws at the smallest length scales. The reductionist school of high-energy physics continues to be a noble intellectual enterprise, but is now virtually decoupled from physics at ångström-unit scales, just as atomic physics is decoupled from the Navier–Stokes equations. New particles discovered in high-energy physics are unlikely to help us understand problems like turbulence or how itinerant magnetism arises from the Hubbard model; neither will they unravel other hard problems like the complexities of reptation dynamics in entangled polymer melts [5].

Although the precise nature of physics at very short length scales need not have a profound impact on deep unresolved questions at much larger scales, knowledge of the correct short-distance theory is of course far from useless in condensed matter physics. A first-principles calculation of the viscosity and density of water, for example, would require a molecular or atomic starting point. Deriving hydrodynamic parameters such as the viscosity from an atomistic framework is the task of kinetic theory, in which significant progress has been made during the last century, at least for weakly interacting gases; and we are impressed when *ab initio* band-structure experts are able to correctly predict the lattice constant and crystal structure of silicon via numerical solutions of Schrödinger's equation. Nevertheless, there will always be important problems that a strict *ab initio* approach based on a more fundamental theory are unlikely to resolve.

The problems discussed in this book are all represented by coarse-grained long-wavelength "hydrodynamic" models, with the detailed physics packaged into a small number of phenomenological coupling constants. To illustrate the approach, we discuss two interesting but unusual situations in which ratios of long-wavelength hydrodynamic parameters are themselves *universal constants*, just like the universal exponents at a critical point. The first is the universal value of the Prandtl number of a two-dimensional incompressible fluid (it equals $(\sqrt{17}-1)/2$) and the second is the universal (negative!) value of Poisson's ratio associated with polymerized membranes or "tethered surfaces" (it is about $-\frac{1}{3}$). We then review the role of topological defects in destabilizing the hydrodynamic *surface* of fixed points associated with two-dimensional crystals. Here, the physics associated with strong thermal disorder leads to models dominated by fluctuations in the *phase* of the translational and orientational order parameters, instead of the usual amplitude fluctuations associated with mean field or Landau theories. Amplitude fluctuations now reside only in the cores of defects such as dislocations and disclinations. We discuss the new fixed point that takes over in membranes characterized by a bending rigidity, where defects such as dislocations can buckle easily out of the plane. This introduction concludes with an overview of the contents of the rest of the book.

1.2 The universal Prandtl number in two-dimensional hydrodynamics

Understanding chaotic fluid flows, particularly those at high Reynolds numbers, remains one of the most challenging problems in theoretical physics and fluid mechanics, despite the concerted efforts of many experts during the past half century [4]. Under many circumstances, it is expected that the fluid velocity field $\mathbf{v}(\mathbf{r}, t)$ and the concentration $\psi(\mathbf{r}, t)$ of tracer particles are described by the equations

$$\partial_t \mathbf{v} + (\mathbf{v}\cdot\nabla)\mathbf{v} = -\frac{1}{\rho}\nabla p + \nu\nabla^2\mathbf{v}, \tag{1.1}$$

$$\nabla\cdot\mathbf{v}=0, \tag{1.2}$$

$$\partial_t\psi+(\mathbf{v}\cdot\nabla)\psi= D\nabla^2\psi, \tag{1.3}$$

where ν is the kinematic viscosity, ρ is the density of the fluid, D is the diffusivity of the tracer particles and the condition of incompressibility (which is valid in the limit of velocities much less than the speed of sound) is enforced by Eq. (1.2). This condition can be used to eliminate the term involving the pressure field, $p(\mathbf{r}, t)$, in Eq. (1.1). The dynamics becomes insensitive to the density. Equations (1.1)–(1.3) then represent a "universal" long-wavelength, low-frequency description of a large number of atomic and molecular fluids, parameterized only by the

Fig. 1.1. The degrees of freedom for a randomly stirred Navier–Stokes fluid, indexed by wavevector and frequency, with wavevectors above a cutoff Λ excluded. A renormalization group transformation focusing on the long-wavelength hydrodynamic behavior can be constructed by iterative elimination of Fourier modes in a shell of wavevectors in the range $\Lambda\exp(-\ell)<k<\Lambda$ indicated by the shaded region.

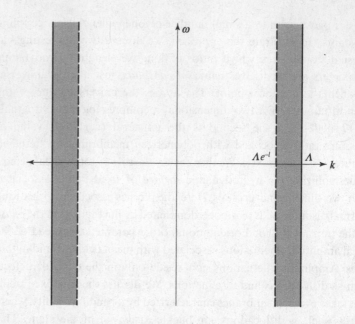

viscosity ν and the diffusivity D. If a colored dye of tracer particles is injected into water, one has $\nu\approx10^{-2}$ cm^2 s^{-1} for the viscosity and $D\approx10^{-6}$ cm^2 s^{-1} for the molecular diffusivity and thus the dimensionless "Prandtl number" is

$$Pr=\nu/D\approx10^4. \tag{1.4}$$

However, in general, one would expect that the Prandtl number could assume any positive value in a three-dimensional fluid. This is not the case in two dimensions!

By including various additive forcing terms on the right-hand side of Eqs. (1.1) and (1.3), one can simulate the effects of thermal fluctuations or even random stirring which can provoke chaos and high-Reynolds-number turbulence. The influence of the nonlinear terms $((\mathbf{v}\cdot\nabla)\mathbf{v}$ and $(\mathbf{v}\cdot\nabla)\psi)$ on the physics at long wavelengths can be assessed by the iterative coarse-graining procedure embodied in the renormalization group. The idea behind this perturbative renormalization procedure is reviewed in Fig. 1.1 [6, 7]. The velocity field of a d-dimensional fluid is first decomposed into Fourier modes according to

$$\mathbf{v}(\mathbf{r},\,t)=\int_{k<\Lambda}\frac{d^d k}{(2\pi)^d}\int_{-\infty}^{\infty}\frac{d\omega}{2\pi}\mathbf{v}(\mathbf{k},\omega)e^{i\mathbf{k}\cdot\mathbf{r}-i\omega t}. \tag{1.5}$$

The spatial modes are cut off above a spatial wavevector $k=\Lambda$ of order the inverse interparticle separation and the frequencies are unconstrained. The equations of motion for modes in the range $\Lambda/e^{\ell}<k<\Lambda$ are formally solved (via a diagrammatic perturbation theory) and these solutions then substituted into the equations of motion for the remaining modes in the range $0 < k < \Lambda/e^\ell$. After averaging over forcing contributions in the range $\Lambda/e^\ell < k < \Lambda$ and rescaling frequencies and wavevectors, one finds a coupled set of equations for the remaining modes which have the same form as the initial set. Thus, this procedure can be iterated indefinitely. By tracking the effect of repeated transformations on the coupling constants and correlation functions, this implementation of the renormalization group often allows a systematic resummation of Feynman diagrams in regimes for which conventional perturbation theory fails. This renormalization procedure also predicts situations in which it, too, fails, i.e., when the perturbative coupling constants iterate outside the range accessible to perturbation theory. Related implementations of this type of renormalization group underpin much of this book.

Although some progress is possible when the fluid is driven far out of equilibrium with appropriate forcing functions [7], we concentrate here on equilibrium forcing representing thermal fluctuations that obey the fluctuation–dissipation theorem. In three dimensions, one finds that the nonlinearities are "irrelevant variables," in that short-distance fluctuations lead only to finite renormalizations of the viscosity and diffusivity. However, in two dimensions the situation is more interesting. Here, thermal fluctuations couple into long-range hydrodynamic back-flow, which causes the renormalized wavevector-dependent kinematic viscosity $\nu_R(k)$ to diverge at small wavevectors k [7],

$$\nu_R(k) = \nu\left[1 + \frac{k_B T}{\rho \nu^2}\ln\left(\frac{\Lambda}{k}\right)\right]^{1/2}, \tag{1.6}$$

where k_B is Boltzmann's constant and T is the temperature. Here, $\nu(k)$ controls the rate of relaxation of a velocity fluctuation on wavelength $2\pi/k$, which decays like $\exp[-\nu(k)k^2 t]$ as $t \to \infty$. Thus, the apparent viscosity will depend (logarithmically) on the length scale $\sim k^{-1}$ of the measurement! The renormalized diffusivity $D_R(k)$ is also affected by the velocity fluctuations and it too diverges as $k \to 0$,

$$D_R(k) \sim \ln^{1/2}(\Lambda/k). \tag{1.7}$$

Although both ν_R and D_R diverge, it can be shown that the Prandtl number Pr remains well defined and approaches a *universal* constant characteristic of all incompressible fluids in the limit $k \to 0$ [8],

$$Pr \equiv \lim_{k \to 0} \frac{\nu_R(k)}{D_R(k)}$$

$$= \tfrac{1}{2}(\sqrt{17} - 1). \tag{1.8}$$

As is evident from Eq. (1.6), the dimensionless coupling constant which controls the strength of nonlinear effects in two-dimensional hydrodynamics is

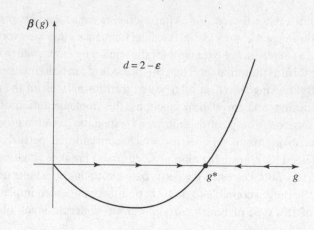

Fig. 1.2. The function $\beta(g)$ which determines the flow of the dimensionless coupling constant $k_B T/(\rho\nu^2)$ describing driven Navier–Stokes fluids in $2-\varepsilon$ dimensions, with $0<\varepsilon \ll 1$. The fixed point at $g=0$ describes conventional linearized hydrodynamics. The stable nontrivial fixed point at $g=g^*=O(\varepsilon)$ indicates a breakdown of conventional hydrodynamics. The two fixed points merge for $d=2$ and the fixed point at the origin dominates the physics for all $d \geq 2$.

$$g=\frac{k_B T}{\rho\nu^2}. \tag{1.9}$$

The effective coupling constant $g(\ell)$ in d dimensions after the short-distance cutoff has been reduced from Λ to Λ/e^ℓ is the solution of the renormalization-group differential equation,

$$\frac{dg(\ell)}{d\ell}=\tfrac{1}{2}(2-d)g-A_d g^2$$

$$\equiv -\beta(g), \tag{1.10}$$

where A_d is a positive constant. As illustrated in Fig. 1.2, conventional linearized hydrodynamics is described by a "Navier–Stokes" fixed point $g=g^*=0$. For $d>2$, $g(\ell)$ decays rapidly to zero and linearized hydrodynamics is stable. In two dimensions, the slow decay to zero of $g(\ell)$ for large ℓ ($g(\ell)\sim 1/\ell$) leads to logarithmic corrections to linearized hydrodynamics discussed above. There is an interesting nontrivial stable fixed point with $g^*>0$ for $d<2$. Unfortunately, incompressible fluid flow is not very meaningful in the physically relevant limit $d=1$. However, nontrivial hydrodynamic fixed points do appear in two and three dimensions for forcing functions that do not obey the fluctuation–dissipation theorem [7]. In the next section, we discuss the nontrivial fixed point which describes the anomalous long-wavelength elastic properties and bending rigidity of a two-dimensional solid membrane, free to fluctate in three dimensions and subject to thermal fluctuations.

1.3 The universal Poisson ratio in fluctuating polymerized membranes

Like fluid mechanics, the continuum elastic description of solids subject to arbitrary deformations has a long and rich history. It reached its

current form early in the twentieth century. The theory of the bending
of thin plates resembles a simplified form of general relativity and both
subjects use the language of differential geometry. A thin elastic plate
can be parameterized by two internal indices, x_1 and x_2, which we can
take as the particle positions in the flat undeformed state. Its deformations in the xy-plane are described by phonon coordinates $u_1(x_1,x_2)$ and
$u_2(x_1,x_2)$. Out-of-plane deformations are parametrized via a perpendicular displacement $f(x_1,x_2)$. The actual position in three dimensions of a
particle labeled by (x_1,x_2) in the deformed membrane is thus given by

$$\mathbf{r}(x_1,x_2) = \begin{pmatrix} x_1 + u_1(x_1,x_2) \\ x_2 + u_2(x_1,x_2) \\ f(x_1,x_2). \end{pmatrix}. \tag{1.11}$$

The energy of a thin plate for small deformations is characterized by a
bending rigidity κ and in-plane elastic constants μ and λ [9],

$$F = \tfrac{1}{2}\kappa \int d^2x (\nabla^2 f)^2 + \tfrac{1}{2} \int d^2x (2\mu u_{ij}^2 + \lambda u_{kk}^2), \tag{1.12}$$

where $u_{ij}(x_1,x_2)$ is the nonlinear strain matrix. (For more details, see
Chapters 5 and 6.) Like general relativity, the theory is intrinsically nonlinear, which shows up at this level in the relation between the strain
matrix and the displacements $\mathbf{u}(x_1,x_2)$ and $f(x_1,x_2)$ [9],

$$u_{ij}(x_1,x_2) = \tfrac{1}{2}(\partial_i u_j + \partial_j u_i) + \tfrac{1}{2}(\partial_i f)(\partial_j f). \tag{1.13}$$

To determine the deformation of plates subjected to edge forces with
various boundary conditions, one must minimize Eq. (1.12). Upon
eliminating $u_{ij}(x_1,x_2)$ in favor of the Airy stress function $\chi(x_1,x_2)$, via
$\sigma_{ij}(x_1,x_2) \equiv 2\mu u_{ij}(x_1,x_2) + \lambda u_{kk}(x_1,x_2)\delta_{ij} \equiv \varepsilon_{im}\varepsilon_{jn}\partial_m\partial_n\chi(x_1,x_2)$, where $\varepsilon_{ij} =$
$\delta_{i1}\delta_{j2} - \delta_{j1}\delta_{i2}$ is the antisymmetric unit matrix in two dimensions, one
obtains the von Karman equations,

$$\kappa\nabla^4 f = \frac{\partial^2\chi}{\partial y^2}\frac{\partial^2 f}{\partial x^2} + \frac{\partial^2\chi}{\partial x^2}\frac{\partial^2 f}{\partial y^2} - 2\frac{\partial^2\chi}{\partial x\,\partial y}\frac{\partial^2 f}{\partial x\,\partial y}, \tag{1.14}$$

$$\frac{2\mu + \lambda}{4\mu(\mu + \lambda)}\nabla^4\chi = -\frac{\partial^2 f}{\partial x^2}\frac{\partial^2 f}{\partial y^2} + \left(\frac{\partial^2 f}{\partial x\,\partial y}\right)^2. \tag{1.15}$$

The negative of the Gaussian curvature $\det(\partial^2 f/\partial x_i\partial x_j)$ of the deformed
surface appears as a source on the right-hand side of (1.15). According
to Landau and Lifshitz [9], these nonlinear equations are "very complicated, and cannot be solved exactly, even in very simple cases." However,
at least the elastic constants κ, μ and λ are indeed *constants* for long-
wavelength deformations in the von Karman theory! When thermal
fluctuations are taken into account, *these parameters depend on the*

wavelength, even more strongly than do the diverging viscosity and diffusivity in Eqs. (1.6) and (1.7).

Equation (1.12) also describes biological sheet polymers that are one molecule thick, such as the spectrin skeleton on the inside walls of red blood cells [10, 11]. When these polymers are extracted from red blood cells and put into water solution, the nodes of this "biological fishnet" are subject to violent thermal fluctuations in the form of Brownian motion [12]. The equilibrium statistical mechanics of this structure (similar to the fluctuations of semiflexible linear polymer chains) can be modelled by assuming that configurations occur with probability proportional to $\exp[-F/(k_B T)]$. Because the out-of-plane displacement $f(x_1, x_2)$ occurs in the generalized strain matrix (1.13), there is a nontrivial coupling between in-plane and out-of-plane displacements.

This nonlinearity can be studied by first expanding u_1, u_2 and f in Fourier modes in the range $0 < k < \Lambda$, where Λ is a suitable microscopic cutoff. Upon integrating Fourier modes in the interval $\Lambda/e^\ell < k < \Lambda$ out of the partition function (defined as a functional integral over u_1, u_2 and f)

$$Z = \int \mathcal{D}\mathbf{u}(x_1, x_2) \int \mathcal{D}f(x_1, x_2) \exp(-F/k_B T), \tag{1.16}$$

and rescaling degrees of freedom and wavevectors, one can implement a renormalization group similar to that for the Navier–Stokes equations discussed in the previous section. Iterating this transformation determines effective dimensionless coupling constants $\bar{\lambda}(\ell) = k_B T \lambda/(\Lambda^2 \kappa^2)$ and $\bar{\mu}(\ell) = k_B T \mu/(\Lambda^2 \kappa^2)$ that describe the statistical mechanics for wavevectors $k < \Lambda/e^\ell$. Figure 1.3 shows schematically the solutions of the renormalization-group differential equations for $\bar{\lambda}(\ell)$ and $\bar{\mu}(\ell)$ for fluctuating membranes embedded in three dimensions [13]. The "von Karman" fixed point at the origin describes the physics at zero temperature, i.e., situations to which Eqs. (1.14) and (1.15) apply with *constant* elastic parameters κ, μ and λ. However, this fixed point is unstable at any finite temperature relative to a fixed point where conventional elasticity theory breaks down at long wavelengths. Indeed, at this fixed point the renormalized wavevector-dependent bending rigidity $\kappa_R(k)$ *diverges* strongly as k goes to zero [14], while the renormalized wavevector-dependent elastic parameters $\mu_R(k)$ and $\lambda_R(k)$ *vanish* in this limit [13]. Among the best current estimates of these singularities are those of Radzihovsky and Le Doussal [15],

$$\kappa_R(k) \sim 1/k^{0.82}, \qquad \mu_R(k) \sim k^{0.36}, \qquad \lambda_R(k) \sim k^{0.36}, \tag{1.17}$$

where these power-law singularities are governed by universal critical exponents.

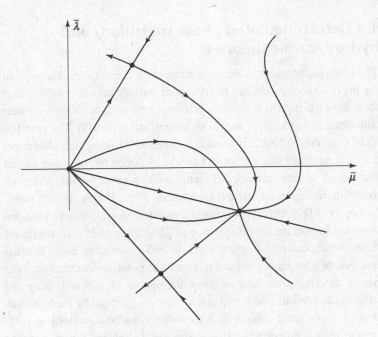

Fig. 1.3. Renormalization-group flows in the space of coupling constants appropriate to a two-dimensional crystalline membrane subject to thermal fluctuations in three dimensions. The (unstable) von Karman fixed point at the origin describes the nonlinear elasticity of bent plates at zero temperature. The stable fixed point in the fourth quadrant leads to elastic parameters that depend strongly on wavevector when thermal fluctuations are important. The runaway trajectories outside the domain of attraction of this fixed point indicate elastic instabilities.

The physical interpretation of these results is that the elastic "constants" associated with the stable finite-temperature fixed point are not in fact constant at all, but depend instead in a universal way, independent of the microscopic details, on the length scale at which the measurement is made. Although the renormalized elastic constants μ_R and λ_R vanish as $k \to 0$, the two-dimensional Poisson ratio $\sigma_R = \lambda_R/(2\mu_R + \lambda_R)$ approaches a *universal* negative value [13, 15],

$$\sigma_R = \lim_{k \to 0} \frac{\lambda_R(k)}{2\mu_R(k) + \lambda_R(k)} \approx -\frac{1}{3}, \qquad (1.18)$$

similar to the universal Prandtl number discussed in the previous section. The Poisson ratio describes the transverse shrinkage (or, in this case, expansion!) of an elastic medium pulled at its edges along a particular direction with a given force [9].

The physical origin of the negative Poisson ratio for polymerized membranes (a very rare phenomenon in conventional materials) can be seen by partially flattening a crumpled piece of paper to simulate a membrane in its flat phase in the presence of thermal fluctuations. Subsequent pulling on the paper in one direction releases some of the stored area remaining in the creases, so the sheet *expands* in the perpendicular direction, corresponding to the negative Poisson ratio

1.4 Defect-mediated phase transitions and hydrodynamic theories

The isotropic continuum elastic treatment of membranes discussed in the previous section applies to crystalline particle arrays, provided that these have the high symmetry of a triangular lattice, with each particle connected in an identical way to six nearest neighbors [9]. The possibility of topological defects in these arrays (leading to near-neighbor coordination numbers other than six) can be excluded by requiring a fixed "tethering" or connectivity between nearest neighbors [16]. Although many interesting experimental realizations of such polymerized membranes exist [17], defects such as vacancies, dislocations and disclinations are important excitations in equilibrated particle arrays without this tethering constraint, particularly in two dimensions. From the perspective of the renormalization group, various hydrodynamic fixed points describing the low-temperature phases of thin-film magnets, superfluids, crystals and liquid crystals are destabilized by the introduction of topological defects such as vortices and dislocations at sufficiently high temperatures. In *membranes*, in which only a bending rigidity (as opposed to a surface tension or binding to a substrate) prevents deformations into the third dimension, buckled dislocations in fact destabilize the crystalline phase at *any* finite temperature, leading to a stable low-temperature hexatic phase. (See Chapter 2 for an introduction to the hexatic phase.) Defects play an important role in this book and we use the example of crystalline monolayers and membranes here to illustrate the effect of defects on statistical mechanics and geometry in condensed matter physics.

Strong thermal fluctuations often drive the ordering temperature for phase transitions well below the mean field value, leading to a description in which amplitude fluctuations can be neglected to a first approximation. Because the order parameters of traditional Landau or mean field theories usually vanish in defect cores, defects can be viewed as vestigial remnants of amplitude fluctuations in this low-temperature description. The familiar quartic Landau theory with an up/down symmetry in two dimensions illustrates this point of view. The Landau energy associated with a scalar order-parameter configuration $M(x)$ with partition function $Z = \int \mathcal{D}M(x)\exp[-E/(k_B T)]$ may be written

$$E[\{M(x)\}] = \int d^2x \left(\frac{1}{2}|\nabla M(x)|^2 + \frac{1}{2}rM^2 + uM^4 \right). \tag{1.19}$$

Well below the mean-field ordering temperature T_c^0, defined by $r \approx a(T - T_c^0) = 0$, this energy can be rewritten (up to an additive constant) as

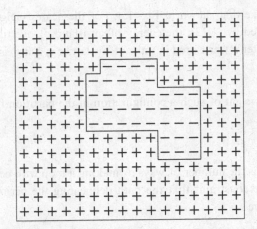

$$E = \int d^2 x \left(\frac{1}{2} |\nabla M(x)|^2 + u(M^2 - M_0^2)^2 \right), \tag{1.20}$$

with $M_0 = \sqrt{-r/(4u)}$. We now assume that u is very large, rescale the order parameter so that the lowest energy state is $M(x) \equiv \sigma(x) = \pm 1$ and impose a microscopic cutoff by putting these Ising-like "spins" on a square lattice. The spins $\sigma(x)$ assume the values ± 1 on these lattice sites and the gradient coupling in (1.20) is represented as a nearest-neighbor exchange energy $J > 0$, leading to the familiar two-dimensional Ising model,

$$E \rightarrow \mathcal{H}_I = -J \sum_{\langle ij \rangle} \sigma_i \sigma_j, \qquad \sigma_i = \pm 1, \tag{1.21}$$

where the sum is over distinct nearest-neighbor pairs. The amplitude fluctuations originally present in (1.19) are now consigned to the *boundaries* of islands of overturned spins, as in Figure 1.4. This boundary is a loop where $M(x)$ would pass from zero while changing from $+1$ to -1 in the original continuum model. The line-like loop excitation in Fig. 1.4 plays the same role for the two-dimensional Ising field theory as point-like defects such as vortices and dislocations play for the problems discussed in this book. When many islands of overturned spins are present, one can reformulate the Ising-model partition function

$$Z_I = \mathop{\mathrm{Tr}}_{\{\sigma_j = \pm 1\}} \exp \left(-\frac{J}{k_B T} \sum_{\langle ij \rangle} \sigma_i \sigma_j \right) \tag{1.22}$$

in terms of the statistical mechanics of a loop gas of defects.

Crude estimates of transition temperatures can often be obtained by balancing the energy and entropy of defect excitations out of an ordered state. One of the first arguments of this kind was applied by Peierls [18] to the two-dimensional Ising model. He focused on a loop excitation like that shown in Fig. 1.4. For the nearest-neighbor Ising model defined by (1.22), the energy of an island of overturned spins with perimeter P (measured in units of the lattice spacing) is

$$E = 2JP. \tag{1.23}$$

The actual weight of such a configuration in the partition function is, however,

$$N(P)e^{-E/(k_{\mathrm{B}}T)}, \tag{1.24}$$

where $N(P)$ is the number of ways of making a loop with perimeter P. Ignoring the constraint that the path delineating the perimeter must close, we have

$$N(P) \sim 3^P \tag{1.25}$$

for a square lattice. A more accurate estimate which includes the constraint is [19]

$$N(P) \sim (2.639)^P. \tag{1.26}$$

From (1.24) we see that the relevant quantity to consider is the island's free energy

$$\begin{aligned}
F &= E - TS \\
&= 2JP - k_{\mathrm{B}}T\ln[N(P)] \\
&\approx [2J - k_{\mathrm{B}}T\ln(2.639)]P.
\end{aligned} \tag{1.27}$$

An island with arbitrarily large perimeter will overwhelm the ordered state when its free energy goes negative, i.e., above a critical temperature

$$T_{\mathrm{c}} \approx 2J/[k_{\mathrm{B}}\ln(2.639)]. \tag{1.28}$$

A succinct derivation of the well-known *exact* critical temperature of the two-dimensional Ising model [20] is given in Appendix B. The result reads

$$\begin{aligned}
k_{\mathrm{B}}T_{\mathrm{c}} &= 2J/\ln(1 + \sqrt{2}) \\
&\approx 2J/\ln(2.414).
\end{aligned} \tag{1.29}$$

The crude entropy argument of Peierls does fairly well, but it is, of course, not exact.

What happens when we introduce defects in an analogous way into the elastic description of a two-dimensional crystalline solid? The Landau theory describing how liquids freeze into a solid, with its translational and orientational order parameters, is discussed in Chapter 2 and leads at low temperatures to Eq. (1.12) with $f = 0$, as would be appropriate if surface tension or binding to a substrate prevented large particle deformations into a third dimension. We shall call such flat

two-dimensional particle configurations "monolayers," to distinguish
them from "membranes," which can buckle easily into the third dimension. The elementary defects which disrupt translational and orientational correlations in monolayers are *point-like* dislocations and
disclinations, instead of the *line-like* boundaries between up and down
spins appropriate to the two-dimensional Ising model [21].

Disclinations are points of local 5- and 7-fold symmetry embedded in a triangular lattice, as illustrated on the right-hand side of Fig.
1.5. Both orientational order and translational order are disrupted by
these defects and the appropriate order parameters vanish at the cores
of these defects. Disclinations can be created from a perfect crystal by
inserting or deleting a 60° wedge of material in an otherwise perfect
crystal. The azimuthal particle displacement u_ϕ varies linearly with
distance r from the center of the resulting disturbance, so the component $u_{\phi r}$ of the strain matrix in polar coordinates is approximately
constant. A more detailed calculation shows that, in a circular sample
of radius R (see, e.g., [16]),

$$E = \frac{1}{2} \int d^2x [2\mu u_{ij}^2(x) + \lambda u_{kk}^2(x)]$$

$$\underset{R \to \infty}{\approx} \frac{Ks^2}{32\pi} R^2, \tag{1.30}$$

where the two-dimensional Young modulus is

$$K = \frac{4\mu(\mu + \lambda)}{2\mu + \lambda}, \tag{1.31}$$

the disclination "charge" associated with the ±60° wedges is given (in
radians) by

$$s = \pm(2\pi/6) \tag{1.32}$$

and the quadratic divergence with R arises from the constant strain
matrix. This calculation shows that isolated disclinations are energetically very costly and unlikely to occur in a low temperature crystal.

Fig. 1.5. Phases and defects involved in the dislocation/disclination-unbinding scenario for two-dimensional melting. Free dislocations proliferate in a hexatic phase above a temperature T_m. The circuit around a dislocation which defines the Burgers vector is shown. Unbound disclinations (points of local 5- and 7-fold symmetry) appear above a temperature T_i. In real monolayer materials, the temperature window which supports the hexatic phase is usually smaller than that indicated here.

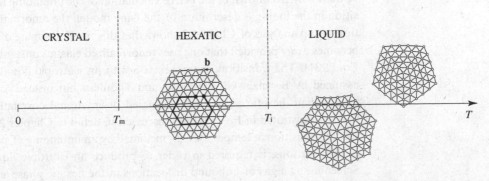

CRYSTAL   HEXATIC   LIQUID

0   T_m T_i T

Dislocations can be viewed as composite defects, composed of a tightly bound pair of neighboring 5- and 7-fold disclinations. As illustrated in Fig. 1.5, a dislocation is characterized by a discrete Burgers vector "charge" **b**, which measures the amount by which a path around the defect fails to close. The translational order parameter vanishes at its core, due to the disruption caused by an extra row of particles relative to a perfect crystal. However, orientational order, measured in terms of the directional *alignment* of distant Bragg rows of atoms on opposite sides of the defect is not much affected. It can be shown that the strain matrix now falls off like $1/r$, leading to a logarithmic divergence in the energy in a circular sample of size R,

$$E = \frac{1}{2} \int d^2x \, [2\mu u_{ij}^2(x) + \lambda u_{kk}^2(x)]$$

$$\underset{R \to \infty}{\approx} \frac{Kb^2}{8\pi} \ln\left(\frac{R}{a}\right), \tag{1.33}$$

where $|\mathbf{b}| = a$ and a is the lattice constant. An energy/entropy argument for dislocations, similar to the Peierls analysis of the two-dimensional Ising model in the 1930s, was constructed by Berenzinskii [22] and by Kosterlitz and Thouless [23]. Because there are of order $(R/a)^2$ places to put a dislocation in a disk of material with radius R, the free energy of an isolated dislocation is

$$F = E - TS$$

$$= \frac{Kb^2}{8\pi} \ln\left(\frac{R}{a}\right) - k_B T \ln\left(\frac{R^2}{a^2}\right), \tag{1.34}$$

which becomes negative for

$$T > T_m = Kb^2/(16\pi k_B), \tag{1.35}$$

suggesting that there is a finite melting temperature above which pairs of dislocations with equal and opposite Burgers vectors unbind. Remarkably, in contrast to the Peierls calculation of the unbinding transition in the loop-gas description of the Ising model, the renormalization-group analysis of Chapter 2 shows that this rough estimate of T_m becomes *exact* provided that one uses renormalized elastic constants in Eq. (1.31). The transition, moreover, is not to an isotropic liquid, as assumed by Berenzinskii, Kosterlitz and Thouless, but instead to an intermediate hexatic phase, with extended orientational correlations [24]. As illustrated in Fig. 1.5 (and discussed in detail in Chapter 2), a second transition at temperature T_i, mediated by an unbinding of pairs of disclinations, is required in order to produce an isotropic liquid. Screening by a gas of unbound dislocations in the hexatic phase leads to a weaker logarithmic divergence in the disclination energy, in contrast to the quadratic divergence displayed in Eq. (1.30).

The long-wavelength "hydrodynamic" description of two-dimensional crystals, with dislocation defects appearing only as tightly bound pairs, is embodied in a *surface* of renormalization-group fixed points, parameterized by the renormalized values of the elastic constants μ and λ. This fixed surface is destabilized above the temperature T_m by unbound dislocations, signaling a breakdown of crystalline hydrodynamics. The renormalization-group recursion relations lead instead to a new stable *line* of fixed points characteristic of the hexatic phase [24]. This hydrodynamic description of hexatics itself becomes unstable relative to an unbinding of pairs of disclination above the temperature T_i (see Chapter 2).

What happens when the monolayer physics described above is replaced by the physics of *membranes*, which can buckle easily into the third dimension? Even in the absence of defects, the usual continuum elastic theory is replaced by one with singular wavevector-dependent elastic constants and a singular wavevector dependent bending rigidity, as we saw in Section 1.3. When defects are allowed, buckling into the third dimension destabilizes the crystal phase completely [14, 16]. Consider a dislocation in a disk of radius R at zero temperature, biased only by a bending rigidity to lie in a plane. As discussed in Chapter 9, this dislocation will buckle for radii larger than [16]

$$R_b \approx 127\kappa/(Ka^2)$$
$$\approx 2.5\kappa/(k_B T_m),$$
(1.36)

where we have inserted the flat-space result (1.35) for the melting temperature. Buckling allows stretching energy to be traded for bending energy and leads to a *finite* dislocation energy,

$$E_D \approx \frac{Kb^2}{8\pi} \ln\left(\frac{R_b}{a}\right)$$
(1.37)

and a nonzero density n_D of these defects

$$n_D \approx a^{-2} e^{-E_D/(k_B T)}.$$
(1.38)

At finite temperatures, one should also incorporate the singular elastic constants discussed in Section 1.3 into this analysis. When the vanishing of the elastic constants predicted by Eq. (1.17) is taken into account, we see that the dislocation-free energy in Eq. (1.34) should be replaced by

$$F(R) \underset{R\to\infty}{\approx} \frac{K(R)b^2}{8\pi} \ln\left(\frac{R}{a}\right) - k_B T \ln\left(\frac{R^2}{a^2}\right),$$
(1.39)

where $K(R) \sim 1/R^{0.36}$, which is always negative for sufficiently large R. We are again led to the conclusion that dislocations are unbound at any finite temperature.

The implications of these results for a membrane confined between flat plates with spacing d [25] is illustrated schematically in Fig. 1.6. When d is small, steric interactions between the membrane and the confining walls suppress the dislocation instabilities discussed above. The monolayer picture of two-dimensional melting works on scales larger than the in-plane distance required for the membrane to "feel" the effect on the confining wall potential. The melting temperature $T_{m}(d)$ drops with increasing d, tending to zero as $d \to \infty$. Disclinations in the hexatic phase can also lower their energy by buckling, but a logarithmically diverging energy remains, leading to a finite-temperature disclination unbinding transition at $T = T_{i}(d)$ even in the limit $d \to \infty$ [14, 26, 27]. If a hexatic phase appears for small $d \lesssim a$ (it could always be preempted by a direct first-order transition from crystal to isotropic liquid), it typically exists over a limited range of temperatures or densities. However, Fig. 1.6 makes it clear that a stable low-temperature hexatic phase is *inevitable* as we approach the limit of an unconfined membrane when $d \to \infty$. Recent computer simulations of the proliferation of defects in vesicles with small numbers of particles seem to support this point of view [28].

1.5 The contents of this book

As discussed earlier, the remainder of this book consists of chapters extracted from various workshops, summer school proceedings and reviews. The order of presentation builds up from ideas about the statistical mechanics of point-like defects in two dimensions to more sophisticated treatments of surfaces and lines in subsequent chapters. A brief commentary on more recent developments is presented at the beginning of each chapter.

Chapter 2 is a broad survey of the statics and dynamics of defect-mediated phase transitions, primarily in two dimensions. The main focus is on two-dimensional XY magnets, films of superfluid helium and two-dimensional melting, although some material on three-dimensional smectics, cubic and icosahedral bond orientational order in liquids and vortex loops in bulk superfluid helium is presented as well. The idea of two-dimensional melting with an intermediate hexatic phase with long-range bond correlations is developed in detail.

The concept of bond-orientational order is applied to two-dimensional glasses in Chapter 3. Although two-dimensional arrays of identical particles pack nicely into a perfect triangular lattice at zero

(a)

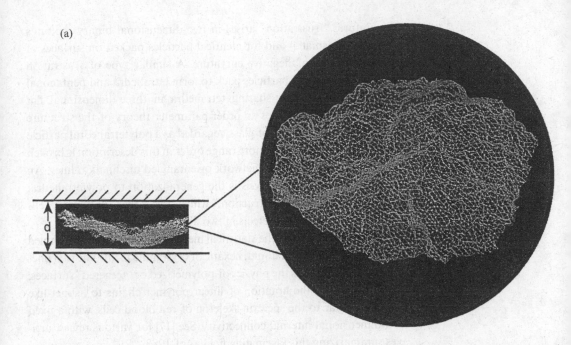

(b)

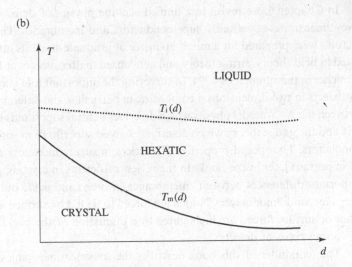

Fig. 1.6. A crystalline array of particles confined between two parallel plates with separation d and the corresponding phase diagram as a function of d and the temperature T. For short distances relative to d, the system behaves like a membrane and the equilibrium melting temperature $T_m(d)$ is suppressed by fluctuations and the buckling of dislocations. On large scales, however, the particles feel the presence of the walls and the system behaves like a monolayer. As $d \to \infty$, $T_m(d)$ goes to zero and the hexatic becomes the dominant low-temperature phase, with true crystalline order only at $T = 0$.

temperature, "frustration" arises in two-dimensional binary mixtures with a size mismatch and for identical particles packed on surfaces of constant positve or negative curvature. A similar type of frustration arises when identical particles pack to form tetrahedra and pentagonal bond spindles of edge-sharing tetrahedra in three-dimensional flat space. Chapter 4 describes an order parameter theory of the structure and statistical mechanics of glass, regarded as a polytetrahedral particle packing. The icosahedral short-range order in this description is broken up by a three-dimensional network of entangled disclination lines. An important tool in the analysis is the generalization to polytetrahedral packings of the "5–7 construction," which uses Voronoi/Dirichlet cells to visualize disclination defects in two dimensions.

Chapter 5 describes the statistical mechanics of surfaces. Crumpled membranes can exist in liquid, hexatic or crystalline form. The rich phenomenology includes the physics of polymerized or "tethered" surfaces, which generalize the notation of linear polymer chains to fishnet-like objects, similar to the spectrin skeleton of red blood cells, with a fixed two-dimensional internal connectivity. See [17] for various review articles summarizing this fascinating field as of 1988.

In Chapter 6, we revisit in a unified way the physics of defects in two-dimensional superfluids, superconductors and membranes. These lectures were prepared for a mixed audience of graduate students interested in field theory, string theory and condensed matter physics at Les Houches in the summer of 1994. In surveying the important role played by defects in two dimensions, I emphasize, in particular, the differences between the X–Y-model physics which describes neutral superfluids like ^{4}He and the gauge theory which describes charged superfluids in superconductors. The special properties of vortices in superconductors will be important later in the book. In the review of defects in crystals, the important differences between "membranes," which can buckle out of the plane, and "monolayers," which are forced to lie flat by surface tensions or surface forces, are highlighted by a discussion of the buckling of various types of defects.

The remainder of this book describes the statistical mechanics of directed lines. The discovery of the new high temperature superconductors in 1988 revitalized the field of "vortex physics" in type-II superconductors, especially regarding the statistical mechanics of many identical directed line-like defects subject to strong thermal excitations and pinning. Chapter 7 shows how many important results regarding vortex fluctuations and pinning can be derived using a mapping (via imaginary time Feynman path integrals) onto elementary quantum mechanics. A more sophisticated version of this mapping, discussed in Chapter 8, shows that the physics of many interacting vortices in a type-II superconductor is intimately related to the physics of superfluid bosons in two dimensions. Superfluidity of the bosons is equivalent to entanglement of the vortex lines. This chapter also discusses nonlocal hydrodynamics of flux-flow resistivity generated by polymer-like vortex lines subject to a Lorentz force in inhomogeneous geometries.

Ideas developed for vortex lines regarded as polymer-like excitations in type-II superconductors have interesting counterparts for various mesophases of real polymers. The final chapter (Chapter 9) discusses the physics of polymer nematics and the hexagonal columnar phase of materials such as DNA. The underlying elasticity theory must be rotationally invariant, which leads to softer restoring forces than for the corresponding description of tangled vortex lines. The equivalent "boson" theory for directed polymer melts plays a role similar to the mapping by de Gennes and by des Cloiseaux of isotropic polymer melts onto the $n \rightarrow 0$ limit of interacting n-component spins. We also discuss the physics and possible proliferation of vacancies and interstitials, which are particularly interesting and unusual line-like excitations when they arise in polymer crystals.

Appendix A. Renormalization

Adapted with permission from *Encyclopedia of Physics*, edited by R. G. Lerner and G. L. Trigg (VCH Publishers, New York, 1991), pp. 1060–1063.

The concept of renormalization in physics has its origins in nineteenth-century hydrodynamics. It was discovered that large objects moving slowly through a viscous fluid behave in some ways as if they possess an enhanced mass due to the fluid particles they drag along. We would now say that the mass of such objects is renormalized away from the "bare" value it has in isolation by interactions with the medium. This idea finds modern expression in Landau's quasiparticle picture of Fermi liquids and in the band theory of metallic conduction. Landau argued that liquid ^{3}He could be described at low temperatures as a dilute gas of quasiparticle excitations near a spherical Fermi surface. The quasiparticles were to be viewed as "dressed" versions of the constituent ^{3}He fermions, with altered masses, interactions, etc. The dynamics of conduction electrons in metals is often parameterized in a similar fashion by invoking an effective electron mass describing interactions with the periodic lattice potential and with other electrons.

In all these examples, the basic idea is to replace a complicated many-body problem by a simpler system in which interactions are absent or negligible. Complicated many-body effects are absorbed into redefinitions of masses and coupling constants. For objects moving slowly through a viscous medium, the mass renormalization can be calculated perturbatively as a power series in the Reynolds number. Even if interactions are strong enough to make perturbation theory intractable, as is the case for liquid ^{3}He, phenomenological renormalized parameters taken from experiments provide a very useful description of the physics.

Problems developed when these ideas were applied to quantum electrodynamics. Calculations of the renormalized mass of the electron due to its interaction with electromagnetic radiation were carried out perturbatively in powers
of the fine-structure constant. At every order in perturbation theory, infinities,
due to interactions at very short distances were discovered. The renormalized
mass and coupling constant, for example, diverged as the cutoff Λ (the largest
allowed inverse length scale) was taken to infinity. Such infinities can be viewed
as arising from equivalent contributions to the renormalized mass and coupling
constant from many different length scales. Although Feynman, Schwinger and
Tomanaga showed how to extract meaningful results from such poorly behaved
series, their renormalization procedure does not give a great deal of physical
insight.

A somewhat different way of approaching such problems was subsequently
explored by Gell-Mann and Low, who reformulated renormalized perturbation
theory in terms of a cutoff-dependent coupling constant, $g(\Lambda)$. This allowed
them to study the gradual evolution of bare, unrenormalized quantities into
dressed, renormalized ones. The set of transformations which gradually change
g from a bare to a dressed quantity is called a renormalization group. These ideas
were developed further in the 1960s by Wilson, Callan, Symanzik and others. An
equation, now called the Callan–Symanzik equation, was derived for the n-point
functions of renormalizable field theories by demanding that the physical values
be asymptotically independent of the cutoff. One is led quite naturally to consider the logarithmic derivative of $g(\Lambda)$, often called the "β function."

$$\beta(g) = \frac{\partial g(\Lambda)}{\partial \ln \Lambda}.$$

(A.1)

A spectacular application of these ideas came with the discovery by Politzer
and by Gross and Wilczek of asymptotic freedom. It was found that the functions $\beta(g)$ for a select set of candidates for strong-interaction field theories
(describing "quantum chromodynamics") were such as to drive the renormalized couplings to zero at short distances, or, equivalently, at very high energies.
The idea of asymptotically free theories of the strong interactions led to an
understanding of experiments on electron–proton scattering, which showed that
protons behave at high energies as if they were composed of weakly interacting
constituents called quarks or partons.

Modern renormalization ideas were generalized and used to construct a
theory of critical-point singularities in 1971 by K. G. Wilson. By building on
work by Kadanoff, Wilson was able to explain the singularities near critical
points and show how to calculate them. In doing so, he showed that renormalization-group ideas had applicability far beyond high-energy physics and placed
a powerful mathematical tool at the disposal of the rest of the scientific community. Critical-point phenomena provide a particularly simple context in which to
explain this new perspective on renormalization theory.

Phase changes of matter in thermodynamic equilibrium are often characterized by sharp discontinuities in quantities such as the magnetization of a
ferromagnet and the density of a liquid-vapor system. Usually, these phase transitions can be made continuous by adjusting an external parameter like the

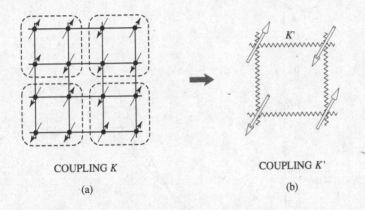

COUNTING K

(a)

COUPLING K'

(b)

Fig. 1.7. The Kadanoff block scheme. Dashed lines enclose groups of spins on a lattice (a), which is characterized by a nearest-neighbor spin coupling K. Out of each block, we construct a collective spin variable by integrating over the internal degrees of freedom. The new spin degrees of freedom populate a lattice (b), with twice the old lattice spacing, and interact as if they had coupling K'.

temperature or pressure. The point at which the discontinuity vanishes is called a critical point.

The behavior of a liquid–gas system near such a singular point is particularly striking. The fluid suddenly becomes milky and opaque and scatters laser light intensely in the forward direction. It seems clear that some sort of cooperative phenomenon is taking place and that the particles of fluid are strongly correlated over distances comparable to the wavelength of the laser light. The range of correlation $\xi(T)$ is a measure of this cooperativity and is expected to diverge strongly at T_c,

$$\xi(T) \sim |T - T_c|^{-\nu}. \tag{A.2}$$

This divergence introduces into the problem a range of length scales not unlike that encountered when the cutoff is sent to infinity in quantum field theory.

The striking thing about this and other anomalies near critical points is that it appears to be universal. A wide variety of liquid–vapor systems and ferromagnets exhibit correlation lengths that appear to diverge in precisely the same way. A second curious fact is that universal numbers like ν are not simply one half or unity, although most simple theories lead inevitably to these values. In fact, it is now believed that ν lies in the range

$$0.64 \leq \nu \leq 0.70 \tag{A.3}$$

in most three-dimensional systems. Modern renormalization theory focuses on the calculation of such numbers. The exponent ν turns out be closely analogous to a number that determines the way masses scale with the cutoff in quantum field theories.

To illustrate the renormalization-group method, we consider a simple model of magnetism, a two-dimensional square lattice of Ising spins $\sigma_i = \pm 1$ (see Fig. 1.7). To each configuration $\{\sigma_i\}$ of spin we assign an energy

$$H[\{\sigma_i\}] = -J \sum_{\langle i, j \rangle} \sigma_i \sigma_j, \tag{A.4}$$

where i and j label lattice sites and the sum is restricted to nearest-neighbor pairs of spins. The calculation of the partition sum associated with Eq. (A.4) as a function of the dimensionless inverse temperature $K = J/(k_B T)$ is a very difficult

Fig. 1.8. An approximate recursion relation for Ising spins, with a fixed point at $K^* = K_c = 0.34$. The "ladder" construction shows successive temperatures produced by repeated iterations of the transformation. The slope of the function $K'(K)$ through its fixed point is related to the critical exponent ν.

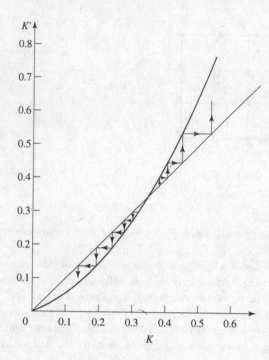

task. It was, in fact, computed exactly for this two-dimensional problem by Onsager in 1944. Onsager's work, which remains today a true mathematical *tour de force*, eventually led to the exponent prediction

$$\nu = 1. \tag{A.5}$$

An analytical evaluation of such partition sums in three dimensions or in much more general situations appears hopeless, however.

Rather than attempting to partition sums exactly, we can, in fact, make progress by merely thinning out the spin degrees of freedom slightly. To do this, we follow the original suggestion of Kadanoff and group the spins on the original lattice (which is at coupling K) into blocks as shown in Fig. 1.7. With some ingenuity, it turns out to be possible to integrate approximately over the internal degrees of freedom within each block. We are then left with a new statistical-mechanical problem with twice the lattice spacing. Provided the interactions between spins in the new lattice are of the same form as those in the old, couplings in the new lattice can be characterized simply by a new temperature K'.

This sort of program was first carried out successfully for Ising spins on a triangular lattice by Niemeyer and Van Leeuwen. A very primitive version of their transformation gives a relationship between the new and old temperatures, namely

$$K' = 2K(e^{3K} + e^{-K})^2/(e^{3K} + 3e^{-K})^2. \tag{A.6}$$

Equation (A.6) is depicted graphically in Fig. 1.8, together with a "ladder" construction showing how the temperature changes with repeated iterations of the transformation. The recursion formula (A.6) has a fixed point at $K^* \approx 0.34$, so it seems natural to identify this isolated point with the critical point coupling K_c of the spin system.

We can see the utility of this procedure of "thinning out" spins as follows: even an approximate calculation of the partition function is difficult near T_c because of the large correlation length $\xi(T)$ in this region. However, note from Fig. 1.8 that the renormalization transformation *increases* temperatures that, initially, are slightly above T_c. By repeatedly iterating the transformation (A.6), we can relate a difficult calculation near T_c to a more tractable high-temperature problem. A very accurate calculation of the partition sum at high temperatures is made possible by expanding in powers of K. In a similar fashion, we see that the transformation (A.6) *decreases* temperatures initially slightly below T_c and eventually produces a more manageable (low-temperature) problem.

The physics behind a renormalization transformation is in the reduction of the correlation length. If the lattice spacing is increased by a factor b by the process of thinning out, it follows that the correlation length transforms according to

$$\xi(K') = \xi(K)/b. \tag{A.7}$$

For the square-lattice spin system shown in Fig. 1.8, $b = 2$, whereas for the triangular lattice of Niemeyer and Van Leeuwen it turns out that $b = \sqrt{3}$. Equation (A.6) can be used in conjunction with (A.7) to produce a critical exponent. Near the fixed point $K^* = K_c = 0.34$, Eq. (A.6) can be written in the approximate form

$$K' - K_c - b^{0.89}(K - K_c), \tag{A.8}$$

which, when it is combined with (A.7), gives a functional equation for $\xi(K - K_c)$,

$$\xi(K - K_c) = b\xi[b^{0.89}(K - K_c)]. \tag{A.9}$$

It is easy to check that the solution of this function equation is $\xi(T) \sim |K - K_c|^{-\nu}$ with $\nu \approx 1.12$, which is not too different from the exact result $\nu = 1.0$.

In more sophisticated calculations, it is not possible to describe a renormalization transformation in terms of a single coupling constant. More complicated couplings such as next-nearest-neighbor and multispin interactions are generated after one iteration. When these extra couplings are taken into account, they not only produce more accurate estimates of critical exponents but also provide a qualitative explanation of universality: If several couplings are included in the calculation, the renormalization transformation can be viewed as a mapping of one Hamiltonian onto another in a multidimensional coupling-constant space. As K is adjusted toward K_c, it turns out that any initial set of couplings "flows" toward a unique fixed point under repeated iterations of the transformation. Just as in the one-interaction-constant example, critical exponents are related to the eigenvalues of the transformation about this fixed point.

From its origins in nineteenth-century hydrodynamics, renormalization theory has developed into a powerful and sophisticated tool for understanding complicated problems in physics. Much of the recent progress in this area has resulted from a fertile interaction of statistical mechanics with quantum field theory. In its modern form, renormalization theory is especially useful for any problem involving multiple length scales. Anderson applied renormalization-group methods to the Kondo problem in condensed matter physics at about the same time as Wilson applied them to critical phenomena. Such methods have also proved useful in solving certain hydrodynamic problems, in studies of electron localization in metals and in problems of insect-population biology. There is some hope that they can be used to understand the multiple length scales encountered in turbulent fluids at high Reynolds numbers.

Suggested reading

N. N. Bogoliubov and D. V. Shirkov. *Introduction to the Theory of Quantized Fields* (Interscience, New York, 1959).

S. Coleman, in *Properties of the Fundamental Interactions, 1971*, edited by A. Zichichi (Editrice Compositori, Bologna, 1972).

M. E. Fisher, *Rev. Mod. Phys.* **42**, 597 (1974).

S.-K. Ma, *Modern Theory of Critical Phenomena* (Benjamin, Reading, MA, 1976).

D. R. Nelson, *Nature* **269**, 379 (1977).

P. Pfeuty and G. Toulouse, *Introduction to the Renormalization Group and Critical Phenomena* (Wiley, New York, 1977).

H. D. Politzer, *Phys. Rev.* C **14**, 131 (1974).

K. G. Wilson and J. Kogut, *Phys. Rep.* C **12**, 77 (1974).

Appendix B. The self-dual point of Ising spins in two dimensions

A compact derivation of the duality transformation for the two-dimensional Ising model follows from the transformations similar to those developed for two-dimensional XY spin systems by José *et al.* [29]. To motivate this approach, note that the partition function of the *one*-dimensional Ising model,

$$Z(K) = \underset{\{\sigma_i = \pm 1\}}{\text{Tr}} \exp\left(K \sum_{\langle ij \rangle} \sigma_i \sigma_j\right),$$

(B.1)

with $K = J/(k_B T)$, may be computed immediately by assigning an Ising spin variable to every nearest-neighbor spin pair $\langle ij \rangle$,

$$\tau_{ij} = \sigma_i \sigma_j = \pm 1.$$

(B.2)

Upon assuming that we have N sites and periodic boundary conditions, this leads to the well-known exact result,

$$Z = \underset{\{\tau_{ij} = \pm 1\}}{\text{Tr}} \exp\left(K \sum_{\langle ij \rangle} \tau_{ij}\right) = (2 \cosh K)^N.$$

(B.3)

If we make the same change of variables for an N-site square lattice in two dimensions, we obtain

$$Z = \underset{\{\tau_{ij} = \pm 1\}}{\text{Tr}} \exp\left(K \sum_{\langle ij \rangle} \tau_{ij}\right) \times \text{constraints},$$

(B.4)

where the "constraints" are required in order to insure that the product of the τ_{ij} around every closed loop vanishes. For example, the product of the bond variables around the elementary plaquette P shown in Fig. 1.9 must equal unity,

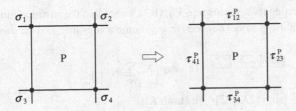

Fig. 1.9. The elementary plaquette of a two-dimensional Ising model. The site variables $\{\sigma_j\}$ on each plaquette P are replaced by bond variables $\{\tau_{ij}^P\}$, subject to the constraint (B.5).

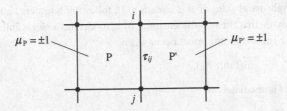

Fig. 1.10. A pair of elementary plaquettes P and P′. These plaquettes are associated with new Ising spins μ_P and $\mu_{P'}$, which enforce the constraint.

$$\tau_{12}^P \tau_{23}^P \tau_{34}^P \tau_{41}^P = \sigma_1 \sigma_2 \sigma_2 \sigma_3 \sigma_3 \sigma_4 \sigma_4 \sigma_1 = 1. \tag{B.5}$$

It is sometimes convenient to label the new spins by their positions around plaquette P as indicated in Fig. 1.9. Explicit introduction of constraint factors which vanish unless (B.5) is satisfied into the partition function leads to

$$Z = \operatorname*{Tr}_{\{\tau_{ij} = +1\}} \prod_{\langle i,j \rangle} e^{K\tau_{ij}} \prod_P \left(\frac{1 + \tau_{12}^P \tau_{23}^P \tau_{34}^P \tau_{41}^P}{2} \right). \tag{B.6}$$

The partition function for N Ising variables on a square lattice with N sites now seems to contain $2N$ degrees of freedom! However, there are N constraints, insuring that those configurations which fail to satisfy (B.5) receive zero weight. It is easy to check that the constraints on plaquettes are sufficient to correctly insure that the product of the τ variables around *any* closed loop equals unity.

We now introduce a new spin variable $\mu_P = \pm 1$ for each plaquette constraint,

$$Z = 2^{-N} \operatorname*{Tr}_{\{\mu_P = \pm 1\}} \operatorname*{Tr}_{\{\tau_{ij} = \pm 1\}} \prod_{\langle i,j \rangle} e^{K\tau_{ij}} \prod_P (\tau_{12}^P \tau_{23}^P \tau_{34}^P \tau_{41}^P)^{(1 - \mu_P)/2}. \tag{B.7}$$

Tracing over the $\{\tau_{ij}\}$ can now be carried out explicitly. We focus on the sum involving a single-bond variable τ_{ij}, which leads to a coupling between two neighboring plaquettes P and P′ of the dual lattice (see Fig. 1.10),

$$\sum_{\tau_{ij} = \pm 1} e^{K\tau_{ij}} \tau_{ij}^{1 - (\mu_P + \mu_{P'})/2} = e^K - e^{-K} (-1)^{-(\mu_P + \mu_{P'})/2}$$
$$\equiv A e^{K' \mu_P \mu_{P'}}. \tag{B.8}$$

The last equivalence will be satisfied for all possible plaquette spin configurations $(\mu_P, \mu_P) = (+,+), (+,-), (-,+)$ and $(-,-)$ provided that

$$A = \sqrt{2 \sinh(2K)} \tag{B.9}$$

and

$$K' = -\tfrac{1}{2} \ln \tanh K. \tag{B.10}$$

Upon inserting these results into Eq. (B.7), we see that the partition function of the original Ising model has been mapped onto a copy living on the dual lattice and that

$$Z(K) = \frac{[2\sinh(2K)]^{2N/2}}{2^N} \operatorname*{Tr}_{\{\mu_P = \pm 1\}} \exp\left(K' \sum_{\langle PP' \rangle} \mu_P \mu_{P'}\right)$$

$$= [\sinh(2K)]^N Z[-\tfrac{1}{2}\ln(\tanh K)].$$

Except for a nonsingular prefactor, the *same* partition function appears on the left- and right-hand sides of this equation. If, following Kramers and Wannier [20], we assume that the two-dimensional Ising model has a *single* finite-temperature phase transition, this must occur when

$$K = K_c = -\tfrac{1}{2}\ln(\tanh K_c), \tag{B.12}$$

which leads immediately to

$$K_c = \frac{J}{k_B T_c} = \tfrac{1}{2}\ln(1 + \sqrt{2}). \tag{B.13}$$

References

[1]   For a more elementary introduction, see P. M. Chaikin and T. C. Lubensky, *Principles of Condensed Matter Physics* (Cambridge University Press, Cambridge, 1995).

[2]   A. Z. Patashinskii and V. L. Pokrovskii, *Fluctuation Theory of Phase Transitions* (Pergamon, New York, 1979), Preface. Similar scaling ideas were developed in the West by B. Widom, L. Kadanoff and M. E. Fisher.

[3]   K. G. Wilson, in *Magnetism and Magnetic Materials – 1972,* edited by C. D. Graham Jr and J. J. Rhyne (American Institute of Physics, New York, 1973), p. 843.

[4]   U. Frisch, *Turbulence: The Legacy of A. N. Kolmogorov* (Cambridge University Press, Cambridge, 1995).

[5]   Of course, research into particle physics (and string theory) still has great intellectual value and many condensed matter physicists including myself have shamelessly exported its beautiful mathematical ideas into our own research. There is wonderful cross-fertilization between the two disciplines, which dates back at least to the work of K. G. Wilson in the early 1970s and continues to this day.

[6]   See, e.g., S. K. Ma and G. Mazenko, *Phys. Rev.* B **11**, 4077 (1975).

[7]   D. Forster, D. R. Nelson and M. J. Stephen, *Phys. Rev.* A **16**, 732 (1977); some of the results for *equilibrium* forcing functions in this paper were derived earlier using a mode-coupling approach: See, e.g., Y. Pomeau and P. Résibois, *Phys. Rev.* C **19**, 64 (1975).

[8]   There is a misprint in Eq. (3.69) of the first part of [7].

[9]   L. D. Landau and E. M. Lifshitz, *Theory of Elasticity* (Pergamon, New York, 1970).

[10] A. Elgsaeter, B. Stokke, A. Mikkelsen and D. Branton, *Science* **234**, 1217 (1986).

[11] F. Abraham and D. R. Nelson, *Science* **249**, 393 (1990); *J. Physique* **51**, 2653 (1990).

[12] C. F. Schmidt, K. Svoboda, N. Lei, I. B. Petsche, L. E. Berman, C. R. Safinya and G. S. Grest, *Science* **259**, 952 (1993).

[13] J. A. Aronovitz, L. Golubovic and T. C. Lubensky, *J. Physique* **50**, 609 (1989).

[14] D. R. Nelson and L. Peliti, *J. Physique* **48**, 1085 (1987).

[15] L. Radzihovsky and P. Le Doussal, *J. Physique* I **2**, 599 (1992).

[16] S. Seung and D. R. Nelson, *Phys. Rev.* A **38**, 10005 (1988).

[17] See the articles in *Statistical Mechanics of Membranes and Surfaces*, edited by D. R. Nelson, T. Piran and S. Weinberg (World Scientific, Singapore, 1989).

[18] R. E. Peierls, *Proc. Cambridge Phil. Soc.* **32**, 477 (1936).

[19] See, e.g., C. Domb in C. Domb and M. S. Green (eds.), *Phase Transitions and Critical Phenomena*, vol. III (Academic, New York, 1974), pp. 357–478.

[20] H. A. Kramers and G. H. Wannier, *Phys. Rev.* **60**, 252 (1941).

[21] Grain boundaries between micro-crystallites with different orientations are an interesting *linear* defect configuration in two-dimensional solids. Grain boundaries, however, can be viewed as rows of dislocations with Burgers-vector orientations strictly perpendicular to the grain direction. A careful estimate of the entropy associated with a linear grain boundary shows that the free energy changes sign at the same point as that for an isolated dislocation. See D. S. Fisher, B. I. Halperin and R. Morf, *Phys. Rev.* B **20**, 4692 (1979).

[22] V. L. Berenzinskii, *Zh. Éksp. Teor. Fiz.* **59**, 907 (1970) [*Sov. Phys. JETP* **32**, 493 (1971)]; *Zh. Éksp. Teor. Fiz.* **61**, 1144 (1972) [*Sov. Phys. JETP* **34**, 493 (1972)].

[23] J. M. Kosterlitz and D. J. Thouless, *J. Phys.* C **5**, L124 (1972); *J. Phys.* C **6**, 1181 (1973).

[24] B. I. Halperin and D. R. Nelson, *Phys. Rev. Lett.* **41**, 121 (1978); *Phys. Rev. Lett.* **41**, 519 (1978).

[25] D. C. Morse and T. C. Lubensky, *J. Physique* II **3**, 531 (1993).

[26] F. David, E. Guitter and L. Peliti, *J. Physique* **48**, 2059 (1987).

[27] J. M. Park and T. C. Lubensky, *Phys. Rev.* E **53**, 2648 (1996); *Phys. Rev.* E **53**, 2665 (1996).

[28] G. Gompper and D. Kroll, *Phys. Rev. Lett.* **78**, 2859 (1997); *J. Physique* I **11**, 1 (1997).

[29] J. José, L. P. Kadanoff, S. Kirkpatrick and D. R. Nelson, *Phys. Rev.* B **16**, 1217 (1977).

Chapter 2
Defect-mediated phase transitions

Preface

Defects play an essential role in many aspects of condensed matter physics. Vacancies and interstitials catalyze diffusion of particles in solids, dislocations determine the strength of crystalline materials and vortex motion controls the resistance of superconductors as well as the capacity of neutral superfluids such as ^{4}He to carry heat currents. These familiar defects are also important excitations in equilibrium statistical mechanics, especially in two dimensions, in which their point-like character makes them relatively easy to generate in the presence of thermal fluctuations. This chapter describes theoretical developments up to 1983. Renormalization-group and dynamical scaling ideas applied to critical phenomena in the 1970s proved to be powerful tools for studying the equilibrium statistical mechanics and dynamics of defect-mediated phase transitions, especially in two dimensions.

Perhaps the most surprising result to emerge from these investigations is the existence of a fourth "hexatic" phase of matter, intermediate between a crystal and a liquid. Most of us were taught in school that matter comes in three basic forms: solid, liquid and gas. This classification scheme goes back at least to the Greeks and had remained unchanged for thousands of years. The important refinements have come only during the last 100 years or so. We now know, for example, that there is no fundamental distinction between a liquid and a gas above the critical point. On the other hand, it is important to distinguish between the repeating arrays of atoms and molecules in a crystalline solid and the nonequilibrium disorderly arrangements in glassy solids. We find additional phases in liquid crystals that, because of the anisotropic shape of their constituent molecules, are in many ways intermediate between liquids and solids. Despite such refinements, there remains a strong basic prejudice that simple point-like atoms or molecules in equilibrium should have just three basic phases.

This prejudice overlooks the fact that two broken symmetries separate low-temperature crystalline solids from high-temperature liquids. The regular rows of atoms in a crystal responsible for Bragg peaks reflect a breakdown in the translational invariance characteristic of a liquid. The *directions* of the crystallographic axes of the solid represent a different, broken orientational symmetry. A broken translational symmetry necessarily implies long-range orientational order. It is possible to imagine phases of matter, however, in which the short-range translational order of a liquid coexists with a broken rotational symmetry. The long-range rotational order of this liquid is associated with "bonds" (determined by, say, a Wigner–Seitz-like construction) joining neighboring atoms or molecules. Such an intermediate phase in fact appeared when a model of two-dimensional melting, proposed independently by Kosterlitz and Thouless [1], by Berezinski [2] and by Feynman [3] was finally solved. The usual latent heat of the crystal-to-liquid melting transition can be spread out over an intermediate hexatic phase, somewhat similar to a nematic liquid crystal. Unlike in a liquid crystal, though, the residual 6-fold orientational order of hexatics is not associated with the anisotropy of the constituent particles. Equilibrium hexatics display a fuzzy 6-fold symmetric diffraction pattern (indicative of extended orientational correlations) accompanied by a vanishing shear modulus, enforced by a finite density of free dislocations.

A fourth, hexatic phase of matter has now been observed in free-standing liquid-crystal films [4], in colloidal crystals [5, 6], in magnetic bubble arrays [7] and, quite extensively, in Langmuir–Blodgett surfactant monolayers [8]. Hexatic order is not confined to two dimensions. It appears in layered smectic liquid crystals [9], in dense solutions of DNA [10] and (at least in glassy form) in the flux-line arrays of high-temperature superconductors [11]. The modest time scales available even on the fastest computers make equilibration in Monte Carlo or molecular-dynamics simulations of two-dimensional melting difficult. However, there is now evidence via computer simulations for continuous melting and a narrow sliver of hexatic phase for hard disks [12] and for particles interacting with a repulsive $1/r^{12}$ potential [13]. For interesting evidence that the hexatic order appears at least in metastable form for the familar Lennard-Jones 6–12 pair potential, see [14]. Although the above references are not exhaustive, they do suggest that the subject matter of this chapter continues to be a lively area of research.

2.1 Introduction

2.1.1 Theoretical background

Fluctuations dominate the modern theory of phase transitions. Fluctuation effects can be neglected near ordinary critical points only above four dimensions, where mean field theory provides a good description.

Below $d = 4$, fluctuations shift critical exponents from their mean field values in a way calculable as a power series in $\varepsilon = 4 - d$ [15]. If the dimensionality is low enough, fluctuations can destroy long-range order entirely and drive the critical temperature to zero. This happens in the limit $d \to 1^+$ for Ising systems with short-range interactions and for other models with discrete symmetry [16–18]. Broken *continuous* symmetries become impossible in two dimensions or fewer [19–22]. The critical temperatures of n-component Heisenberg ferromagnets, with $n \geqslant 3$, for example, go continuously to zero as d approaches two from above [23–25].

If the continuous symmetry is *Abelian*, it turns out that a finite-temperature two-dimensional phase transition is possible, despite the lack of a genuine broken symmetry. Abelian symmetries occur in superfluid ^4He, in two-dimensional crystalline solids and in XY magnets. Phase transitions in such systems can occur via an unbinding of point defects such as vortices or dislocations, as pointed out by Berezinskii [26] and by Kosterlitz and Thouless [27, 28].

In this review, we shall attempt to survey what is known about these defect-mediated phase transitions. Although defects are sometimes unimportant in experiments on bulk materials, they can provide crucial insights into a wide variety of experiments on thin films. In some cases, an understanding of two-dimensional defect-controlled phase transitions actually lends insight into unsolved three-dimensional problems. As will be discussed further in Section 2.7, fluctuations that destroy long-range order also occur in bulk smectic-A liquid crystals. *Line* defects may be important in understanding the smectic-A-to-nematic transition in *three* dimensions.

To see more quantitatively the special role played by two dimensions in superfluids, crystals and magnets, we first review the behavior of equilibrium order-parameter correlation functions in these systems. Consider, in particular, the equilibrium correlation functions

$$G(\mathbf{r}) = \langle \psi(\mathbf{r})\psi^*(\mathbf{0}) \rangle \quad (^4\text{He superfluids}) \tag{2.1a}$$

$$= \langle \hat{\rho}_G(\mathbf{r})\hat{\rho}_G^*(\mathbf{0}) \rangle \quad (\text{crystals}) \tag{2.1b}$$

$$= \langle \mathbf{S}(\mathbf{r}) \cdot \mathbf{S}(\mathbf{0}) \rangle \quad (\text{magnets}). \tag{2.1c}$$

Here, $\psi(\mathbf{r})$ is the complex condensate wave function of a superfluid, while $\mathbf{S}(\mathbf{r})$ is a two-component spin at a site $\mathbf{r}$ in a magnet. The quantity $\hat{\rho}_G(\mathbf{r})$ is the local Fourier component of the mass or number density of a crystal at a reciprocal lattice vector $\mathbf{G}$. The average displayed in (2.1b) is related to the Debye–Waller factor; its Fourier transform gives the X-ray structure factor.

In *three* dimensions, correlation functions like those displayed in

Eqs. (2.1) can behave in three distinct ways at large distances. If there is no long-range order, as in the normal phase of a superfluid, in a liquid and in a paramagnet, $G(r)$ decays exponentially,

$$G(r) \sim e^{-r/\xi(T)}, \qquad (2.2)$$

where $\xi(T)$ is a temperature-dependent correlation length. A broken symmetry is signaled by the decay of $G(\mathbf{r})$ to a nonzero constant at large r:

$$\lim_{r \to \infty} G(r) = \text{constant} \neq 0. \qquad (2.3)$$

In a superfluid, this kind of long-range order means that the phases of $\psi(\mathbf{r})$ add coherently from point to point, whereas, in a magnet, the magnetization points in a definite direction. Long-range order in crystals represents broken continuous translational symmetry and shows up as delta-function Bragg peaks at the reciprocal-lattice vectors in the X-ray structure factor. Of course, there is also broken *orientational* symmetry in crystals, represented by singled-out crystallographic axes. A third, more exceptional, possibility is that $G(r)$ decays algebraically to zero, taking the form

$$G(\mathbf{r}) \sim 1/r^{d-2+\eta}, \qquad (2.4)$$

where d is the spatial dimension. This power-law fall-off occurs only at isolated second-order critical points; the quantity η is expected to be a universal critical exponent for systems with the same symmetry. Algebraic decay does not usually occur in three-dimensional melting, since this transition is almost always of first order.

Two-dimensional systems are intriguing because the behavior (2.3), indicating long-range order, is impossible. Long before rigorous proofs were constructed in the 1960s, Bloch [29], Peierls [30] and Landau [31] observed that any hypothetical broken magnetic or translational symmetry would be overwhelmed by spin-wave or phonon excitations. The key point is the reduced phase space available for these "Goldstone modes" in two dimensions. Third sound plays an analogous role in helium films. Two dimensions seems to be a borderline where fluctuations are just strong enough to prevent conventional ordering with an *Abelian* continuous symmetry like superfluids [32–34], crystals [35–37] and XY magnets [38, 39]; one expects low-temperature phases in which correlations fall off everywhere as in (2.4):

$$G(r) \sim 1/r^{\eta(T)}. \qquad (2.5)$$

Analogous power-law decays occur in bulk smectic liquid crystals [40]. The exponent $\eta(T)$ is no longer universal, but turns out to depend continuously on temperature! In this sense, we must deal with whole *phases* of critical points. One consequence is that these systems are particularly susceptible to perturbations. When correlations decay as in Eq. (2.5), we shall say that the system possesses "quasi-long-range order."

The conventional continuum elastic description [41] of isotropic two-dimensional solids provides a nice illustration of a "phase of critical points." A quantity of immediate experimental interest is the X-ray structure factor, which may be written as a sum over a discrete set of possible lattice sites $\{\mathbf{r}\}$,

$$S(\mathbf{q}) = \sum_{\mathbf{r}} e^{i\mathbf{q}\cdot\mathbf{r}} \langle e^{i\mathbf{q}\cdot[\mathbf{u}(\mathbf{r})-\mathbf{u}(0)]}\rangle, \tag{2.6}$$

where $\mathbf{u}(\mathbf{r})$ is the displacement field associated with the site $\mathbf{r}$. On writing the density $\rho(\mathbf{r})$ of a collection of point particles at positions

$$\mathbf{R}(\mathbf{r}) \equiv \mathbf{r} + \mathbf{u}(\mathbf{r}) \tag{2.7}$$

as

$$\rho(\mathbf{r}) = \sum_{\mathbf{r}} \delta[\mathbf{r} - \mathbf{R} - \mathbf{u}(\mathbf{r})], \tag{2.8}$$

we recognize $S(\mathbf{q})$ as the Fourier transform of the density–density correlation function. Moreover, it is easily seen that

$$\hat{\rho}_{\mathbf{G}}(\mathbf{r}) = e^{i\mathbf{G}\cdot\mathbf{u}(\mathbf{r})}, \tag{2.9}$$

so $S(\mathbf{q})$ is also the Fourier transform of the Debye–Waller correlation (2.1b) over the discrete set of lattice sites $\mathbf{r}$ for $\mathbf{q} \approx \mathbf{G}$.

Correlations in $\hat{\rho}_{\mathbf{G}}(\mathbf{r})$ may be determined with the aid of the continuum elastic-free-energy functional [41],

$$F = \frac{1}{2} \int d^2r \, [2\mu u_{ij}^2(\mathbf{r}) + \lambda u_{kk}^2(\mathbf{r})]. \tag{2.10}$$

Here, u_{ij} is the symmetric strain tensor,

$$u_{ij} = \frac{1}{2} \left(\frac{\partial u_i}{\partial r_j} + \frac{\partial u_j}{\partial r_i} \right) \tag{2.11}$$

and we have employed the summation convention. The parameters μ and λ are called Lamé coefficients; only two such elastic constants are needed to describe close-packed triangular solids. Rotational invariance implies that F cannot depend on the antisymmetric part of $\partial_i u_j$, namely

$$\theta = \tfrac{1}{2}(\partial_x u_y - \partial_y u_x). \tag{2.12}$$

Of course, gradients of θ would be allowed as higher-order terms in (2.10).

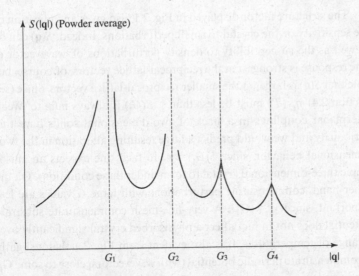

Since F is quadratic in the displacement field, it is a simple matter to compute the Debye–Waller average at larger r. One easily finds

$$\langle \rho_G(\mathbf{r}) \rho_G^*(0) \rangle \sim r^{-\eta_G(T)}, \tag{2.13}$$

where

$$\eta_G(T) = \frac{k_B T |\mathbf{G}|^2 (3\mu + \lambda)}{4\pi\mu(2\mu + \lambda)}. \tag{2.14}$$

Note that $\eta_G(T)$ depends on temperature, on the Lamé coefficients and on the particular reciprocal-lattice vector involved. This power-law decay leads to power-law singularities in $S(\mathbf{q})$ at the reciprocal-lattice points $\{\mathbf{G}\}$. For $\mathbf{q} \approx \mathbf{G}$, we find

$$S(\mathbf{q}) \sim \frac{1}{|\mathbf{q} - \mathbf{G}|^{2 - \eta_G(T)}}. \tag{2.15}$$

A sketch of a powder average of $S(\mathbf{q})$ is shown in Fig. 2.1. The structure factor diverges at the smaller reciprocal-lattice vectors and displays cusps at larger ones. Figure 2.1 is quite different from the sequence of delta-function Bragg peaks superimposed on a diffuse background which we would expect for a three-dimensional solid. On the other hand, it bears no resemblance to a liquid structure factor either! Note that power-law Bragg peaks require only nonzero elastic constants, which we take as our definition of a two-dimensional crystal. It is easy to show that

$$\langle \rho_G(\mathbf{r}) \rangle = 0, \tag{2.16}$$

so that there is indeed no conventional long-range order.

The structure factor displayed in Fig. 2.1 also gives an indication of the sensitivity of the crystal to small perturbations. Indeed, $S(\mathbf{q})$ can be viewed as the susceptibility to density perturbations of wavevector $\mathbf{q}$. The response is strongest at the reciprocal-lattice vectors, of course, but note that $S(\mathbf{q})$ is larger at the smaller reciprocal-lattice vectors. Since (see Section 2.4) $\eta_{G_1}(T)$ must be less than $\frac{1}{3}$, $S(G_1)$ is always infinite. Weak interplanar couplings in a stack of two-dimensional solids have this periodicity and we would predict a large resulting alteration in the two-dimensional behavior, since $S(G_1) = \infty$. In fact, one expects an aniso-tropic three-dimensional crystal to form under these conditions. On the other hand, commensurate perturbations with large G values are less important, since $\eta_G \propto |\mathbf{G}|^2$. A very-fine-mesh commensurate substrate potential does not in fact affect a physisorbed crystal significantly over a range of temperatures. It is also evident from Fig. 2.1 that a slightly incommensurate periodic potential (with wavevector $\mathbf{q}$ close to some G_i) produces less of a response than does a corresponding commensurate one with the same strength.

Since the correlations discussed above decay exponentially at suffi-ciently high temperatures, there must be a temperature T_c above which power-law decay is no longer possible. All these systems can be described by two-component order parameters, which we parameterize by an amplitude and a phase. By considering the relative importance of amplitude versus phase fluctuations, one can see why defects are likely to be important near T_c: We have just seen that phase fluctuations sup-press genuine long-range order. One would certainly expect T_c to be depressed by these fluctuations far below the mean-field or Landau crit-ical temperature. Amplitude fluctuations near the actual critical tem-perature are then constrained by steep minima in the corresponding local Landau free energy. There is, however, no such constraint on phase fluctuations, which are associated with a locally broken continuous symmetry. One is led naturally to "low-temperature" models in which phase fluctuations dominate and amplitude fluctuations play a secon-dary role. This point of view takes us some distance from traditional Landau theories, which are expansions in a small order-parameter amplitude.

Amplitude fluctuations, which are always present at sufficiently high temperatures, appear in the models we wish to investigate in the form of defects. Vortices in two-dimensional superfluids and XY magnets and dislocations in two-dimensional crystals may be viewed as isolated points where the amplitude of the order parameter vanishes. Dislocation lines play a similar role in bulk smectic-A liquid crystals. One can account approximately for amplitude variations by allowing for a finite density of such defects. If the defects become more numerous with increasing tem-perature, it is at least plausible that they eventually produce a transition

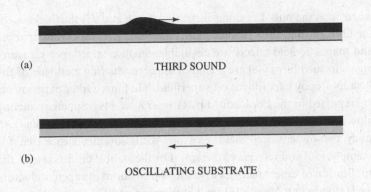

(a)

THIRD SOUND

(b)

OSCILLATING SUBSTRATE

Fig. 2.2. Experiments on superfluid helium films. (a) the superfluid density can be extracted from the speed of a film-height deformation called third sound. (b) The superfluid density can be determined directly from the moment of inertia of an oscillating substrate. The superfluid fraction does not participate in the motion at low frequencies.

into a phase with exponentially decaying correlations. The detailed justification for these ideas will emerge later in this review.

2.1.2 Experiments

The dimension in which fluctuations first prevent long-range order is often called the "lower critical dimension." There is a surprisingly rich variety of experimental systems that are precisely at this borderline dimensionality. Before describing the defect-mediation theories, it seems worth reviewing the experiments which allow them to be tested.

Very precise experiments are possible on superfluid ^{4}He films, which have been under careful investigation for many years. There have been many observations of a propagating wave of film height and temperature called "third sound" [42, 43] (see Fig. 2.2(a)), following the pioneering work of Atkins [44]. Since the velocity of propagation is related to the fraction of the film which is superfluid, one can determine the superfluid density. This quantity plays a role analogous to that of the shear modulus in the theory of two-dimensional superfluidity. A direct measure of the superfluid density is provided by the oscillating-substrate experiment of Bishop and Reppy [45]. A roll of flexible substrate coated with helium is wrapped around a torsional oscillator (see Fig. 2.2(b)) and the moment of inertia and absorption of energy are determined. The assembly of substrate and helium film seems "lighter" than it should be, since the superfluid fraction does not participate in the torsional oscillations at low frequencies.

Superconducting films have some similarities with films of superfluid ^{4}He, with the added complication that the particles exhibiting superfluidity (electrons) are charged. As emphasized by Beasley *et al.* [46], however, magnetic fields due to the motion of a charged condensate can be neglected on scales less than a magnetic "screening length" given [47] by

$$L_s = 2\lambda_L^2/d, \qquad (2.17)$$

where λ_L is the bulk London penetration depth and d is the thickness of the film. In thin, dirty, superconducting films, L_s can be 1 cm or more and magnetic-field effects are negligible. Surface-impedance measurements in such films [48] are a kind of superconducting analogue of the Bishop–Reppy experiment on superfluid ^{4}He films. Other experiments are reported in the book edited by Gubser *et al.* [49]. Superconducting films in a perpendicular magnetic field may provide an opportunity to study two-dimensional melting of an Abrikosov flux lattice (see, for example, [50] and references therein). For theoretical discussions of the application of defect theories to resistive transitions in superconductors see Halperin and Nelson [51] and Huberman and Doniach [52].

A particularly pristine example of a superconducting XY model has been created and studied by Resnick *et al.* [52]. Photolithography was used to create an extremely regular lattice of over a million tiny lead disks, embedded in a tin matrix. The disks were 1.3 μm apart at closest approach, and each was 150 nm thick and 13 μm in diameter! Below the bulk superconducting transition of the lead, but above that of the tin, the condensate phases within each lead disk are coupled just as in the two-dimensional XY model.

Free-standing smectic liquid-crystal films [53, 54] may provide some of the best realizations of isolated, two-dimensional liquid and solid phases in which to study melting. As few as two layers of a bulk smectic liquid crystal can be suspended across a mount (Fig. 2.3(a)) and studied with light and X-ray [55] scattering. More recently, Pindak *et al.* [56] attached a torsional oscillator to the bottom of "soap-bubble" films as few as four molecular layers thick and measured their shear moduli!

Grimes and Adams [57] made measurements on electrons pinned to the surface of helium by a positively charged capacitor plate (Fig. 2.3(b)). The measurements are consistent [58] with a kind of Wigner crystallization of the electrons into a triangular lattice with a lattice constant of about 500 nm at 0.5 K. The electrons interact classically with a repulsive $1/r$ potential in the plane of the surface and the capacitor plate effectively provides a neutralizing background of positive charge. Pieranski [59] has made colloidal crystals with a comparable lattice constant by trapping polystyrene spheres at a water/air interface. Related two-dimensional colloids have provided a powerful environment in which to investigate two-dimensional melting [5, 6].

Lipid monolayers floating on water (Fig. 2.3(c)) provide another two-dimensional system in which to investigate melting [8, 60–64]. To make contact with the theories discussed in this review, however, one must first exclude transitions triggered by cooperative deformations of the lipid hydrocarbon chains. Such deformations change the density and

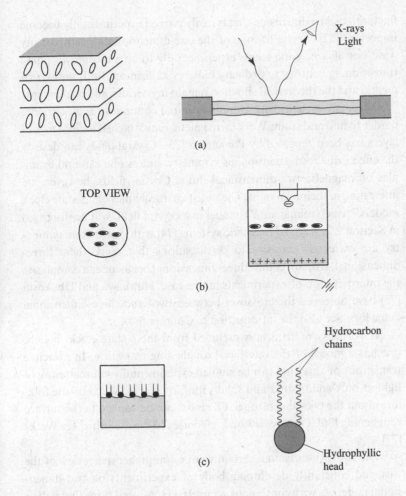

Fig. 2.3. Experiments that probe two-dimensional melting. (a) A few layers of a bulk smectic liquid crystal (shown on the left-hand side) are suspended across a hole in a glass slide, like a flat soap bubble. Order in the orientations and centers of mass of the rod-like nematogen molecules can be probed by scattering of light and X-rays. (b) Electrons can be trapped on the surface of liquid helium by a positively charged capacitor plate. (c) Monolayer films of amphiphillic or fat molecules trapped at an air/water interface.

could in the process melt a crystal in a way we do not wish to consider here.

There is a vast literature [65] on the properties of, say, rare-gas atoms physisorbed onto substrates such as graphite. Incommensurate solids of adsorbate atoms can exist at submonolayer coverages, which appear to melt via continuous transitions [66, 67] into a liquid-like phase. Intriguing questions arise when one considers the transition from incommensurate solids to commensurate crystalline phases, which are in registry with the periodic graphite substrate. Early low-energy electron-diffraction studies of the commensurate–incommensurate transition [68] have been supplemented by high-resolution synchrotron-X-ray-radiation experiments [69, 70]. High-quality synchrotron-X-ray studies of incommensurate melting have also been performed [71].

Melting of smectic liquid crystals into nematic liquids is interesting for a number of reasons. The lower critical dimension is three, so fluctuations in a dimension that is easily probed experimentally become important [72]. Crystallization of the one-dimensional smectic density wave out of a nematic seems experimentally to be a *continuous* phase transition, in contrast to ordinary bulk crystallization. Both the experimental and the theoretical situation remain mysterious and challenging.

Magnetic systems with two-dimensional continuous symmetry are harder to find and study. Work on magnetic ions attached to lipid monolayers has been reported by Pomerantz [73]. Crystal fields can destroy the effects due to the continuous symmetry, unless one can find examples of magnetic two-dimensional fluids. Crystal fields, however, are interesting in their own right. The statistical mechanics of "p-state clock models," which mimic an XY model in a crystal field, will be discussed in Section 2.2. *Layered* magnetic systems [74] with continuous symmetry are extremely sensitive to perturbations that make them three-dimensional. Crossover into three-dimensional behavior can complicate the interpretation of experiments in this case. Hirakawa and Ubukoshi [75] have observed the crossover between two- and three-dimensional behaviors. See also the introduction to Chapter 5.

If vortices are artificially excluded from the p-state clock models, one has a model of the interfacial roughening transition. In principle transitions of this kind can be studied experimentally via scattering of light at bulk solid/liquid and solid/vapor interfaces. Much of the folklore about the two-dimensional XY model can be applied to the surface roughening [76]. For a review and references to the literature, see Weeks [77].

The above discussion is certainly not a comprehensive review of the vast and constantly developing body of experiments on two-dimensional materials with continuous symmetry. However, I hope that it does give non-experts a feel for the flavor and variety of the experiments which stimulate and motivate much of the theory.

2.1.3 Bibliographic note

There are a number of reviews on systems that may exhibit defect-mediated phase transitions. The book by Dash [65] and the review by Kosterlitz and Thouless [78] set the stage for most of the work described here. This survey is itself a revised and expanded version of lectures first presented at the 1980 Enschede Summer School on Fundamental Problems in Statistical Mechanics [79]. Discussions of related experimental and theoretical developments may be found in the proceedings of the NATO Advanced Study Institute on "Ordering in Strongly Fluctuating Condensed Matter Systems" [80], the Ettore Majorana Summer School on "Phase Transitions in Surface Films" [81] and the conference on "Ordering in Two Dimensions" [82]. The material covered in the 1979 reference by Halperin and Nelson [51] has some overlap with this review, although our presentation is different. Halperin has also reviewed melting and liquid-crystal phases in two dimensions [83]. Two-dimensional superfluidity and *XY* magnetism, as well as two-dimensional phase transitions with *discrete* symmetry, have been surveyed by Barber [84]. For reviews of commensurate–incommensurate transitions in physisorbed monolayers, see Villain [85, 86] and Pokrovsky and Talapov [87].

2.2 The *XY* model and superfluidity in two dimensions

2.2.1 Model systems

A collection of n-component, classical spins ($n \geqslant 2$) interacting on a two-dimensional lattice provides a standard example of statistical mechanics with a continuous symmetry. The Hamiltonian for these fixed-length spins ($|\mathbf{S}_i| = 1$) is given by

$$\frac{\mathcal{H}}{k_{\mathrm{B}}T} = -K \sum_{\langle i,j \rangle} \mathbf{S}_i \cdot \mathbf{S}_j, \tag{2.18}$$

where the sum is over nearest-neighbor lattice sites. Analysis of high-temperature series expansions [88–90] suggested that finite-temperature phase transitions occur for these systems, the arguments and rigorous proofs [91] that conventional long-range order is absent notwithstanding. This work did not accurately distinguish between systems with $n \geqslant 3$ and the special case $n = 2$, however.

Such a distinction is clearly indicated in the work of Polyakov [23, 92], who constructed a low-temperature renormalization-group recursion relation for the Hamiltonian (2.18). When it is generalized to $d = 2 + \varepsilon$ dimensions [24, 25], Polyakov's recursion relation becomes, to lowest order in ε and the temperature-like parameter K^{-1},

$$\frac{dK^{-1}(\ell)}{d\ell} = -\varepsilon K^{-1}(\ell) + \frac{n-2}{2\pi} K^{-2}(\ell). \tag{2.19}$$

In contrast to the recursion relation (1.10) for thermally agitated fluids, one now has an upwards parabola in K^{-1} and an *unstable* fixed point at nonzero coupling. Here, $K^{-1}(\ell)$ is the effective temperature after fluctuations with wavelengths shorter than $e^{\ell}a$ (a is the lattice constant and a^{-1} is the ultraviolet cutoff) have been eliminated. For $n \geqslant 3$, one finds a finite-temperature critical fixed point (at $(K^*)^{-1} = 2\pi\varepsilon/(n-2)$), which tends to zero temperature as the dimensionality approaches $d = 2$. In precisely two dimensions, $K^{-1}(\ell)$ always increases toward larger values, indicating that the spins behave paramagnetically at long wavelengths. This is another way of saying that fluctuations have suppressed T_c to zero. Note, however, that $n = 2$ is special, since $dK^{-1}(\ell)/d\ell$ is identically zero in this case. This result, which holds to all orders in K^{-1} [24, 25], suggests that a line of fixed points is necessary to describe spins with an XY internal symmetry.

Focusing in on $n = 2$, it is convenient to rewrite (2.18) as

$$\frac{\mathcal{H}}{k_B T} = -K \sum_{\langle i,j \rangle} \cos(\theta_i - \theta_j), \tag{2.20}$$

where θ_i is the angle which the ith spin makes with some reference axis. A useful approximation is obtained by expanding $\cos(\theta_i - \theta_j)$ to second order in $\theta_i - \theta_j$:

$$\frac{\mathcal{H}}{k_B T} \rightarrow \frac{\mathcal{H}_0}{k_B T} \equiv \text{constant} + \frac{1}{2} K \sum_{\langle i,j \rangle} (\theta_i - \theta_j)^2. \tag{2.21}$$

As shown, for example, by José et al. [93], the spin correlation function (for a square lattice) is then

$$C(\mathbf{r}_i - \mathbf{r}_j) = \langle \mathbf{S}_i \cdot \mathbf{S}_j \rangle$$

$$= \exp\left(-\frac{1}{K} \int_{-\pi}^{\pi} \frac{dq_x}{2\pi} \int_{-\pi}^{\pi} \frac{dq_y}{2\pi} \frac{(1 - e^{i q \cdot (\mathbf{r}_i - \mathbf{r}_j)/a_0})}{(4 - 2\cos q_x - 2\cos q_y)}\right). \tag{2.22}$$

For large $\mathbf{r}_{ij} = \mathbf{r}_i - \mathbf{r}_j$, one finds at any temperature the algebraic decay advertised in Section 2.1:

$$C(\mathbf{r}_{ij}) \sim \mathbf{r}_{ij}^{-\eta(T)}, \tag{2.23}$$

where

$$\eta(T) = 1/(2\pi K). \tag{2.24}$$

Of course, important elements of the physics may have been lost in the passage to (2.20) from (2.21)!

A very analogous picture emerges from a simple description of superfluid ^{4}He films. Assuming that the normal component of the film is pinned to the substrate, we write the energy of a superfluid-velocity fluctuation as

$$\mathcal{H} = \frac{1}{2} \rho_s^0(T) \int d^2r \, |\mathbf{v}_s(\mathbf{r})|^2, \tag{2.25}$$

where $\rho_s^0(T)$ is the superfluid density.

Galilean invariance of the film-and-substrate system requires that $\mathbf{v}_s(\mathbf{r})$ be simply related to the gradient of the phase of the wave function of the condensate for slowly varying disturbances,

$$\mathbf{v}_s(\mathbf{r}) = \frac{\hbar}{m} \nabla \theta(\mathbf{r}), \tag{2.26}$$

where m is the mass of a ^{4}He atom. Note that the true mass of helium, not some effective mass due to the substrate, appears in (2.26). We may then write

$$\frac{\mathcal{H}}{k_B T} = \frac{1}{2} K \int d^2 \mathbf{r} (\nabla \theta)^2, \tag{2.27}$$

where

$$K = \frac{\hbar^2}{m^2} \frac{\rho_s^0(T)}{k_B T}. \tag{2.28}$$

Evidently, this is a continuum version of (2.21), provided that we identify the phase of the condensate with the angle a spin makes at site i. When it is combined with Eq. (2.28), this identification allows results for superfluids to be transcribed into corresponding conclusions for magnetic systems. As we shall see, vortices can have important thermodynamic consequences at finite temperatures. Vortices in superfluids are isolated points where the superflow is not strictly irrotational,

$$\nabla \times \mathbf{v}_s(\mathbf{r}) \neq 0. \tag{2.29}$$

They are characterized by a net circulation, which according to (2.26) must be quantized in units of $2\pi \hbar / m$. Thus,

$$\oint \mathbf{v}_s \cdot d\ell = \frac{2\pi \hbar}{m} s, \qquad s = 0, \pm 1, \ldots, \tag{2.30}$$

where the contour surrounds a vortex of charge s. It is easy to see from (2.30) that $\mathbf{v}_s(\mathbf{r})$ falls off like $1/r$ far from a vortex. One sees then from Eq. (2.25) that the energy of an isolated vortex diverges logarithmically with the size of the system. There is also a logarithmic cost for vortex configurations of spins in the XY model, which are shown for $s = \pm 1$ in Fig. 2.4.

To determine the effect of an assembly of vortices on the statistical mechanics, we decompose $\mathbf{v}_s(\mathbf{r})$ into two parts [28]:

$$\mathbf{v}_s(\mathbf{r}) = \frac{\hbar}{m} \nabla \phi(\mathbf{r}) + \frac{2\pi \hbar}{m} (\hat{\mathbf{z}} \times \nabla) \int d^2 r' \, n(\mathbf{r}') G(\mathbf{r}, \mathbf{r}'), \tag{2.31}$$

where $\phi(\mathbf{r})$ is a nonsingular phase function and $\hat{\mathbf{z}}$ is a unit vector perpendicular to the plane of the film. The quantity $n(\mathbf{r})$ is the vortex charge density, while the Green function $G(\mathbf{r}, \mathbf{r}')$ satisfies

$$\nabla^2 G(\mathbf{r}, \mathbf{r}') = \delta(\mathbf{r} - \mathbf{r}'). \tag{2.32}$$

For $|\mathbf{r} - \mathbf{r}'|$ large and both points far from boundaries, we have

$$G(\mathbf{r}, \mathbf{r}') = \frac{1}{2\pi} \ln\left(\frac{|\mathbf{r} - \mathbf{r}'|}{a}\right) + C, \tag{2.33}$$

Fig. 2.4. Vortex configurations of spins in a two-dimensional *XY* model on a square lattice. The upper configurations have charge + 1, while the lower spin textures have charge − 1. The configurations on the right-hand side are related to those on the left-hand side by a 90° rotation.

$S = +1$

$S = -1$

where a is an ultraviolet cutoff of order the vortex-core diameter and C is a constant associated with the core energy. It is easily checked from (2.31) and (2.32) that

$$\nabla \times \boldsymbol{v}_{s}(\mathbf{r}) = \frac{2\pi\hbar}{m} n(\mathbf{r})\hat{\mathbf{z}},\tag{2.34}$$

as required by (2.30).

On inserting Eq. (2.31) into (2.25), we find that the reduced Hamiltonian $\mathcal{H}/(k_{B}T)$ breaks into two pieces,

$$\frac{\mathcal{H}}{k_{B}T} = \frac{1}{2}K \int d^{2}r (\nabla\phi)^{2} + \frac{\mathcal{H}_{v}}{k_{B}T},\tag{2.35}$$

with

$$K = \hbar^{2}\rho_{s}^{0}(T)/(m^{2}k_{B}T),\tag{2.36}$$

and where the vortex part is

$$\frac{\mathcal{H}_{v}}{k_{B}T} = 2\pi^{2}K \int d^{2}r \int d^{2}r_{1} \int d^{2}r_{2}\, n(\mathbf{r}_{1}) n(\mathbf{r}_{2}) \nabla G(\mathbf{r}, \mathbf{r}_{1}) \cdot \nabla G(\mathbf{r}, \mathbf{r}_{2}).\tag{2.37}$$

There is no term linear in $\phi(\mathbf{r})$, since Eq. (2.31) with $\nabla\phi = 0$ is in fact a functional minimum of the Hamiltonian (2.25), with the constraint of a given number of vortices at fixed positions. Upon integrating by parts in (2.37) and using (2.32), one finds a logarithmic interaction between the vortex charges,

$$\frac{\mathcal{H}_v}{k_B T} = -\pi K \int\!\!\int_{|r-r'|>a} d^2r\, d^2r'\, n(\mathbf{r}) n(\mathbf{r}') \ln\left(\frac{|\mathbf{r}-\mathbf{r}'|}{a}\right)$$

$$+ \frac{E_c}{k_B T} \int d^2r\, n^2(\mathbf{r}), \tag{2.38}$$

where the core energy E_c is a positive constant proportional to C. In order to integrate by parts in (2.37) we have assumed that $\nabla G(\mathbf{r}, \mathbf{r})_2$ vanishes sufficiently rapidly at infinity. This will certainly be the case for a charge-neutral collection of vortices confined to a region of size L, which will eventually be taken arbitrarily large.

To calculate averages with (2.35), one must average over smoothly varying velocity fields determined by $\phi(\mathbf{r})$ and over distinct vorticity complexions that satisfy a charge-neutrality condition,

$$\int d^2r\, n(r) = 0.$$

If vortices can be neglected, the first term of (2.35) gives rise to the same algebraic decay of correlations as was found from (2.22).

In the work of José *et al.* [93], duality transformations and an approximate renormalization procedure due to Migdal [16] justified an analogous lattice vortex Hamiltonian for the *XY* model of magnetism, namely

$$\frac{\mathcal{H}_v}{k_B T} = -\pi K \sum_{r\neq r'} s(\mathbf{r}) s(\mathbf{r}') \ln(|\mathbf{r}-\mathbf{r}'|/a) + \frac{E_c}{k_B T} \sum_r s^2(\mathbf{r}). \tag{2.39}$$

Here, the summations extend over a lattice of sites dual to the lattice which defines (2.21) and the $s(\mathbf{r})$ assume integer values, $s(\mathbf{r}) = 0, \pm 1, \ldots$. The "spin-wave" approximation (2.22), of course, neglects vortices in the spin field. The lattice-Coulomb-gas Hamiltonian (2.39) is often used to impose a convenient ultraviolet cutoff on (2.38).

The combination of phase and vortex degrees of freedom embodied in Eq. (2.35) is the starting point for further analysis of two-dimensional superfluidity and *XY* magnetism.

2.2.2 Landau theory and amplitude fluctuations

It is important to understand the connection between phase-fluctuation–vortex models of phase transitions and the more traditional Landau approach, which is often successful near four dimensions. In Landau theory, one starts with a coarse-grained free energy of the form [94]

$$F = \int d\mathbf{r} \left(\frac{1}{2} |\nabla \psi|^2 + \frac{1}{2} a |\psi|^2 + b |\psi|^4 \right), \tag{2.40}$$

where a changes sign at the Landau critical temperature and b is a positive constant. In superfluids, the complex field $\psi(\mathbf{r})$ is related to the wave function of the condensate; in magnets, one identifies $\psi(\mathbf{r})$ with a two-dimensional magnetization vector $\mathbf{S}(\mathbf{r})$ via

$$\psi(\mathbf{r}) = S_x(\mathbf{r}) + i S_y(\mathbf{r}). \tag{2.41}$$

The probability of a particular configuration $\{\psi(\mathbf{r})\}$ is taken to be proportional to $e^{-F/(k_B T)}$.

In two dimensions, one expects that T_c will be depressed well below its Landau value, leading to large, negative values of a, even above the true critical temperature. It is then convenient to rewrite (2.40) in the form

$$F = \text{constant} + \int d^2 r \left[\frac{1}{2} |\nabla \psi|^2 + b \left(|\psi|^2 - \frac{|a|}{4b} \right)^2 \right], \tag{2.42}$$

which shows that $|\psi|^2$ becomes trapped in a steep minimum at $|a|/(4b)$ in the limit of large b. In the limit $b \to \infty$, with $|a|/b$ fixed, we can set

$$\psi(\mathbf{r}) = \frac{1}{2} \sqrt{|a|/b} \, e^{i\theta(\mathbf{r})}$$

and recover a free energy that resembles the Hamiltonian (2.27) with

$$K = |a|/(4b k_B T). \tag{2.43}$$

The role of amplitude fluctuations for large but *finite* b can be studied perturbatively in a low-temperature expansion [95, 96]. One finds that the dominant effect at all temperatures below T_c is a slight renormalization of the exponent $\eta(T)$ appearing in Eq. (2.24). In this sense, perturbative effects due to amplitude variations can be absorbed into a redefinition of K in analyses of the long-wavelength physics. Similar conclusions were reached by Amit *et al.* [97] for n-vector spin models with $n \geq 3$ near two dimensions.

The vortices introduced in the previous subsection may be viewed as a nonperturbative manifestation of amplitude fluctuations. Since the phase is undefined at the center of a vortex, it is as if the amplitude of the order parameter vanishes there. Vortices do not appear in a systematic power-series expansion in T, since they occur with a probability proportional to $e^{-E_c/(k_B T)}$, where E_c is the core energy.

Kosterlitz and Thouless [27] identified vortex-pair excitations as the important fluctuations out of a state with quasi-long-range order in two-dimensional superfluids and XY models (see Fig. 2.5). If it becomes preferable for such a pair to separate on average, one expects algebraic decay to be replaced by the exponential decay of correlations present in a normal fluid or paramagnet. The energy of an unbound isolated vortex occupying a region of size L is easily found to be

$$E = \pi k_B T K \ln(L/a). \tag{2.44}$$

There are $(L/a)^2$ places to put such a vortex, so the entropy is

$$S \approx 2k_B \ln(L/a). \tag{2.45}$$

The free energy $F = E - TS$ changes sign at a temperature such that

$$K(T_c) \approx 2/\pi. \tag{2.46}$$

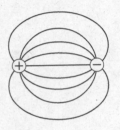

Fig. 2.5. Vortex-pair excitation, which upon unbinding would break up order in superfluid films. Lines of constant condensate phase are shown.

Equation (2.46) is often stated as an *inequality* since other mechanisms (perhaps leading to a first-order transition) could preempt the vortex unbinding at a lower temperature. In terms of the coupling K appropriate to a two-dimensional superfluid (Eq. (2.36)), we then have the condition

$$\frac{\hbar^2 \rho_s^0(T)}{m^2 k_B T} \gtrsim \frac{2}{\pi}. \tag{2.47}$$

Kosterlitz and Thouless [28] argued that interactions with the substrate could be accounted for by replacing the mass of the helium atom in (2.47) by an effective mass m^*. As we shall see, Eq. (2.47) is in fact an exact *equality* at the vortex-unbinding temperature, provided that one uses instead a renormalized superfluid density $\rho_s^R(T)$ [98].

Landau theories of two-dimensional melting will be discussed in Section 2.4. One again finds a description in terms of phase fluctuations, with amplitude fluctuations appearing only in the form of singularities called dislocations. An entropy argument similar to that sketched above for vortices was applied by Kosterlitz and Thouless [27] to dislocations in a two-dimensional solid. It, too, predicts a finite transition temperature and turns out to be an exact consequence of more sophisticated theories.

2.2.3 The Kosterlitz recursion relations and the universal jump

Renormalization-group recursion relations are often constructed by working directly with the partition function [15]. However, any convenient correlation function can in fact be used. Here we focus explicitly on the correlation which gives the superfluid density $\rho_s^R(T)$. This provides a particularly transparent derivation of recursion relations first derived by Kosterlitz [99] and demonstrates that there is a universal jump discontinuity in the superfluid density [98].

To define $\rho_s(T)$, we follow Hohenberg and Martin [100] and consider autocorrelations of a superfluid momentum current,

$$g_s(r) = \frac{\delta \mathcal{H}}{\delta v_s} = \rho_s^0 v_s(r), \tag{2.48}$$

where $\mathcal{H}$ is given by (2.25). Recall that ρ_s^0 is a "bare" superfluid density unrenormalized by vortex excitations. The quantity we want to consider is

$$C_{\alpha\beta}(\mathbf{q}, K, y) \equiv \left(\frac{\hbar}{mk_B T}\right)^2 \langle \hat{g}_s^{\alpha}(\mathbf{q}) \hat{g}_s^{\beta}(-\mathbf{q}) \rangle, \tag{2.49}$$

where $\hat{g}_s^{\alpha}(\mathbf{q})$ is the Fourier transform of the αth component of $g_s(r)$. To define the renormalized superfluid density properly, it is important to decompose $C_{\alpha\beta}$ into tranverse and longitudinal parts:

$$C_{\alpha\beta} = A(q) \frac{q_\alpha q_\beta}{q^2} + B(q) \left(\delta_{\alpha\beta} - \frac{q_\alpha q_\beta}{q^2} \right), \tag{2.50}$$

where $A(q)$ and $B(q)$ depend on the thermodynamic parameters K and y. In an isotropic normal liquid, one would have $A(q) = B(q)$ at long wavelengths. The renormalized superfluid density $\rho_s^R(T)$ is given by the difference between these quantities as q tends to zero [100],

$$K_R \equiv \frac{\hbar^2 \rho_s^R(T)}{m^2 k_B T} = \lim_{q \to 0} [A(q) - B(q)]. \tag{2.51}$$

See Appendix A of Chapter 6 for a derivation of this important result. The velocity field (2.31) is already decomposed into longitudinal and transverse parts. Upon Fourier transforming and averaging over the smoothly varying phase function $\phi(r)$, we find that $C_{\alpha\beta}$ takes the form (2.50), with

$$A(q) = K, \tag{2.52a}$$

$$B(q) = \frac{4\pi^2 K^2}{q^2} \langle \hat{n}(\mathbf{q}) \hat{n}(-\mathbf{q}) \rangle. \tag{2.52b}$$

Here, $\hat{n}(\mathbf{q})$ is the Fourier transform of $n(r)$ and the average in (2.52b) is to be carried out over the vortex part of the reduced Hamiltonian (2.35). Although $\langle n(r)n(0) \rangle$ falls off like a power law when vortices are bound in pairs, the decay is always rapid enough that the second moment exists. Thus we can perform the moment expansion out to order q^2,

$$\langle n(\mathbf{q})n(-\mathbf{q}) \rangle = C_0 + C_2 q^2 + O(q^2). \tag{2.53}$$

The charge-neutrality condition means that $C_0 = 0$, while, assuming that the charge–charge correlation function is isotropic, we have

$$C_2 = \frac{1}{4} \int d^2 r \, r^2 \langle n(\mathbf{r})n(0) \rangle. \tag{2.54}$$

Combining everything together, we obtain from (2.51) the basic formula

$$K_R = K - \pi^2 K^2 \int d^2 r \, r^2 \langle n(\mathbf{r})n(0) \rangle. \tag{2.55}$$

Equation (2.55) can be used as the basis for a perturbative calculation of K_R, in powers of the "vortex fugacity"

$$y = e^{-E_c/(k_B T)}. \tag{2.56}$$

Using the lattice-Coulomb-gas Hamiltonian (2.39) for convenience and taking an angular average, we find

$$K_R^{-1} = K^{-1} + 4\pi^3 y^2 \int_a^\infty \frac{dr}{a} \left(\frac{r}{a}\right)^{3-2\pi K} + O(y^4). \tag{2.57}$$

At low temperatures, Eq. (2.57) provides an exponentially small $(y^2 \sim e^{-2E_c/(k_B T)})$ correction to the value of K_R^{-1}. When $K \leqslant 2/\pi$, however, the integral becomes infrared divergent and perturbation theory breaks down.

We can overcome this difficulty with a renormalization procedure first used by José *et al.* [93]. The idea is to break the integral in (2.57) into two parts,

$$\int_a^\infty \to \int_a^{ae^\delta} + \int_{ae^\delta}^\infty, \tag{2.58}$$

and evaluate only the nonsingular contribution of small r. This can be done order by order in y, even if the perturbation theory formally diverges. Equation (2.57) becomes

$$K_R^{-1} = (K')^{-1} + 4\pi^3 y^2 \int_{ae^\delta}^\infty \frac{dr}{a} \left(\frac{r}{a}\right)^{3-2\pi K} + \cdots, \tag{2.59a}$$

where

$$(K')^{-1} = K^{-1} + 4\pi^3 y^2 \int_a^{ae^\delta} \frac{dr}{a} \left(\frac{r}{a}\right)^{3-2\pi K} + O(y^4). \tag{2.59b}$$

In Section IV.A of [101], it is shown explicitly that an $O(y^4)$ contribution to the vortex-charge correlation function appearing in (2.54) causes replacement of K by K' in (2.59a). This replacement is correct to $O(y^2)$ in any case. Upon rescaling the integral to restore the original cutoff, we find a perturbative expression for K_R^{-1} of precisely the same form as (2.57),

Fig. 2.6. Renormalization flows arising from the Kosterlitz recursion relations. The shaded domain of attraction of the fixed line at $y(\ell) \equiv 0$ is a superfluid. A locus of initial conditions is shown as a dotted line. The superfluid is bounded by an incoming separatrix terminating at $K^{-1}(\ell) = 2/\pi$. Quantities above T_c can be calculated by first evaluating them on the vertical solid line at high temperature, using a kind of Debye–Hückel theory. One then integrates the recursion relations backward, with this information as a boundary condition.

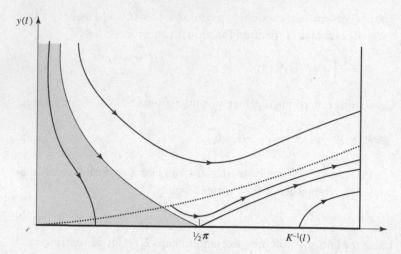

$$K_R^{-1} = (K')^{-1} + 4\pi^3(y')^2 \int_a^\infty \frac{dr}{a} \left(\frac{r}{a}\right)^{3-2\pi K'} + O(y'^4), \tag{2.60a}$$

with

$$y' = e^{(2-\pi K)\delta} y. \tag{2.60b}$$

It is convenient to build up a large increase in the core diameter from a to ae^ℓ by repeating this transformation many times with δ infinitesimal. In this way, one arrives at differential renormalization-group equations for effective couplings $K(\ell)$ and $y(\ell)$:

$$\frac{dK^{-1}(\ell)}{d\ell} = 4\pi^2 y^2(\ell) + O[y^4(\ell)], \tag{2.61a}$$

$$\frac{dy(\ell)}{d\ell} = [2 - \pi K(\ell)] y(\ell) + O[y^3(\ell)], \tag{2.61b}$$

The invariance of the perturbation series for K_R^{-1} may now be written [98]

$$K_R(K,y) = K_R(K(\ell), y(\ell)). \tag{2.62}$$

Equations (2.61) are the famous Kosterlitz recursion relations, originally derived by him using a method developed by Anderson and Yuval for the Kondo problem [102]. The Hamiltonian trajectories they generate in the (K^{-1}, y)-plane are shown in Fig. 2.6, together with a temperature-dependent locus of initial conditions. Hamiltonians to the left of the incoming separatrix renormalize into the line of fixed points at $y = 0$, which describes the low-temperature phase. At higher temperatures, $y(\ell)$ eventually becomes large, indicating that vortices are important even at very long wavelengths. One then expects that order-parameter correlations decay exponentially,

$$\langle \psi^*(\mathbf{r}) \psi(0) \rangle \sim e^{-r/\xi_+}, \tag{2.63}$$

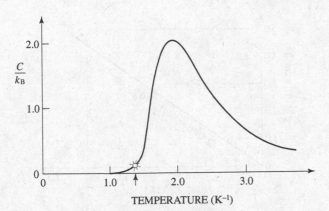

TEMPERATURE (K^{-1})

Fig. 2.7. The approximate contribution of vortices to the specific heat of an XY model. There is only an unobservable essential singularity at T_c (open circle) and a maximum above this temperature. The precise height, shape and position of this maximum (which occurs above T_c due to the entropy liberated by an increasing number of unbound vortices) is *nonuniversal*.

where the correlation length ξ_+ is related to the density of free vortices n_f:

$$n_f(T) \sim \xi_+^{-2}(T). \tag{2.64}$$

The behavior of the correlation length above T_c can be extracted straightforwardly from the recursion relations (2.61) and turns out [99] to be

$$\xi_+(T) \sim e^{b'/|T-T_c|^{1/2}}, \tag{2.65}$$

where b' is a nonuniversal constant. The superfluid density $\rho_s(T)$ approaches a finite value as T goes to T_c from below, preceded by a square-root cusp [78, 99],

$$\rho_s^R(T) \approx \rho_s^R(T_c^-)[1 + b(T - T_c)^{1/2}], \tag{2.66}$$

with b nonuniversal. As observed by Ambegaokar *et al.* [103], there is a universal relation between b and b':

$$b' = 2\pi/b. \tag{2.67}$$

The singular part of the specific heat has only an essential singularity,

$$C_{\text{sing}}(T) \sim \xi_+^{-2}, \tag{2.68}$$

which will be undetectable when it is measured against a regular background. The qualitative behavior of the specific heat was determined by Berker and Nelson [104], who also studied superfluidity in ^{4}He films with finite concentrations of ^{3}He impurities and vacancies. As shown in Fig. 2.7, there is a maximum above T_c, due to the entropy liberated by a gradual dissociation of pairs of vortices. Only pairs with very large separations become unbound just above T_c. This calculation has been improved by Solla and Riedel [105]. A direct Monte Carlo simulation of the specific heat of an XY model has been carried out by Tobochnik and Chester [106].

Fig. 2.8. Renormalized superfluid density $\rho_s^R(T)$ curves simulating the effect of varying film height, substrate, etc., in a real experiment. Although the curves are different, they all terminate in a line with a *universal* slope, 3.491×10^{-9} g cm^{-2} K^{-1}.

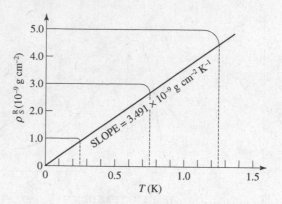

Since vortices become unimportant at long wavelengths below T_c ($y(\ell) \rightarrow 0$), there is algebraic decay of correlations of the form (2.24), with K replaced by K_R. Precisely at T_c, there is a logarithmic correction to the result $\eta(T_c) = \frac{1}{4}$ [99],

$$\langle \psi^*(\mathbf{r})\psi(\mathbf{r})\rangle \sim \frac{\ln^{1/8}(r/a)}{r^{1/4}}. \tag{2.69}$$

Perhaps the most important experimental consequence of the Kosterlitz–Thouless theory is the prediction of a *universal* jump discontinuity in the superfluid density. This is a direct consequence of the relation (2.62) and the renormalization-group flows shown in Fig. 2.6. Below T_c, $y(\ell)$ tends to zero for large ℓ and we have

$$K_R(K,y) = \lim_{\ell \to \infty} K_R(K(\ell), y(\ell))$$
$$= \lim_{\ell \to \infty} K(\ell) \qquad (T \leqslant T_c). \tag{2.70}$$

Since it is clear from Fig. 2.6 that this limit is just $2/\pi$ at T_c, we have

$$\lim_{T \to T_C^-} K_R(K,y) = \lim_{T \to T_C^-} \frac{\hbar^2 \rho_s^R(T)}{m^2 k_B T} = \frac{2}{\pi}, \tag{2.71}$$

independent of the way in which the locus of initial conditions crosses the incoming separatrix. It is instructive to produce plots of $\rho_s(T)$ against T for various loci of initial conditions, as shown in Fig. 2.8. Quantities such as the core energy and the unrenormalized value of ρ_s could be varied experimentally by changing the film thickness or substrate. One obtains a sequence of curves with jump discontinuities and T_c values all falling on a line with slope

$$\frac{\rho_s(T_c^-)}{T_c^-} = \frac{2m^2}{\pi \hbar^2 k_B T} \approx 3.491 \times 10^{-9} \text{ g cm}^{-2}\text{K}^{-1}. \tag{2.72}$$

At this point, we have demonstrated the universality of $\rho_s(T_c^-)/T_c^-$ only to order $y^2(\ell)$. Higher-order corrections to Eq. (2.61) will not change this result, provided that (2.62) is preserved. However, this relation between K_R evaluated for bare and renormalized parameters is just the Josephson relation [107], specialized to two dimensions. To see this, note that the Josephson relation amounts to a connection between $\rho_s^R(T)$ and a suitable defined transverse correlation length $\xi_\perp$ in d dimensions [108],

$$K_R = \frac{\hbar^2 \rho_s^R(T)}{m^2 k_B T} \equiv \xi_\perp^{2-d}. \tag{2.73}$$

Denoting by $\{y_i\}$ a collection of Hamiltonian parameters in addition to K and y, which are "irrelevant" along the line of fixed points in Fig. 2.6, we have

$$K_R(K, y, \{y_i\}) = e^{(2-d)\ell} K_R(K(\ell), y(\ell), \{y_i(\ell)\}). \tag{2.74}$$

This generalization of Eq. (2.62) follows from (2.73) and the way in which correlation lengths transform under a renormalization group [15]. It may also be derived by dimensional analysis from (2.48), for any renormalization procedure that does not eliminate the zero-wavevector component of $v_s(\mathbf{r})$. Since $v_s(\mathbf{r})$ is expressible at the gradient of a phase at long wavelengths, it must scale like an inverse length. José et al. [109] have argued that large numbers of Hamiltonian couplings consistent with an XY symmetry are irrelevant below the vortex-unbinding temperature. If all such y_i are irrelevant in this way, Fig. 2.6 accurately describes the Hamiltonian flows for large ℓ and one is led quite generally to the universal prediction (2.72). This prediction has been confirmed experimentally [43, 45, 110, 111] for a variety of thicknesses of ^{4}He film, substrates and concentrations of ^{3}He impurity.

Nelson and Kosterlitz [98] actually used a different definition of $K_R(T)$, one that does not distinguish between the longitudinal and transverse parts of the correlation function. The same definition is quoted in the review article by Kosterlitz and Thouless [78]. Although this formula is correct when it is used to obtain $K_R(T)$ to $O(y^2)$, it breaks down at higher orders. Since the general arguments for the universal jump given in the previous paragraph apply to the correctly defined superfluid density $\rho_s^R(T)$, the results are unchanged. Nevertheless, it seems worth emphasizing that it is Eq. (2.51) – as originally given by Hohenberg and Martin [100] – which should be used in higher-order calculations. Similar conclusions have been reached by Minnhagen and Warren [112].

2.2.4 The wavevector-dependent superfluid density above T_c

The unstable Hamiltonian flows to the right of the incoming separatrix in Fig. 2.6 are strongly suggestive of a transition to a normal fluid. To see that this is really the case, it is necessary to deal at least approximately with effects due to free vortices. Here, we determine how free vortices affect the superfluid density at finite wavelengths above T_c. Unbound vortices cause this quantity to vanish at long wavelengths in a normal fluid. The method of calculation gives results analogous to those obtained in the Debye–Hückel theory of charged plasmas and was used to calculate the specific heat above T_c by Berker and Nelson [104].

Above T_c, one expects small patches of superfluid helium broken up by a background of normal material. It should be possible to define a wavevector-dependent superfluid density $\rho_s^R(q, T)$ that is finite for wavelengths smaller than a typical blob of superfluid, but vanishes at small q. To define $\rho_s^R(q, T)$, we simply use a finite-wavevector version of (2.54):

$$K_R(q, K, y) \equiv \frac{\hbar^2 \rho_s^R(q, T)}{m^2 k_B T} = A(q) - B(q), \tag{2.75}$$

where $A(q)$ and $B(q)$ are given by (2.52).

To calculate $\rho_s^R(q, T)$, we need a q-dependent generalization of the matching relation (2.62), which is easily found to be

$$K_R(q, K, y) = K_R(e^\ell q, K(\ell), y(\ell)). \tag{2.76}$$

Effects of *bound* vortices can be taken into account by integrating the recursion relations (2.61) until $\ell = \ell^*$ such that the renormalized correlation length equals the core diameter a. Since the transformation law for the correlation length [15] is

$$\xi_+(K, y) = e^\ell \xi_+(K(\ell), y(\ell)), \tag{2.77}$$

the required value $\ell = \ell^*$ is

$$\ell^* = \ln(\xi_+/a), \tag{2.78}$$

and Eq. (2.75) becomes

$$K_R(q, K, y) = K_R(q\xi_+/a, K(\ell^*), y(\ell^*)). \tag{2.79}$$

The renormalized couplings $K(\ell^*)$ and $y(\ell^*)$ are finite constants of order unity somewhere in the right-hand half of Fig. 2.6. All remaining vortices are unbound and can be treated by the Debye–Hückel theory. Upon evaluating the right-hand side of Eq. (2.79), we have

$$K_R(q, K, y) = K(\ell^*) - \frac{4\pi^2 K^2(\ell^*)}{(q\xi_+/a)^2} \left\langle \hat{n}\left(\frac{q\xi_+}{a}\right) \hat{n}\left(\frac{-q\xi_+}{a}\right) \right\rangle_{\ell^*}, \tag{2.80}$$

where the average is to be evaluated in a Coulomb-gas ensemble with renormalized couplings $K(\ell^*)$ and $y(\ell^*)$.

To evaluate the average in (2.80) we integrate, rather than sum over the vortex charges. This is a reasonable approximation when there is a high density of free vortex charges. Since the Coulomb-gas Hamiltonian (2.38) in Fourier space is

$$\frac{\mathcal{H}_v(\ell^*)}{k_B T} = \frac{1}{2} \int d^2k \left(\frac{4\pi^2 K(\ell^*)}{k^2} + B(\ell^*) \right) \hat{n}(k)\hat{n}(-k), \tag{2.81}$$

the desired average is

$$\left\langle \hat{n}\left(\frac{q\xi_+}{a}\right) \hat{n}\left(\frac{-q\xi_+}{a}\right) \right\rangle_{\ell^*} = \frac{1}{(4\pi^2 K(\ell^*)/q^2) + B(\ell^*)}. \tag{2.82}$$

The quantity $B(\ell^*)$ is related to $y(\ell^*)$; its precise value is unimportant. Our final result for the wavevector-dependent superfluid density above T_c is then

$$K_R(q, T) = \frac{B(\ell^*)K(\ell^*)}{B(\ell^*) + 4\pi^2 K(\ell^*)a^2/(q^2\xi_+^2)}. \tag{2.83}$$

Note that $K_R(q, T)$ does indeed vanish for small wavevectors, as one would expect in a normal fluid:

$$K_R(q, T) \sim (q\xi_+)^2. \tag{2.84}$$

It is sometimes useful to cast the theory of two-dimensional superfluidity into a dielectric formalism [28, 103, 112, 113]. The dielectric function $\varepsilon(q, T)$ of the vortices is then defined in terms of the ratio of K_R and K:

$$\varepsilon(q, T) \equiv \frac{K}{K_R(q, T)}. \tag{2.85}$$

From Eq. (2.85) we see that $\varepsilon(q, T)$ takes the well-known "Thomas–Fermi" or "Debye–Hückel" form,

$$\varepsilon(q, T) = \frac{1}{1 + (q_s/q)^2} \tag{2.86}$$

with

$$q_s \sim \xi_+^{-1}. \tag{2.87}$$

2.2.5 Symmetry-breaking fields and clock models

Two-dimensional XY spins on periodic lattices are subject to symmetry-breaking crystal fields. In the continuum limit, we can model crystal field effects in the limit of classical, fixed-length spins by the dimensionless Hamiltonian,

$$\frac{\mathcal{H}}{k_B T} = \frac{1}{2} K \int d^2r |\nabla\theta|^2 - h_p \int d^2r \cos(p\theta). \tag{2.88}$$

The first term represents the usual, rotationally invariant, ferromagnetic nearest-neighbor exchange coupling; the strength h_p of the anisotropic part is determined by much weaker dipole–dipole interactions. An Ising-like anisotropy would be represented by $p = 2$, whereas the case $p = 3$ has the symmetry of the three-state Potts model. The case $p = 1$ simply represents a uniform magnetic field. XY magnets on square and triangular lattices should be characterized by $p = 4$ and 6, respectively. Two sublattice *antiferromagnets* in a uniform p'-fold crystal field will be described at long wavelengths by (2.88) with $p = p'$ for p' even and $p = 2p'$ for p' odd. The Hamiltonian (2.88) is often called a "clock model" in the rather unphysical limit $h_p \to \infty$, since the only allowed states consist of spins pointing in p discrete directions symmetrically distributed around a circle.

Physically, one might expect p-fold anisotropies to become unimportant for fixed h_p in the limit of large p. At sufficiently low temperatures, however, the system will surely lock into one of the p directions favored by the symmetry-breaking perturbations. More detailed information can be extracted from a duality theorem [93]. The theorem holds provided that we replace $\exp[h_p \cos(p\theta)]$ in the partition function by a periodic function with the same symmetry [114]:

$$e^{h_p \cos[p\theta(r)]} \to \sum_{m(r)=-\infty}^{\infty} e^{ipm(r)\theta(r)} e^{(\ln y_p)m^2(r)}. \tag{2.89}$$

Since terms with $|m(r)| > 1$ contribute negligibly for small y_p, one has the identification

$$y_p \approx \frac{1}{2} h_p \tag{2.90}$$

in this limit. According to the hypothesis of "universality," the precise form of the interaction should be immaterial, provided that the symmetry is preserved. The XY partition function with a p-fold symmetry-breaking field becomes

$$Z = \sum_{\{m(r)\}} \int \mathcal{D}\theta \exp\left[\int d^2r\left(-\frac{1}{2} K|\nabla\theta|^2 + ipm\theta + (\ln y_p)m^2\right)\right], \tag{2.91}$$

where, in addition to a functional integration over $\theta(\mathbf{r})$, we must carry out summations over sets of integers $\{m(\mathbf{r})\}$ associated with each point $\mathbf{r}$.

For small y_p, excitations with $m \neq 0$ become very unlikely and vortices can be inserted explicitly as if there were no symmetry-breaking field. By performing a decomposition that is equivalent to Eq. (2.31),

$$\theta(\mathbf{r}) = \phi(\mathbf{r}) + \sum_{\mathbf{r}' \neq \mathbf{r}} n(\mathbf{r}') \tan^{-1} \left(\frac{y - y'}{x - x'} \right), \tag{2.92}$$

where $n(\mathbf{r}')$ is the vortex charge, we can integrate out the smoothly varying phase function $\phi(\mathbf{r})$. It is convenient to consider a lattice of possible positions for vortices and to perform a similar discretization of the functional integral in Eq. (2.91). One finds that Z is proportional to the partition function for two coupled Coulomb gases:

$$Z \propto Z_c(K, y, y_p) = \sum_{\{n(\mathbf{r})\}}' \sum_{\{m(\mathbf{r})\}}' e^{-\mathcal{H}_c/(k_B T)}, \tag{2.93a}$$

where

$$
\begin{aligned}
\frac{-\mathcal{H}_c}{k_B T} &= \pi K \sum_{\mathbf{r} \neq \mathbf{r}'} n(\mathbf{r}) n(\mathbf{r}') \ln \left(\frac{|\mathbf{r} - \mathbf{r}'|}{a} \right) + (\ln y) \sum_{\mathbf{r}} n^2(\mathbf{r}) \\
&\quad + \frac{p^2}{4\pi K} \sum_{\mathbf{r} \neq \mathbf{r}'} m(\mathbf{r}) m(\mathbf{r}') \ln \left(\frac{|\mathbf{r} - \mathbf{r}'|}{a} \right) + (\ln y_p) \sum_{\mathbf{r}} m^2(\mathbf{r}) \\
&\quad + ip \sum_{\mathbf{r} \neq \mathbf{r}'} m(\mathbf{r}) n(\mathbf{r}') \tan^{-1} \left(\frac{y - y'}{x - x'} \right).
\end{aligned}
\tag{2.93b}
$$

The summations in Eqs. (2.93) are constrained by charge neutrality of vortices, $\sum_{\mathbf{r}} n(r) = 0$, and by "neutrality" of the integers $\{m(r)\}$, $\sum_{\mathbf{r}} m(\mathbf{r}) = 0$. The second constraint arises from the functional integral over $\phi(\mathbf{r})$. The contribution of the core energy to the Coulomb energy of the vortex has been parameterized by $y = e^{-E_c/(k_B T)}$. Note the peculiar arctangent coupling between the two scalar Coulomb gases [115]. This representation of the statistical mechanics makes manifest a duality theorem, namely

$$Z_c(K, y, y_p) = Z_c \left(\frac{p^2}{4\pi^2 K}, y_p, y \right). \tag{2.94}$$

The proportionality factors relating Z_c to Z in Eq. (2.93a) are non-singular, so the two partition functions should exhibit identical critical behavior.

Below the vortex-unbinding temperature ($K_c \approx 2/\pi$), vortices are negligible at long wavelengths, so we can set $y \approx 0$. According to the duality relation (2.94), the behavior is then just that of an unperturbed *XY* model with K replaced by $p^2/(4\pi^2 K)$ and y replaced by y_p. Consequently, y_p will grow under a renormalization-group transformation *below* a temperature given by $p^2/(4\pi^2 K_c') \approx 2/\pi$. This instability in

Fig. 2.9. The schematic behaviors of the susceptibility and magnetization of $\cos(p\theta)$ *XY* models with $p>4$. The susceptibility diverges on either side of a band of temperatures ($T_1 \le T \le T_2$) characterized by algebraically decaying correlations with continuously variable exponents. The susceptibility is infinite in the starred intermediate phase. The spontaneous magnetization $M_0(T) = |\langle e^{ip\theta}\rangle|$ is zero above T_2 and vanishes very rapidly upon approaching this temperature from below. C_1, C_2 and C_3 are positive constants.

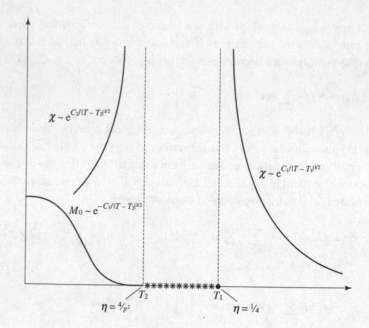

$$\chi \sim e^{C_2/|T-T_2|^{1/2}}$$

$$\chi \sim e^{C_1/|T-T_1|^{1/2}}$$

$$M_0 \sim e^{-C_3/|T-T_2|^{1/2}}$$

$$\eta = 4/p^2 \qquad\qquad T_2 \qquad\qquad\qquad T_1 \qquad\qquad \eta = 1/4$$

$y_p \approx \frac{1}{2} h_p$ presumably means that the magnetization locks into one of the p directions favored by the perturbation. In general, there will be a band of temperatures, determined by

$$\frac{2}{\pi} \le K_R(T) \le \frac{p^2}{8\pi}, \tag{2.95}$$

such that *both* the symmetry-breaking perturbation and vortex excitations are unimportant in this range [93, 116]. Since only Gaussian spin-wave fluctuations remain in this limit, one expects this intermediate phase to be characterized by algebraically decaying order-parameter correlations. On substituting Eq. (2.95) into (2.24), we see that $\eta(T)$ ranges from $\eta = \frac{1}{4}$ at the high-temperature end of this phase to $\eta = 4/p^2$ just above the lock-in transition. Evidently this kind of three-phase behavior is possible only for $p>4$. The temperature dependences of the susceptibility and magnetization in this case are illustrated in Fig. 2.9.

The system is always unstable with respect to either vortex unbinding or the symmetry-breaking perturbation when $p=1, 2,$ or 3. One then expects, respectively, no transition, an Ising-like transition or a Potts transition as a function of temperature. The case $p=4$ requires more study. The only renormalization-group recursion relations to leading order in y and y_p consistent with symmetry, duality and what we know already about the unperturbed *XY* model are [93]

$$\frac{dK^{-1}}{d\ell} = 4\pi^3 y^2 - K^2 p^2 y_p^2, \tag{2.96a}$$

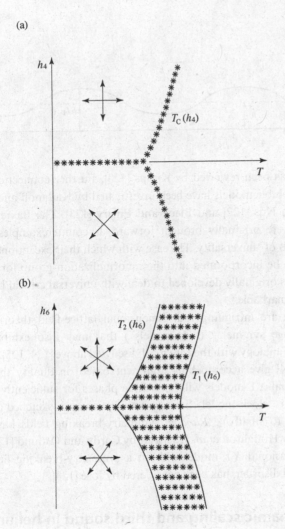

Fig. 2.10. Phase diagrams in the (h_p, T)-plane for (a) $p = 4$ and (b) $p = 6$. The asterisks denote critical points with continuously variable critical exponents.

$$\frac{dy}{d\ell} = (2 - \pi K)y, \tag{2.96b}$$

$$\frac{dy_p}{d\ell} = \left(2 - \frac{p^2}{4\pi K}\right)y_p. \tag{2.96c}$$

For the borderline case $p = 4$, one finds three lines of fixed points, each with its own set of critical exponents! Phase diagrams as a function of temperature and h_p are shown in Fig. 2.10 for $p = 4$ and $p = 6$.

The critical properties along the lines of fixed points for $p = 4$ turn out to be remarkably similar to those of the Baxter model [117]. The behavior near the special point $K = 2/\pi$ and $y = y_4 = 0$ is identical to that of the F model, which was solved exactly by Lieb [118]. A detailed correspondence has been proposed and tested by Kadanoff [119]; the

Fig. 2.11. A third-sound wave in a helium film of average height h_0 and local height $h(\mathbf{r}, t)$.

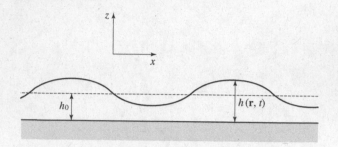

subject has been reviewed by Knops [120]. Further connections with exactly soluble models have been investigated by Kadanoff and Brown [121], den Nijs [122] and Black and Emery [123]. The Baxter and F models were originally brought forward as counterexamples to the hypothesis of universality. The ease with which their exceptional properties can be incorporated into the renormalization-group formalism, which was originally developed to deal with universal critical behavior [15], is remarkable.

There are intriguing four-dimensional lattice-field theories with local gauge symmetry ("Z_p models") that may also exhibit three phases, in analogy with the case $p > 4$ discussed above [124, 125]. Elitzur *et al.* [124] give a compelling argument based on duality that two-dimensional XY models will have three phases for sufficiently large p even in the "clock-model" limit $h_p \to \infty$. The analysis sketched above is restricted to small h_p. *Random* symmetry-breaking fields have been studied by Houghton *et al.* [126] and by Cardy and Ostlund [127]. The two-dimensional XY model without a symmetry-breaking field, but with bond disorder, has been considered by José [128].

2.3 Dynamic scaling and third sound in helium films

Third sound is a peculiar wave-like excitation present even in very thin helium films. Experimental studies of velocities and damping of third sound [42, 43] provide an important probe of vortex dynamics near the Kosterlitz–Thouless transition. The theory described here is due to Ambegaokar *et al.* [103]. Its connection with dynamic scaling as applied to, say, second sound in bulk helium has been elucidated by Hohenberg *et al.* [129].

2.3.1 The hydrodynamics of helium films

A third-sound wave is shown schematically in Fig. 2.11, following the original presentation of Atkins [44]. This excitation consists of a

wave of helium-film height accompanied by periodic variations in the
in-plane superfluid velocity. Such a wave is possible even in films
which are, say, 20 atomic layers thick, because there is no viscous
interaction of the superfluid with the substrate. Third sound is not
unlike "gravity waves" in a conventional inviscid liquid film [130],
except that the van der Waals attraction of helium atoms to the substrate replaces the gravitational force. The equations for third sound
are particularly simple in the limit of thin films at low temperatures,
for which one can neglect mass transport through the vapor phase in
equilibrium with the film. We also assume that there is good thermal
contact with the substrate, so variations in temperature are negligible;
and we neglect irreversible dissipative couplings as well. A more
general treatment of third-sound hydrodynamics has been given by
Bergman [131, 132].

With the simplifying assumptions proposed above, the equations of
motion for the superfluid velocity $v_s(r, t)$ and film height $h(\mathbf{r}, t)$ take the
forms

$$\frac{\partial v_s}{\partial t} = -f \nabla h, \tag{2.97}$$

$$\frac{\partial(\bar{\rho} h)}{\partial t} = -\rho_s^0 \nabla \cdot v_s. \tag{2.98}$$

Here, f is a van der Waals constant, measuring the attraction of the film
to the substrate, and $\bar{\rho} = \bar{\rho}(h)$ is the mean mass density of a film of thickness h. The product $\bar{\rho} h$ is then an *areal* mass density. Upon linearizing
Eq. (2.98) about an equilibrium film height h_0, it is convenient to define

$$m(\mathbf{r}, t) \equiv [h_0 - h(\mathbf{r}, t)] f/g, \tag{2.99}$$

where

$$g^2 = f \left/ \left[\bar{\rho}(h_0) + h_0 \left(\frac{\partial \bar{\rho}}{\partial h} \right) \bigg|_{h=h_0} \right] \right. . \tag{2.100}$$

Equations (2.97) and (2.98) then become

$$\frac{\partial v_s}{\partial t} = g \nabla m, \tag{2.101}$$

$$\frac{\partial m}{\partial t} = \rho_s^0 \nabla \cdot v_s. \tag{2.102}$$

It is easily seen that these equations support waves proportional to
$e^{ik \cdot r - i\omega(k)t}$, with

$$\omega(k) = \pm c_3 k, \tag{2.103}$$

where the third-sound velocity is

$$c_3 = g \sqrt{\rho_s^0}. \tag{2.104}$$

Simple models of dissipation in films produce an imaginary contribution to (2.103) proportional to k^2.

Above the superfluid-transition temperature, $\rho_s^R(T)$ is renormalized to zero and we do not expect third sound to propagate. Instead, height deformations relax via a diffusive process, which can be written

$$\frac{\partial(\bar{\rho}h)}{\partial t} = \lambda_h f \nabla^2 h, \tag{2.105}$$

where λ_h is a transport coefficient. Again linearizing in $h(\mathbf{r}, t) - h_0$, we arrive at a diffusion equation

$$\frac{\partial m(\mathbf{r}, t)}{\partial t} = \lambda_h g \nabla^2 m(r, t). \tag{2.106}$$

The characteristic frequency of this process is just

$$\omega(k) = -i\lambda_h g k^2. \tag{2.107}$$

The main question I would like to address in this section is that of how the vortex-unbinding transition connects these hydrodynamic behaviors of helium films above and below T_c.

2.3.2 Dynamic scaling

A problem in some ways very similar to the behavior of third sound near the Kosterlitz–Thouless transition arises in bulk helium. Heat propagates as a second-sound wave below the λ transition, but relaxes via ordinary thermal diffusion above T_λ. How the dynamics changes near the critical point can be understood in terms of a dynamic-scaling hypothesis [133, 134]. Here, we review briefly the application of this idea to bulk helium and show what it predicts for third sound in helium films. As we shall see, dynamic scaling is wrong in this case, despite its impressive success in three dimensions!

The ideas behind dynamic scaling are illustrated in Fig. 2.12(a). The characteristic frequencies $\omega(k, \xi)$ for thermal excitations should exhibit three distinct regimes as a function of wavevector and temperature, for k and $\xi_\pm^{-1}$ small. When $k\xi_\pm \lesssim 1$, one must recover the hydrodynamics of either second sound or heat diffusion, depending on the temperature. New behavior is expected when $k\xi_\pm \gtrsim 1$. Dynamic scaling asserts that the changeover from one regime to another can be described by a scaling function,

$$\omega(k, \xi_+) = k^z \Omega_\pm(k\xi_\pm), \tag{2.108}$$

(a)

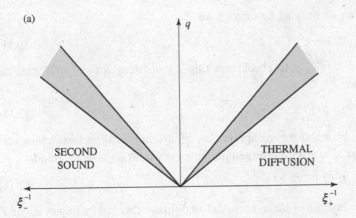

(b)

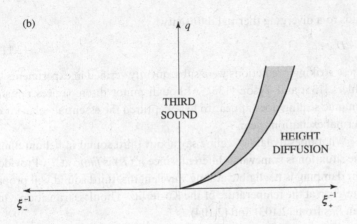

Fig. 2.12. (a) The standard dynamic-scaling picture of excitations in the plane of wavevector q and inverse correlation length. Heat propagates via second sound for small q below T_λ and by ordinary thermal diffusion in the hydrodynamic limit above this temperature. New behavior is expected for finite q precisely at T_λ. Crossover between different behaviors occurs in the shaded regions. (b) An analogous plot for helium films. Third sound propagates even above T_c at finite wavevectors, since $\rho_s^R(T)$ is nonvanishing at T_c^-. If dynamic scaling held, the crossover between wave excitations and height diffusion would occur for $q\xi_+ \approx 1$. Instead, dynamic scaling breaks down and the change occurs for $q \sim \xi_+^{-2}$.

where z is to be determined.

Below T_c, one has

$$\omega(k, \xi_-) = \pm c_2 k - \tfrac{1}{2} D_2 k^2 i, \tag{2.109}$$

where D_2 is the damping constant for second sound and the second-sound velocity is

$$c_2 \sim \sqrt{\rho_s(T)}, \tag{2.110}$$

near T. The exponent z now follows from the Josephson relation (2.73) for $\rho_s(T)$,

$$\rho_s(T) \propto \xi_\perp^{2-d}, \tag{2.111}$$

where $\xi_\perp$ diverges near T_λ in three dimensions,

$$\xi_\perp \approx \xi_0 \left| \frac{T - T_\lambda}{T_\lambda} \right|^{-2/3}. \tag{2.112}$$

After identifying $\xi_\perp$ with ξ_-, we have

$$\mathrm{Re}\,\omega \sim \xi_-^{1-d/2} k \tag{2.113}$$

and it is easy to check that this is consistent with (2.107), provided that

$$z = d/2. \tag{2.114}$$

Straightforward application of dynamic scaling to the imaginary part of ω gives a divergent damping constant for second sound

$$D_2(T) \sim \xi_\perp^{2-d/2}. \tag{2.115}$$

Application of the hypothesis to a diffusive thermal frequency above T_λ,

$$\omega(k, \xi_+) \approx -D_T k^2 i, \tag{2.116}$$

leads to a diverging thermal diffusivity,

$$D_T \sim \xi_+^{2-d/2}. \tag{2.117}$$

These striking predictions were subsequently verified in experiments by Ahlers [135] and Tyson [136]. Although minor discrepanices remain, dynamic scaling does appear to have captured the essential behavior of thermal excitations near T_λ.

What does dynamic scaling say about third sound in helium films? The situation is somewhat different, since $\rho_s^R(T)$ is *finite* at T_c. Provided that damping is negligible at long wavelengths, third sound will propagate, even at the temperature of the Kosterlitz–Thouless transition. One then has from (2.103) and (2.104)

$$\omega \sim k^z, \qquad z = 1, \tag{2.118}$$

in agreement with (2.114). At finite wavelengths, one would expect third sound to propagate even *above* T_c, provided that $k\xi_+ \gtrsim 1$: A third-sound wave with wavevector $k \gtrsim \xi_+^{-1}$ should be insensitive to widely separated unbound vortices. Application of the scaling to the height-diffusion frequency (2.107) gives a diverging diffusivity

$$\lambda_h \sim \xi_+(T) \sim \exp(b/|T - T_c|^{1/2}), \tag{2.119}$$

where we have used Kosterlitz's result for $\xi_+(T)$. As we shall see, λ_h actually diverges as a *different* power of ξ_+, although the temperature dependence displayed in (2.119) is qualitatively correct.

2.3.3 "Maxwell's equations" for third sound

The effect of vortices on third sound near the vortex-unbinding transition can be determined analytically [103]. To do this, it is convenient to define a quantity $N(\mathbf{r}, t)$, which is proportional to the number density of vortex charges,

$$N(\mathbf{r}, t) = \frac{2\pi\hbar}{m} n(\mathbf{r}, t). \tag{2.120}$$

In terms of this quantity, the fundamental relation (2.34) becomes

$$\nabla \times \mathbf{v}_\mathrm{s} = N\hat{\mathbf{z}}. \tag{2.121}$$

Since vortex charge is conserved, there is a vortex current $\mathbf{J}(\mathbf{r}, t)$ corresponding to (2.120), such that

$$\frac{\partial N}{\partial t} + \nabla \cdot \mathbf{J} = 0. \tag{2.122}$$

Because Eqs. (2.121) and (2.122) can be combined to read

$$\nabla \times \left(\frac{\partial \mathbf{v}_\mathrm{s}}{\partial t} + \hat{\mathbf{z}} \times \mathbf{J} \right) = 0, \tag{2.123}$$

we must have

$$\frac{\partial \mathbf{v}_\mathrm{s}}{\partial t} + \hat{\mathbf{z}} \times \mathbf{J} = \nabla \Xi, \tag{2.124}$$

where Ξ is some smooth scalar function. To obtain agreement with hydrodynamics (Eq. (2.101)) in the *absence* of vortex currents, we must in fact have

$$\frac{\partial \mathbf{v}_\mathrm{s}}{\partial t} = g\nabla m - \hat{\mathbf{z}} \times \mathbf{J}. \tag{2.125}$$

The hydrodynamic relation (2.102),

$$\frac{\partial m}{\partial t} = \rho_\mathrm{s}^0 g \nabla \cdot \mathbf{v}_\mathrm{s}. \tag{2.126}$$

is unaffected by vortex currents.

Equations (2.121), (2.122), (2.125) and (2.126) bear a striking resemblance to Maxwell's equations for electrodynamics [137] specialized to two dimensions. Equation (2.121) plays the role of Gauss's law, while (2.122) is just the equation of charge continuity. Equation (2.125) is like Ampère's law with a displacement current; (2.126) is Faraday's law. The zero-divergence condition on the magnetic field here amounts to the trivial statement

$$\left(\frac{\partial}{\partial x}, \frac{\partial}{\partial y}, \frac{\partial}{\partial z} \right) \cdot (0, 0, m(x, y)) = 0. \tag{2.127}$$

This correspondence is summarized in Table 2.1.

Table 2.1 *Maxwell's equations – the analogy with third sound*

Maxwell equation	Third-sound equation
$\nabla \cdot \mathbf{E} = 4\pi\rho$	$\nabla \times \mathbf{v}_s = N\hat{\mathbf{z}}$
$\nabla \times \mathbf{B} = \dfrac{4\pi}{c}J + \dfrac{1}{c}\dfrac{\partial \mathbf{E}}{\partial t}$	$g\nabla m = \hat{\mathbf{z}} \times \mathbf{J}_v + \dfrac{\partial \mathbf{v}_s}{\partial t}$
$\nabla \times \mathbf{E} + \dfrac{1}{c}\dfrac{\partial \mathbf{B}}{\partial t} = 0$	$r_s^0 g\nabla \cdot \mathbf{v}_s - \dfrac{\partial m}{\partial t} = 0$
$\nabla \cdot \mathbf{B} = 0$	$\left(\dfrac{\partial}{\partial x}, \dfrac{\partial}{\partial y}, \dfrac{\partial}{\partial z}\right) \cdot (0, 0, m(x, y)) = 0$
$\dfrac{\partial \rho}{\partial t} + \nabla \cdot \mathbf{J} = 0$	$\dfrac{\partial N}{\partial t} + \nabla \cdot \mathbf{J}_v = 0$

In the absence of inhomogeneities in the film height ($\nabla m = 0$), Eq. (2.125) becomes

$$\frac{\partial \mathbf{v}_s}{\partial t} = -\hat{\mathbf{z}} \times \mathbf{J} \qquad (2.128)$$

and has the physical interpretation shown in Fig. 2.13. Equation (2.128) implies that uniform supercurrents relax by motion of vortices across the flow [138–140]. The phase is slipped by 2π, and the superfluid velocity reduced correspondingly, each time a vortex goes by.

To study the dynamics above T_c, we first account for bound pairs of vortices by a kind of dielectric theory. This can be done in complete analogy with ordinary electrodynamics. The equations which describe effects of the remaining free-vortex charge are

$$\varepsilon_c \nabla \times \mathbf{v}_s = N_{\text{free}}\hat{\mathbf{z}}, \qquad (2.129)$$

$$\varepsilon_c \frac{\partial \mathbf{v}_s}{\partial t} = g\nabla m - \hat{\mathbf{z}} \times \mathbf{J}_{\text{free}}, \qquad (2.130)$$

$$\frac{\partial m}{\partial t} = \rho_s^0 g\nabla \cdot \mathbf{v}_s, \qquad (2.131)$$

$$\frac{\partial N_{\text{free}}}{\partial t} + \nabla \cdot \mathbf{J}_{\text{free}} = 0. \qquad (2.132)$$

Here N_{free} and $\mathbf{J}_{\text{free}}$ are the density and current of free vortices, while ε_c is the dielectric constant of the bound charges. Ambegaokar *et al.* [103] estimate that

$$\varepsilon_c \approx \rho_s^0(T_c^-)/\rho_s^R(T_c^-).$$

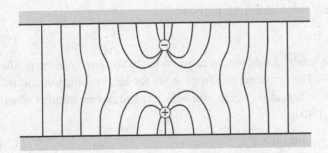

Fig. 2.13. Phase slippage in a superfluid film. A constant superfluid velocity $v_s = (\hbar/m)\nabla\theta$ corresponds to equally spaced, vertical lines of constant phase. This metastable state relaxes to a lower velocity via nucleation of a pair of vortices at the center of the channel. If these vortices escape to the edges, the phase is slipped by 2π.

To close these equations, we use a constitutive relation expressing J_{free} in terms of v_s and ∇N_{free},

$$J_{free} = \gamma\hat{z}\times v_s + D\nabla N_{free}. \tag{2.133}$$

This resembles the relation among current, electric-field and charge-density gradients in a metal. From this point of view, γ is proportional to a vortex mobility, while D is a vortex diffusion constant. As usual, there is an Einstein relation between γ and D, namely [103]

$$\gamma = \frac{2\pi\hbar D\rho_s^0}{mk_BT}n_f, \tag{2.134}$$

where n_f is the equilibrium density of free vortices. The dominant temperature dependence in γ comes from $n_f \sim \xi_+^{-2}$. The vortex diffusion constant has been studied by Petschek and Zippelius [141], who conclude that it remains finite near T_c.

After making use of (2.133), we arrive at three independent equations for v_s, m and N_{free}:

$$\varepsilon_c\frac{\partial v_s}{\partial t} = g\nabla m - \gamma v_s + D(\hat{z}\times\nabla)N_{free}, \tag{2.135}$$

$$\frac{\partial m}{\partial t} = \rho_s^0 g\nabla\cdot v_s, \tag{2.136}$$

$$\frac{\partial N_{free}}{\partial t} = -\gamma\hat{z}\cdot(\nabla\times v_s) + D\nabla^2 N_{free}. \tag{2.137}$$

Upon solving these equations, one finds eigenvectors describing both the relaxation of the vortex charge and "critical slowing down" of v_s above T_c. These characteristic frequencies are imaginary, asymptotically independent of wavevector and proportional to ξ_+^{-2}. Of particular interest to us is the third characteristic frequency, corresponding to height diffusion,

$$\omega_h(k) \approx \frac{\rho_s^0 g^2}{\gamma}k^2 i \tag{2.138}$$

$$\sim \xi_+^2 k^2 i. \tag{2.139}$$

Comparison with Eq. (2.107) gives

$$\lambda_h \sim \xi_+^2 \tag{2.140}$$

for the height diffusivity, in contrast to the dynamic-scaling prediction (2.119). The characteristic frequencies for height diffusion and relaxation of v_s actually change over into third-sound excitations when (see Fig. 2.12(b))

$$k \approx \text{constant} \times \xi_+^{-2}, \tag{2.141}$$

instead of for $k\xi_+ = O(1)$. Transport coefficients for a diverging helium film have been reported by Ratnam and Mochel [142] and by Maps and Hallock [143] and Agnolet and Reppy [144]. Such experiments provide important tests of the dynamical theory and a direct measure of the superfluid correlation length.

2.4 Statistical mechanics of two-dimensional melting

2.4.1 Translational order, orientational order and Landau theory

Two broken symmetries distinguish solids from liquids in three dimensions. X-ray diffraction in solids reveals extended order in the translational order parameter

$$\rho_G(\mathbf{r}) = e^{i\mathbf{G}\cdot\mathbf{r}}, \tag{2.142}$$

where $\mathbf{G}$ is a reciprocal-lattice vector and $\mathbf{r}$ is the position of a particle. Although there is no long-range order in two dimensions, we saw in Section 2.1 that there are quite extended translational correlations even in this case. In addition to anomalous translational correlations, crystals also display a broken orientational symmetry associated with a singled-out set of crystallographic axes. These two symmetries are of course not completely independent: Rotating one patch of perfect crystal relative to another necessarily disrupts translational order as well (see Fig. 2.14(a)). This kind of disruption occurs, for example, across grain boundaries. As shown in Fig. 2.14(b), however, *translating* one patch of crystal relative to another (as might occur, for example, across a stacking fault), keeps the orientational correlations intact. One can imagine equilibrium phases of matter with extended correlations in local crystallographic axes, but *without* any concomitant translational order. A remarkable phase of this kind, displaying a symmetry intermediate between that of a liquid and that of a solid, arises naturally in theories of defect-mediated melting in two dimensions [145, 146].

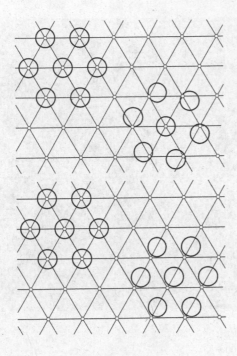

Fig. 2.14. Disruption of crystalline order by rotations and translations. (a) The hexagonal patch of atoms on the upper left-hand side defines a triangular lattice of sites corresponding to a crystal with perfect translational and orientational order. Rotating the patch on the lower right-hand side relative to the reference patch decorrelates translational order as well. (b) A relative translation, however, leaves the orientational correlations intact.

Orientational order in two dimensions can be measured via a complex local order parameter [145],

$$\psi_6(\mathbf{r}) = e^{6i\theta(\mathbf{r})}. \tag{2.143}$$

As illustrated in Fig. 2.15, θ is the angle made by the line joining two neighboring atoms relative to some reference axis, while $\mathbf{r}$ locates the midpoint of this "bond". By a "bond," we mean not a chemical bond, but merely an assignation of near neighbors based on, say, the construction of Dirichlet polygons [147]. The quantity $e^{6i\theta}$ is appropriate for studying melting of triangular lattices, since one is interested only in bond-order modulo 60° rotations. It is also a natural measure of bond order in liquids, since the mean coordination number of any two-dimensional fluid is six. Long-wavelength hydrodynamic theories should be insensitive to the precise microscopic definition of $\psi_6(\mathbf{r})$.

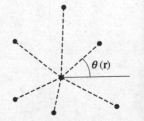

Fig. 2.15. Construction of the "bond-angle field" from the angles which lines joining nearest-neighbor atoms make with a reference axis.

A mechanical model with anomalous orientational correlations is shown in Fig. 2.16 [148]. Here, order in a dense matrix of small ball bearings has been disturbed by a dilute concentration of larger spheres. Although the material seems locally crystalline, the periodic modulations in density are dephased on a scale set by the spacing of the randomly distributed large spheres. Orientational order, on the other hand, persists across the entire sample. A quantitative measure of translational and orientational order is provided by the correlation functions

Fig. 2.16 (a) A mechanical model of extended orientational order in the presence of relatively short-ranged translational order. Translation correlations are disrupted by a dilute concentration of large ball bearings [148]. (b) Translational and orientational correlation functions for a system like that shown in (a). The functions $\bar{G}_T(r)$ and $\bar{G}_6(r)$ are proportional to the quantities $G_T(r)$ and $G_6(r)$ defined in the text [148].

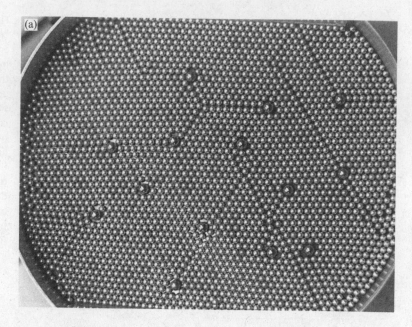

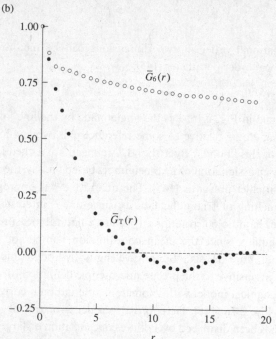

$$G_T(\mathbf{r}) = \langle \rho_G^*(\mathbf{r}) \rho_G(0) \rangle, \tag{2.144a}$$

$$G_6(\mathbf{r}) = \langle \psi_6^*(\mathbf{r}) \psi_6(0) \rangle, \tag{2.144b}$$

which are shown for a ball-bearing system like that in Fig. 2.16(a). Systems that exhibit extended 6-fold orientational correlations but only short-range translational correlations are called "hexatics."

Figure 2.16 is of course a non-equilibrium system with quenched-in disorder. To discuss translational and orientational order in *equilibrium* single-component materials, we start with a Landau theory in terms of the order parameters ρ_G and ψ_6. Although Landau theory is unreliable in the presence of large fluctuations, it is nevertheless instructive to review its predictions for two-dimensional melting and crystallization. Landau himself originally considered the liquid–solid transition in terms of translational order alone [31, 149, 150]. Specializing to $d=2$ and including coupling to an orientational order parameter, we find a coarse-grained free-energy density:

$$\mathscr{F} = \frac{1}{2} r_T \sum_G |\rho_G|^2 + w_T \sum_{G_1+G_2+G_3=0} \rho_{G_1} \rho_{G_2} \rho_{G_3} + \cdots$$

$$+ \frac{1}{2} r_6 |\psi_6|^2 + u_6 |\psi_6|^4 + \cdots$$

$$+ \gamma \sum_G |\rho_G|^2 [\psi_6 (G_x - iG_y)^6 + \psi_6^* (G_x + iG_y)^6]. \tag{2.145}$$

The summations encompass all wavevectors $\mathbf{G}$ lying on a circle located at the first maximum of the liquid X-ray structure function in reciprocal space. Gradient terms have been neglected entirely. Both r_T and r_6 are positive at high temperatures and eventually change sign as T decreases. If r_6 is large and positive when r_T changes sign, one expects a transition directly into a solid. Since a term cubic in ρ_G appears for triplets of $\mathbf{G}$ values forming an equilateral triangle, this theory predicts a first-order liquid–solid transition [31]. Crystallization into a triangular solid means that fluctuations become large at six $\mathbf{G}$ values symmetrically distributed around the circle. The transition is unusual in that there is a continuum of $\mathbf{G}$-sextets related by rotations that can order. This degeneracy (first studied by Brasovskii [151] in another context) could make the transition be of first order even in the absence of a cubic term [150]. When the ρ_G values order, the γ coupling in Eq. (2.145) means that ψ_6 locks into alignment with the singled-out set of $\mathbf{G}$ values.

If r_6 changes sign before r_T, Landau theory predicts a *continuous* transition into an orientationally ordered hexatic phase. Since ψ_6 is a two-component (i.e. complex) order parameter, this transition would be in the universality class of the superfluids and XY models discussed in Section 2.2. Note that ordering in ψ_6 does *not* imply translational order in the ρ_G values. A more sophisticated Landau theory can be constructed to treat a subsequent hexatic–solid transition. When ψ_6 orders, six of the continuum of ρ_G values in Eq. (2.145) will become much larger than all others. Focusing attention on the corresponding $\{G_\alpha\}$, $\alpha = 1$, ..., 6, symmetrically distributed around a circle, we define

$$\rho_\alpha(\mathbf{r}) \equiv \rho_{G_\alpha}(\mathbf{r}), \qquad \alpha = 1, \ldots, 6. \tag{2.146}$$

A coarse-grained free-energy density can be constructed in close analogy with de Gennes' [152] analysis of the nematic–smectic-A transition in liquid crystals (B. I. Halperin, 1979, unpublished):

$$\mathcal{F} = \frac{1}{2} A \sum_{\alpha=1}^{6} |\mathbf{G} \cdot [\nabla - i(\hat{\mathbf{z}} \times \mathbf{G})\theta]\rho_\alpha|^2$$

$$+ \frac{1}{2} B \sum_{\alpha=1}^{6} |\mathbf{G}[\nabla - i(\hat{\mathbf{z}} \times \mathbf{G})\theta]\rho_\alpha|^2$$

$$+ \frac{1}{2} r \sum_{\alpha=1}^{6} |\rho_\alpha|^2 + w[\rho_1\rho_3\rho_5 + \rho_2\rho_4\rho_6] + \frac{1}{2} K_A(\nabla\theta)^2. \tag{2.147}$$

Here, $\theta(\mathbf{r})$ is the phase of the bond-orientational order parameter, while $\hat{\mathbf{z}}$ is a unit vector perpendicular to the xy-plane. The third-order term (again corresponding to equilateral triplets of $\mathbf{G}$ values) implies a first-order transition within this theory. The quantity K_A is a stiffness like that appearing in the low-temperature description of the XY model. The form of the gradient terms involving the $\{\rho_\alpha\}$ is determined by their transformation properties under rotations. Under the rotation

$$\theta(\mathbf{r}) \rightarrow \theta(\mathbf{r}) + \theta_0, \tag{2.148a}$$

the definition of the $\{\rho_\alpha\}$ as Fourier transforms of the density means that

$$\rho_\alpha \rightarrow e^{i(G_\alpha \times \mathbf{r}) \cdot \hat{\mathbf{z}}\theta_0} \rho_\alpha. \tag{2.148b}$$

The form of (2.147) insures that F is invariant under this transformation.

As discussed in Section 2.1, fluctuations in thin films will depress actual transition temperatures well below their Landau-theory values. Since variations in the amplitudes of ρ_G and ψ will then be suppressed relative to phase fluctuations, one has reason to doubt the predictions of Landau theory, which is essentially an expansion in small amplitudes. In particular, the hexatic–solid transition need not be of first order in two dimensions.

The free energy (2.147) simplifies with the neglect of amplitude fluctuations. Assuming that the parameter r is large and negative (although still possibly above the true transition temperature), we can write

$$\rho_\alpha(\mathbf{r}) = \rho_0 e^{iG_\alpha \cdot \mathbf{u}(\mathbf{r})}, \tag{2.149}$$

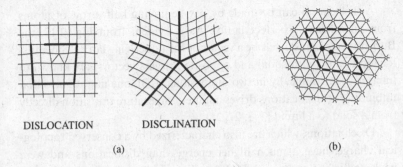

DISLOCATION   DISCLINATION

(a)   (b)

Fig. 2.17. Topological excitations in square and triangular lattices. (a) shows a dislocation (with the associated Burgers circuit) and a disclination embedded in a square lattice. A dislocation in a triangular lattice is exhibited in (b). Note the "disclination pair" – points of local 5- and 7-fold symmetry – buried in the core of the dislocation.

where ρ_0 is constant and $\mathbf{u}(\mathbf{r})$ is a local phonon displacement. The free energy is then minimized when $\theta(\mathbf{r})$ is locked to the curl of the phonon-displacement field,

$$\theta(\mathbf{r}) = \tfrac{1}{2}[\partial_x u_y(\mathbf{r}) - \partial_y u_x(\mathbf{r})]. \tag{2.150}$$

After eliminating $\theta(\mathbf{r})$, one finds that the integrated free energy density F reduces to the usual continuum elastic free energy

$$F = \text{constant} + \frac{1}{2} \int d^2r \, [2\mu u_{ij}^2(\mathbf{r}) + \lambda u_{kk}^2(\mathbf{r})] \tag{2.151a}$$

with

$$\mu = \tfrac{3}{4}\rho_0^2(A+B)G^4, \qquad \lambda = \tfrac{3}{4}\rho_0^2(A-B)G^4. \tag{2.151b}$$

Perturbative effects of amplitude fluctuations in this limit have been discussed by Toner [96].

2.4.2 Melting on a smooth substrate

The role of defects in melting of an isotropic triangular solid on a smooth substrate can be assessed just as in our discussion of superfluids in Section 2.2. Systems on "smooth substrates" include electrons on the surface of helium, liquid-crystal films in equilibrium with a vapor, lipid monolayers on water and colloidal crystals at interfaces or trapped between two glass plates.

The defects associated with the continuum elastic theory of a solid are dislocations and disclinations [153], which are shown for simplicity on a square lattice in Fig. 2.17(a). Dislocations are characterized by a Burgers vector $a_0\mathbf{b}$, defined as the amount by which a contour integral of the displacement field around the dislocation fails to close. Thus,

$$\oint d\mathbf{u} = a_0\mathbf{b}(\mathbf{r}), \tag{2.152}$$

where a_0 is the lattice spacing.

A dislocation can be made by, say, removing half a row of atoms from an otherwise perfect lattice. As is evident from Fig. 2.17, the Burgers vector must itself be a vector of the underlying lattice. Since dislocation pairs (with equal and opposite Burgers vectors) are known to interact logarithmically in two dimensions [153], one might guess that unbinding of dislocations drives a finite-temperature transition directly from a solid to a liquid [2, 3, 27, 28].

Disclinations, which are also characterized by a conserved topological charge, have a much higher energy than dislocations and were neglected in the early work by Kosterlitz and Thouless and others. As illustrated in Fig. 2.17(b), it is impossible to neglect them entirely. A dislocation on a triangular lattice is shown, together with dots indicating two isolated points of 5- and 7-fold symmetry embedded in its core. The dislocation in Fig. 2.17(a) can be regarded as an isolated point of 5-fold symmetry, separated by about half a lattice constant from an interstitial point of 3-fold symmetry. In this sense, every dislocation contains an embryonic pair of disclinations!

Because dislocations are associated with an additional half row of atoms, they can be quite effective at breaking up translational order. As we shall see, dislocations are less disruptive of orientational correlations. Disclinations, on the other hand, can have a profound influence on the decay of orientational order. It is important to study orientational order, because solids display conventional long-range order in $\psi_6(\mathbf{r})$, even in two dimensions [21, 94]. Indeed, it can easily be checked within a continuum elastic theory that the fluctuations in the local orientation angle $\theta(\mathbf{r})$ are *finite*,

$$\langle |\theta(\mathbf{r})|^2 \rangle = \tfrac{1}{4} \langle |\nabla \times \mathbf{u}(\mathbf{r})|^2 \rangle < \infty, \tag{2.153}$$

in contrast to the diverging displacement-field fluctuations noticed by Peierls and Landau. As we shall see, disclinations play an important role above a dislocation-unbinding temperature.

The free energy for interacting dislocations can be worked out just as in our discussion of vortices in superfluids (see, for example, [101]). Upon decomposing the matrix of displacement derivatives into a smooth part and a part due to dislocations,

$$\partial_i u_j = \partial_i \phi_j + (\partial_i u_j)_{\text{sing}}, \tag{2.154}$$

the free energy (2.151) breaks into two parts:

$$F = F_0 + F_D. \tag{2.155}$$

The free energy F_0 is given by Eq. (2.151) with $\mathbf{u}(\mathbf{r})$ replaced by the smooth displacement field $\phi(\mathbf{r})$, while

$$\frac{F_D}{k_B T} = -\frac{1}{8\pi} \sum_{\mathbf{r} \neq \mathbf{r}'} \left(K_1 \mathbf{b}(\mathbf{r}) \cdot \mathbf{b}(\mathbf{r}') \ln(|\mathbf{r} - \mathbf{r}'|/a) \right.$$

$$\left. - K_2 \frac{[\mathbf{b}(\mathbf{r}) \cdot (\mathbf{r} - \mathbf{r}')][\mathbf{b}(\mathbf{r}') \cdot (\mathbf{r} - \mathbf{r}')]}{|\mathbf{r} - \mathbf{r}'|^2} \right)$$

$$+ \frac{E_c}{k_B T} \sum_{\mathbf{r}} |\mathbf{b}(\mathbf{r})|^2, \tag{2.156}$$

with

$$K_1 = K_2 = \frac{4a_0^2}{k_B T} \frac{\mu(\mu + \lambda)}{2\mu + \lambda}. \tag{2.157}$$

The $\{\mathbf{b}(\mathbf{r})\}$ are dimensionless Burgers vectors of the form

$$\mathbf{b}(\mathbf{r}) = m(\mathbf{r})\mathbf{e}_1 + n(\mathbf{r})\mathbf{e}_2, \tag{2.158}$$

where $m(\mathbf{r})$ and $n(\mathbf{r})$ are integers and $\mathbf{e}_1$ and $\mathbf{e}_2$ are unit vectors spanning an underlying triangular lattice.

To determine how dislocations affect elastic constants, one now focuses on the renormalized elastic tensor. (Disclinations will be neglected because of their significantly higher energy.) With suitable boundary conditions [146], one has

$$C_{R,ijk\ell}^{-1} = \langle U_{ij} U_{k\ell} \rangle / (\Omega a_0^2), \tag{2.159}$$

where Ω is the area of the system, a_0 its lattice spacing and

$$U_{ij} = -\frac{1}{2} \oint_P (u_i n_j + u_j n_i) \, d\ell. \tag{2.160}$$

The integration in (2.160) is around the perimeter of the solid, where n is a unit normal. Equation (2.159) can be written

$$C_{R,ijk\ell}^{-1} = C_{ijk\ell}^{-1} + \frac{\langle U_{ij}^{\text{sing}} U_{k\ell}^{\text{sing}} \rangle}{\Omega a_0^2}, \tag{2.161}$$

where $C_{ijk\ell}$ is the matrix of elastic constants unrenormalized by dislocations,

$$C_{ijk\ell}^{-1} = \frac{1}{4\bar{\mu}} (\delta_{ij}\delta_{j\ell} + \delta_{i\ell}\delta_{jk}) - \frac{\bar{\lambda}}{4\bar{\mu}(\bar{\mu} + \bar{\lambda})} \delta_{ij}\delta_{k\ell}, \tag{2.162}$$

and

$$U_{ij}^{\text{sing}} = \frac{1}{2} a_0 \sum_{\mathbf{R}} [b_i(\mathbf{R})\varepsilon_{j\ell}R_\ell + b_j(\mathbf{R})\varepsilon_{i\ell}R_\ell]. \tag{2.163}$$

Here $\bar{\mu}$ and $\bar{\lambda}$ are dimensionless elastic constants,

$$\bar{\mu} \equiv \frac{\mu a_0^2}{k_B T} \qquad \bar{\lambda} \equiv \frac{\lambda a_0^2}{k_B T}, \tag{2.164}$$

Fig. 2.18. Charge-neutral (a) pair and (b) triplet excitations of Burgers vectors that renormalize the elastic constants of a triangular lattice. Note that the triplet in (c) effectively renormalizes the probability distribution of the third pair in part (a).

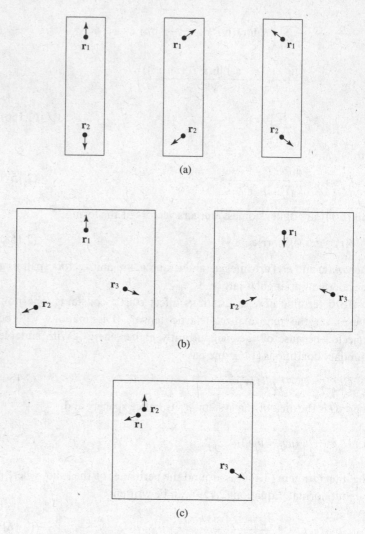

and the summation in (2.163) runs over a lattice of possible sites for dislocations with Burgers vectors **b(R)**.

Renormalization-group recursion relations may now be constructed perturbatively in

$$y = e^{-E_c/(k_{\mathrm{B}}T)}, \tag{2.165}$$

just as for XY models and superfluids. Configurations of Burgers vectors contributing to the renormalized elastic constants are shown in Fig. 2.18. A new feature is the charge-neutral triplet shown in Fig. 2.18(b). As illustrated in Fig. 2.18(c), triplets renormalize the probability of a pair excitation when two members merge together. The recursion relations for the parameters in the Coulomb free energy (2.156) are, to order y^2 [146, 154],

$$\frac{dK^{-1}(\ell)}{d\ell} = \frac{3}{2}\pi y^2(\ell)e^{K(\ell)/(8\pi)}I_0\left(\frac{K(\ell)}{8\pi}\right) - \frac{3}{4}\pi y^2(\ell)e^{K(\ell)/(8\pi)}I_1\left(\frac{K(\ell)}{8\pi}\right),$$

$$\tag{2.166a}$$

$$\frac{dy(\ell)}{d\ell} = \left(2 - \frac{K(\ell)}{8\pi}\right)y(\ell) + 2\pi y^2(\ell)e^{K(\ell)/(16\pi)}I_0\left(\frac{K(\ell)}{8\pi}\right), \tag{2.166b}$$

where $I_0(x)$ and $I_1(x)$ are Bessel functions and

$$K(\ell) = \frac{4\bar{\mu}(\ell)[\bar{\mu}(\ell) + \bar{\lambda}(\ell)]}{2\bar{\mu}(\ell) + \bar{\lambda}(\ell)}. \tag{2.167}$$

These recursion relations produce Hamiltonian flows very much like those shown for superfluids in Fig. 2.6. There is an instability toward large $y(\ell)$, signaling an unbinding of dislocations above a temperature T_m. Free dislocations are separated by a diverging translational correlation length $\xi_+(T)$,

$$\xi_T(T) \sim \exp[b'/|T - T_m|^{\bar{\nu}}], \tag{2.168a}$$

where

$$\bar{\nu} = 0.369\,634\,77\ldots, \tag{2.168b}$$

in contrast to the superfluid result $\bar{\nu} = 0.5$. There are only essential singularities superimposed on an analytical background in the specific heat and compressibility. The renormalized elastic constants below T_m are given by

$$\bar{\mu}_R(\bar{\mu},\bar{\lambda},y) \equiv \lim_{l\to\infty} \bar{\mu}(\ell), \tag{2.169a}$$

$$\bar{\lambda}_R(\bar{\mu},\bar{\lambda},y) \equiv \lim_{l\to\infty} \bar{\lambda}(\ell), \tag{2.169b}$$

where $\bar{\mu}(l)$ and $\bar{\lambda}(l)$ obey the recursion formulas [146]

$$\frac{d\bar{\mu}^{-1}(\ell)}{d\ell} = 3\pi y^2(\ell)e^{K(\ell)/(8\pi)}I_0\left(\frac{K(\ell)}{8\pi}\right), \tag{2.170}$$

$$\frac{d[\bar{\mu}(\ell) + \bar{\lambda}(\ell)]^{-1}}{d\ell} = 3\pi y^2(\ell)e^{K(\ell)/(8\pi)}\left[I_0\left(\frac{K(\ell)}{8\pi}\right) - I_1\left(\frac{K(\ell)}{8\pi}\right)\right]. \tag{2.171}$$

One finds cusp-like singularities just below T_m, e.g.,

$$\bar{\mu}_R(T) \approx \bar{\mu}_R(T_m^-)[1 + b|T - T_m|^{\bar{\nu}}]. \tag{2.172}$$

There is a universal relation between the dimensionless elastic constants, at T_m^-,

$$\lim_{T\to T_m^-} \frac{\bar{\mu}_R(T)[\bar{\mu}_R(T) + \bar{\lambda}_R(T)]}{2\bar{\mu}_R(T) + \bar{\lambda}_R(T)} = 4\pi. \tag{2.173}$$

This universal relation agrees with an "entropy-argument" result originally proposed by Kosterlitz and Thouless [28]. Finally, the structure factor $S(\mathbf{q})$ is singular at the reciprocal-lattice points $\{\mathbf{G}\}$ above T_m,

$$S(\mathbf{G}) \sim \xi_+^{2-\eta_G}, \tag{2.174}$$

where

$$\eta_G = \frac{|\mathbf{G}|^2 k_B T}{4\pi} \frac{3\mu_R + \lambda_R}{\mu_R \, 2\mu_R + \lambda_R}. \tag{2.175}$$

Although translational order decays exponentially above T_m, the behavior of orientational order is more subtle. If there is residual orientational order, even in the presence of unbound dislocations, one might expect it to be controlled by an anisotropic effective free energy,

$$F_A = \frac{1}{2} K_A(T) \int d^2r (\nabla\theta)^2. \tag{2.176}$$

Indeed, this is the form of the hexatic free-energy density (2.147) with all $\rho_\alpha \equiv 0$. It is easily checked that the stiffness $K_A(T)$ is infinite in a genuine solid. To calculate $K_A(T)$ above T_m, we write it as a correlation function,

$$\frac{k_B T}{K_A(T)} = \lim_{q\to 0} \frac{q^2 \langle \hat{\theta}_q \hat{\theta}_{-q} \rangle}{\Omega_0}, \tag{2.177}$$

where $\hat{\theta}_q$ is the Fourier transform of $\theta(\mathbf{r})$ and Ω_0 is the area of the system. The contribution to (2.177) from the smooth part of the phonon field drops out. The distribution of bond angles associated with a set of dislocations is [146]

$$\theta_{\text{sing}}(\mathbf{r}) = \frac{-a_0}{2\pi} \sum_{\mathbf{r}'} \frac{\mathbf{b}(\mathbf{r}') \cdot (\mathbf{r} - \mathbf{r}')}{|\mathbf{r} - \mathbf{r}'|^2}. \tag{2.178}$$

From this we obtain

$$\frac{k_B T}{K_A(T)} = \lim_{q\to 0} \left(\frac{a_0^2}{\Omega_0} \right) \frac{q_i q_j}{q^2} \langle \hat{b}_i(\mathbf{q}) \hat{b}_j(-\mathbf{q}) \rangle. \tag{2.179}$$

Assume for simplicity that we are far enough above T_m that most dislocations are unbound. The average in (2.179) can then be evaluated in a Debye–Hückel approximation, similar to that used in Section 2.2.4. If the charges $\mathbf{b}(\mathbf{r})$ are treated as a continuous vector field, the dislocation free energy (2.156), in Fourier space (with $K_1 = K_2$), becomes

$$\frac{F_D}{k_B T} = \frac{1}{2\Omega_0} \sum_q \left[\frac{K}{q^2} \left(\delta_{ij} - \frac{q_i q_j}{q^2} \right) + \frac{2E_c a^2}{k_B T} \delta_{ij} \right] \hat{b}_i(\mathbf{q}) \hat{b}_j(-\mathbf{q}), \tag{2.180}$$

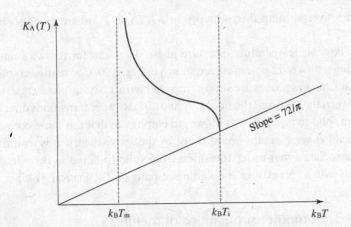

Fig. 2.19. The Frank constant $K_A(T)$ measuring the free-energy cost of bond-angle inhomogeneities in the hexatic phase. This quantity diverges like ξ_+^2 near T_m and exhibits a universal jump discontinuity with slope $72/\pi$ at T_i.

where a is the diameter of a dislocation core and a cutoff-dependent constant has been absorbed into E_c. The average in (2.179) is now easily evaluated and leads to a finite result,

$$\frac{k_B T}{K_A(T)} \approx \frac{k_B T}{2 E_c a^2}. \tag{2.181}$$

The longitudinal projector operator in (2.179) causes the part of F_D proportional to $\delta_{ij} - q_i q_j / q^2$ to drop out. A finite stiffness $K_A(T)$ means that orientational order decays algebraically above T_m,

$$\langle \psi^*(\mathbf{r})\psi(0)\rangle, \sim r^{-\eta_6(T)} \tag{2.182}$$

with

$$\eta_6(T) = 18 k_B T / [\pi K_A(T)], \tag{2.183}$$

in contrast to the exponential decay one expects in a liquid. To calculate $K_A(T)$ at temperatures just above T_m, we first integrate the recursion relations out to higher temperatures and then use (2.181). In this way, one finds that $K_A(T)$ diverges near T_m [146],

$$K_A(T) \sim \xi_T^2(T). \tag{2.184}$$

Free dislocations screen interactions between pairs of *disclinations*, which are tightly bound in the solid. Indeed, a free energy of the form (2.176) gives rise to a weak *logarithmic* interaction between pairs of equal and opposite "disclinicity." It is perhaps not surprising that disclinations, too, eventually unbind at a higher temperature T_i, producing exponential decay of orientational order in a liquid phase. This second transition is very much like the unbinding of vortices discussed for superfluids in Section 2.2. The phase between T_m and T_i, with its residual 6-fold anisotropy, is an equilibrium "hexatic liquid crystal." The behavior of $K_A(T)$ between T_m and T_i is shown in Fig. 2.19. There

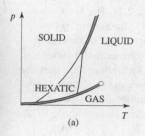

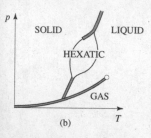

Fig. 2.20. Possible pressure–temperature phase diagrams (a) and (b) for two-dimensional matter interacting via an attractive potential with a repulsive core. Solid, liquid, gas and hexatic phases are shown, together with heavy hatched lines indicating first-order transitions. The loci of continuous dislocation- and disclination-unbinding transitions are shown as light lines. The detailed features of these phase diagrams are pure speculation.

is a universal jump discontinuity in $K_A(T)$ at T_i, where one also has

$$\eta_6 = \tfrac{1}{4}$$

Possible temperature–pressure phase diagrams, for particles interacting with a 6–12 potential, are shown in Fig. 2.20. Conventional solid, liquid and gas phases are shown, together with a hexatic phase, inserted between the solid and the liquid. A line of first-order transitions directly from solid to liquid is also shown, to emphasize that this more conventional behavior cannot be ruled out by the above theory. It is tempting to associate a first-order transition from solid to liquid in two dimensions with a "premature" unbinding of pairs of disclinations [145].

2.4.3. Computer simulations of melting

Few, if any, computer simulations provided evidence for anything other than a first-order melting transition in two dimensions prior to 1979. The most extensive work has been done on hard disks, starting with pioneering work by Alder and Wainwright [155]. There is some early evidence that melting is also of first order, at least at some densities, for particles interacting with a Lennard-Jones 6–12 potential [156]. Simulations must be viewed with caution, however, because of the small system sizes and run times available even on the most modern computers. If melting occurs via two second-order transitions, one would expect very long relaxation times ("critical slowing down") at two distinct temperatures or densities. Simulations that do not heat and cool extremely slowly through these transitions would display erroneous "hysteresis loops."

Particularly severe equilibration difficulties would occur just above a continuous hexatic–solid transition, where one expects large solid-like patches of material with dimensions of the order of the diverging translation correlation length ξ_T. An important equilibration parameter in this case is the time it takes a dislocation to diffuse a translational correlation length,

$$\tau_{\text{climb}} \approx \xi_T^2 / D_{\text{climb}}, \tag{2.185}$$

where D_{climb} is a diffusion constant characterizing "climbing" of a dislocation, i.e. motion perpendicular to the Burgers vector. Climbing takes considerably longer than does "gliding" motion parallel to the Burgers vector, since it requires cooperative motion of vacancies and interstitials [153]. Bruinsma et al. [157] estimate that D_{climb} is of order 10^{-10} cm^2 s^{-1} just below the melting temperature in a two-dimensional solid. Even if this estimate were several orders of magnitude too small, molecular-dynamics simulations (unless maximum run times greatly exceeded of order 10^{-9} s in argon-like units [158]) would be incapable of following an equilibrium melting transition when ξ_T exceeded, say, five to ten interparticle spacings. Correlation lengths of this magnitude are not uncommon in two dimensions, even in those simulations which purport to observe first-order liquid–solid phase transitions.

Despite such limitations, computer studies are an important supplement to the experiments described in Section 2.1. Although equilibration times are many orders of magnitude longer in real experiments, computer studies can probe theoretically important quantities such as the orientational order parameter, which is difficult to measure directly in experiments.

The density–temperature phase diagram for repulsive "soft-disk" pair potentials $V(\mathbf{r})$ of the form

$$V(\mathbf{r}) = C_n r^{-n} \tag{2.186}$$

is quite simple. Because the potential is a simple power law at all distances, one can scale out the number density ρ in the configurational partition sum and find that all thermodynamic quantities must be functions of a single dimensionless parameter, say

$$\Gamma_n \equiv \frac{C_n(\pi\rho)^{1/2n}}{k_B T}. \tag{2.187}$$

The complete phase diagram can then be deduced from a single path in the (ρ, T)-plane. The limit $n \to \infty$ at a temperature T produces a system of hard disks with core diameter

$$a = \lim_{n \to \infty} [C_n/(k_B T)]^{1/n}. \tag{2.188}$$

An intermediate hexatic phase is a possibility, although this could of course be preempted by a first-order transition for any or all values of n.

The case $n = 1$, e.g., a repulsive Coulomb interaction between electrons on the surface of helium, has been studied via a molecular-dynamics simulation by Morf [159]. By measuring the characteristic frequency for the relaxation of shear distortions, it was possible to determine the temperature-dependent shear modulus. On warming the system up, Morf observed a sharp decrease in the shear modulus accompanied by a rise in particle diffusion in the range $120 < \Gamma < 140$, where the dimensionless parameter for this problem is

$$\Gamma \equiv e^2(\pi\rho)^{1/2}/(k_B T). \tag{2.189}$$

Presumably, the electron crystal melts at a value $\Gamma_m \approx 130$. This estimate of Γ_m agrees with the experimental result $\Gamma_m \approx 137 \pm 15$ of Grimes and Adams [57] and with the value $\Gamma_m \approx 125 \pm 15$ inferred from an earlier Monte Carlo simulation by Gann et al. [160]. Morf was able to explain his results by a careful application of the detailed dislocation-unbinding

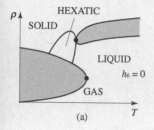

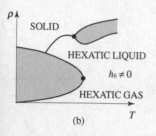

Fig. 2.21. Speculative temperature–density phase diagrams for two-dimensional matter with (a) and without (b) a hexatic field. The shaded regions indicate coexistence of two phases.

theory. A linear depression of $\mu_R(T)$ was used to estimate a "bare" temperature-dependent shear modulus, unrenormalized by dislocations. Combining this information with an evaluation of the zero-temperature dislocation-core energy due to Fisher et al. [161], the recursion relations (2.166) and (2.170) can be integrated numerically to obtain the true shear modulus $\mu_R(T)$ from (2.169a). Morf found a theoretical prediction that not only agrees quantitatively with his data but also gives $\Gamma_m = 128.2$, in good agreement with experiment and recent simulations.

McTague et al. [66] have presented interesting work on soft-disk $1/r^6$ potentials. Using a clever microscopic definition of disclinations and dislocations, they find evidence for melting mediated by a two-stage dislocation–disclination-unbinding process. They also find a profusion of grain boundaries and other complicated structures. It is not yet clear whether these structures represent true equilibrium effects, or are a result of heating and cooling the system too rapidly. A theory of grain-boundary-mediated melting has been studied by Chui [162], who predicts a first-order liquid–solid transition. A study of repulsive $1/r^{12}$ potentials has been reported by Broughton et al. [163]. They find evidence for a first-order phase transition, with properties consistent with extrapolations of liquid- and solid-phase free energies. Kalia et al. [164] have argued that melting of $1/r^3$ soft-disk potentials may be a first-order process.

Recent simulations of particles interacting with a 6–12 potential have produced conflicting evidence for an equilibrium hexatic phase. A density–temperature phase diagram consistent with the pressure–temperature sketch in Fig. 2.20(a) is shown in Fig. 2.21(a). As indicated in Fig. 2.21(a), a hexatic phase could exist over a range of densities. Early evidence that pairs of dislocations might be important in 6–12 melting was found by Cotterill and Peterson [165]. They observed a first-order transition, however, which is inconsistent with the detailed theory. Frenkel and McTague [166] found evidence for a hexatic phase by monitoring the orientational correlation function and viscosity. An orientational correlation length $\xi_6(T)$ became very large with decreasing temperature at a temperature T_i, while the viscosity increased rapidly at a lower temperature T_m. This is what one would expect if a hexatic phase were interposed between these temperatures. They also studied the specific heat and found no observable singularities. The Frenkel–McTague results have been questioned by a number of authors [167, 168], all of whom find first-order transitions. Barker et al. [169] find that the first-order transition observed by Abraham [167] is consistent with extrapolated liquid and solid free energies. Extensive Monte Carlo simulations with long run times by Tobochnik and Chester [106] suggest that a hexatic phase is possible at moderate densities, but that a first-order transition occurs at high densities. These authors conclude, however, that no definite conclusion can be drawn.

As discussed in the preface to this chapter, there is recent evidence for continuous melting and a hexatic phase for hard disks [12] (the classic problem studied by Alder and Wainright [155]) and repulsive $1/r^{12}$ potentials [13]. Two-stage melting may also be a feature of 6–12 Lennard-Jones pair potentials [14].

In constant-density simulations, it is possible to confuse coexistence of two phases with an equilibrium hexatic phase. One way around this difficulty is to introduce a hexatic orienting field of strength h_6. For example, one might simulate a hexatic Lennard-Jones potential,

$$V(\mathbf{r}) = 4\varepsilon\left[\left(\frac{\sigma}{r}\right)^{12} - \left(\frac{\sigma}{r}\right)^{6}\right] + h_6 f(r)\cos(6\theta_r), \tag{2.190}$$

where θ_r is the angle made by $\mathbf{r}$ relative to some reference axis. The function $f(r)$ is unity for small r and drops to zero beyond one or two interparticle spacings. A small, nonzero h_6 would wash out a continuous liquid–hexatic transition but leave a hexatic–solid transition intact (see Section 2.4.4). First-order liquid–solid transitions should be unaltered by a small hexatic orienting field. Thus, if Fig. 2.21(a) is correct for $h_6 = 0$, the phase diagram in an orienting field should appear as shown in Fig. 2.21(b). Observation of *two* distinct transition temperatures at a fixed density with $h_6 \neq 0$ would be strong evidence for a first-order transition in the limit $h_6 \rightarrow 0$. A nonzero hexatic orienting field would also suppress the grain-boundary structures observed by McTague *et al.* [66]. Morf [170] has simulated the repulsive $1/r$ system with $h_6 \neq 0$ and finds evidence for a single, continuous, dislocation-mediated melting transition. An orienting field with an *icosahedral* symmetry has been used in simulations of *three*-dimensional supercooled liquids by Steinhardt *et al.* [171]

2.4.4 Incommensurate melting on periodic substrates

Many melting experiments study rare-gas atoms, or molecules like methane, physisorbed onto a periodic substrate [63]. The two kinds of solids which have been observed experimentally are shown in Fig. 2.22. In Fig. 2.22(a), the adatoms are in registry with the minima of the substrate potential, forming a commensurate solid. Registered solids, which are fascinating in their own right, have induced long-range translational order [84]. An incommensurate solid, whose periodicity varies smoothly with coverage and temperature, is shown in Fig. 2.22(b). Translational order decays algebraically in such solids, just as on a smooth substrate. As will be apparent later, phases of this kind can exist over a range of temperatures, even if their periodicity "accidentally" matches that of the substrate.

In an incommensurate solid, the translational order of the substrate can be disregarded at long wavelengths. There is, however, an *orientational* epitaxy, as was first pointed out by Novaco and McTague [172].

Fig. 2.22. Two kinds of physisorbed solid monolayers. (a) A commensurate structure that localizes adatoms in minima provided by the substrate potential of graphite. There are three equivalent sublattices that will accomodate this adatom lattice. (b) An incommensurate solid, with a spacing that varies continuously with density and temperature. In the absence of substrate inhomogeneities, there is an energy cost to infinitesimal rotations, but not to translations of this "floating solid."

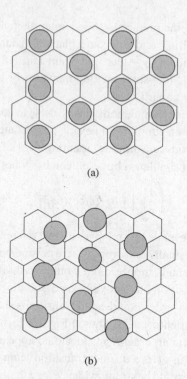

(a)

(b)

To model induced orientational order in isotropic incommensurate solids, consider an elastic free energy of the form

$$F = \frac{1}{2} \int d^2r \, \{2\mu u_{ij}^2 + \lambda u_{kk}^2 - h_p \cos[p\theta(\mathbf{r})]\}. \qquad (2.191)$$

This is just the isotropic continuum elastic theory of a two-dimensional crystal, with a p-fold-symmetric term breaking the rotational invariance. The local orientation angle

$$\theta(\mathbf{r}) \equiv \tfrac{1}{2}[\partial_x u_y(\mathbf{r}) - \partial_y u_x(r)] \qquad (2.192)$$

is measured from the direction of optimal alignment, which need not coincide with a crystallographic axis of the substrate. For triangular solids, resting on hexagonal substrates like graphite, one has $p = 6$. Triangular solids on substrates with a *square* symmetry experience a 12-fold-symmetric orientational perturbation, however (see below) [146].

It is easily checked that fluctuations in $\theta(\mathbf{r})$ are finite, as was already pointed out in the smooth-substrate case $h = 0$. The boundedness of orientation fluctuations allows us to expand the cosine in (2.191). Neglecting an unimportant constant, we find

$$F \approx \frac{1}{2} \int d^2r \, [2\mu u_{ij}^2 + \lambda u_{kk}^2 + \gamma(\partial_x u_y - \partial_y u_x)^2], \qquad (2.193)$$

where the orientational elastic constant is

$$\gamma = \tfrac{1}{8} h p^2. \tag{2.194}$$

As can be easily shown, the diverging Landau–Peierls fluctuations in the displacement field persist and the free energy (2.193) preserves the algebraic decay of translational order present when $\gamma = 0$. The basic results (2.13) and (2.14) still hold provided that we make the replacements

$$\mu \rightarrow \mu + \gamma, \tag{2.195a}$$

$$\lambda \rightarrow \lambda - 2\gamma. \tag{2.195b}$$

The behavior of dislocations interacting on a periodic substrate via Eq. (2.193) was worked out in [146]. The contribution of dislocations to the free energy has the vector Coulomb-gas form of Eq. (2.156), but with different logarithmic and angular couplings, namely

$$K_1 = \frac{4a_0^2}{k_B T}\left(\frac{\mu(\mu+\lambda)}{2\mu+\lambda} + \frac{\mu\gamma}{\mu+\gamma}\right), \tag{2.196a}$$

$$K_2 = \frac{4a_0^2}{k_B T}\left(\frac{\mu(\mu+\lambda)}{2\mu+\lambda} - \frac{\mu\gamma}{\mu+\gamma}\right). \tag{2.196b}$$

This model of dislocation statistical mechanics is particularly easy to solve when $K_2 = 0$. In the absence of this angular coupling, one finds that dislocations unbind at a melting temperature T_m as usual, but with a diverging correlation length [101],

$$\xi_T \sim \exp(b'/|T - T_M|^{2/5}). \tag{2.197}$$

Recursion relations describing the dislocation-unbinding transition for the general situation $K_1 \neq K_2$ have been constructed by Peter Young [154]. In the notation of this review, his recursion relations read

$$\frac{dK_1}{d\ell} = -\frac{3}{4}\pi y^2 e^{K_2/(8\pi)}\left[(K_1^2 + K_2^2)I_0\left(\frac{K_2}{8\pi}\right) - K_1 K_2 I_1\left(\frac{K_2}{8\pi}\right)\right], \tag{2.198a}$$

$$\frac{dK_2}{d\ell} = -\frac{3}{4}\pi y^2 e^{K_2/(8\pi)}\left[2K_1 K_2 I_0\left(\frac{K_2}{8\pi}\right) - \frac{1}{2}(K_1^2 + K_2^2)I_0\left(\frac{K_2}{8\pi}\right)\right], \tag{2.198b}$$

$$\frac{dy}{d\ell} = \left(2 - \frac{K_1}{8\pi}\right)y + 2\pi y^2 e^{K_1/(16\pi)}I_0\left(\frac{K_2}{8\pi}\right). \tag{2.198c}$$

These reduce to (2.166) when $K_1 = K_2$. As shown by Young, these recursion relations display a dislocation-unbinding transition when $K_1^R = 16\pi$, with a diverging correlation length above T_m of the form

(2.168a). Now, however, the exponent $\bar{\nu}$ depends continuously on the ratio

$$\sigma = K_2^R(T_m^-)/K_1^R(T_m^-). \tag{2.199}$$

The exponent $\bar{\nu}$ varies smoothly from $\bar{\nu} = \frac{2}{3}$ when $\sigma = 0$ to $\bar{\nu} = 0.36963\ldots$ when $|\sigma| = 1$.

Although melting of incommensurate solids into fluids proceeds just as in the smooth-substrate case, the situation above the melting temperature is rather different. A possible effective free energy above T_m might be

$$F_A = \int d^2r \left(\frac{1}{2} K_A(T)(\nabla\theta)^2 - h_p\cos(p\theta) \right), \tag{2.200}$$

where the $h_p\cos(p\theta)$ term models the orientational bias of the substrate. For melting of triangular lattices on a hexagonal substrate ($p = 6$), long-range orientational order would be present even above T_m because of this term. All fluids on periodic substrates are trivial examples of the hexatic phase in this sense. The disclination-driven hexatic–liquid transition discussed in the smooth-substrate case will be washed out.

The true situation may be slightly more complicated. As was observed by Novaco and McTague [172], there are typically *two* symmetric degenerate orientational minima, a few degrees off perfect alignment. In thermal equilibrium, the orientation of an incommensurate solid will be locked to one minimum or the other. Above T_m, there will eventually be an Ising phase transition at a temperature T_I, above which the two minima are populated equally [146]. The orientational order parameter $\langle e^{6i\theta}\rangle$ will of course be nonzero even above this transition (see Fig. 2.23(a)). For melting of incommensurate triangular lattices on *square* substrates, there will be an Ising transition above T_m, even without the Novaco–McTague effect. The substrate then presents a 12-fold-symmetric potential, which acts like an *Ising* perturbation on $e^{6i\theta(r)}$. It follows that there will be an Ising-like phase transition, above which $\langle e^{6i\theta}\rangle$ vanishes (Fig. 2.23(b)). Of course, $\langle e^{4i\theta}\rangle$ is nonzero at all temperatures in this case. Note that, although the orientational order parameter drops discontinuously to zero at T_m on a smooth substrate, it has only a shoulder (with an essential singularity) at T_m on a periodic substrate. The Ising transitions discussed above can be viewed as "ghosts" of the hexatic–liquid transition which would occur in the absence of a periodic substrate.

The angular dependence of the X-ray structure function is particularly intriguing just above T_m. In monodomain samples, one expects the 6-fold symmetric pattern of intensity maxima and saddle points shown in Fig. 2.24. The radial width of the pattern is a direct measure of the inverse translational correlation length. The *angular* width of each

(a)

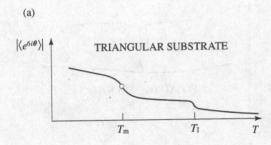

Fig. 2.23. (a) The orientational order parameter for incommensurate triangular lattices melting on a triangular substrate. This quantity decreases sharply at T_m and exhibits an Ising-like singularity at a higher temperature T_I, owing to the Novaco–McTague effect. (b) There are two distinct changes in orientational order for incommensurate triangular lattices melting on a *square* substrate, even without the Novaco–McTague effect. The orientational order parameter now vanishes completely in an Ising-like phase transition at T_I.

(b)

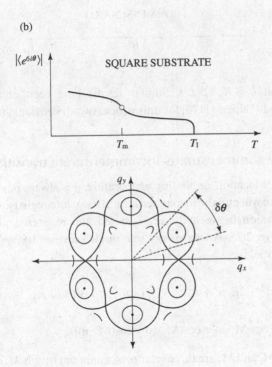

Fig. 2.24. A monodomain X-ray scattering pattern just above the melting temperature of an incommensurate solid on a periodic substrate. Lines of constant intensity are shown.

maximum should be comparable to the root-mean-square fluctuations in the bond angle $\theta(\mathbf{r})$. Upon expanding the free energy (2.200) to quadratic order in $\theta(\mathbf{r})$, a straightforward calculation gives

$$\langle \theta^2(\mathbf{r}) \rangle = \frac{k_B T}{2\pi} \int_{L^{-1}}^{a^{-1}} \frac{q \, dq}{K_A q^2 + \frac{1}{2}p^2 h_p}, \tag{2.201}$$

where a is a microscopic length scale and L is comparable to the linear dimensions of the system. Since it can be shown by the methods used in the smooth-substrate case that K_A diverges like ξ_T^2 near T_m, one finds that

$$\delta\theta \equiv \sqrt{\langle \theta^2(r) \rangle} \sim \xi_T^{-1} \ln^{1/2}(L/a) \tag{2.202}$$

Fig. 2.25. A portion of the pressure–temperature phase diagram for incommensurate melting on a periodic substrate. The dashed line joining A and B denotes floating solids that are "accidentally" commensurate with the large-lattice-constant commensurate solid occurring at lower temperatures. The floating-solid–commensurate transition at B is very similar to the unbinding of dislocations at A, except that reciprocal-lattice vectors of the substrate play the role of Burgers vectors. Note the peculiar shape of the floating-solid–commensurate-transition curves $\delta p_c^{\pm}(T)$. The dotted line is a locus of the Ising-like phase transitions shown in Fig. 2.23.

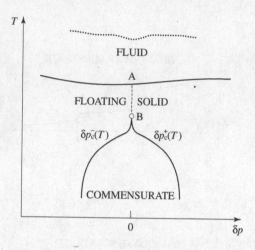

in the limit $h_p \lesssim K_A/(p^2 L^2)$. Similar results have been obtained by Ostlund and Halperin [173] for anisotropic two-dimensional melting.

2.4.5 The commensurate–incommensurate transition

Below the dislocation-unbinding temperature, it is always possible that the substrate will pull the incommensurate lattice into registry. One then has a transition between the unregistered and registered phases, as shown in Fig. 2.25. A free energy that models a small hexagonal substrate potential is

$$F = \frac{1}{2} \int d^2 r \, \{ [2\mu_R u_{ij}^2 + \lambda_R u_{kk}^2 + \gamma_R (\partial_x u_y - \partial_y u_x)^2] + \delta p u_{kk}$$

$$+ h[\cos(\mathbf{M}_1 \cdot \mathbf{u}) + \cos(\mathbf{M}_2 \cdot \mathbf{u}) + \cos(\mathbf{M}_3 \cdot \mathbf{u})] \}, \qquad (2.203)$$

where $\mathbf{M}_1$, $\mathbf{M}_2$ and $\mathbf{M}_3$ are three vectors of minimum length M_0 common to the reciprocal lattices both of the substrate and of the adsorbate, at 120° angles to one another. Elastic constants μ_R, λ_R and γ_R renormalized by dislocations have been used. The parameter δp represents the deviation from a pressure that allows perfect registry with the substrate. Upon making the change of variable

$$\mathbf{u}(\mathbf{r}) = \boldsymbol{\phi}(\mathbf{r}) + \alpha \mathbf{r}, \qquad (2.204a)$$

where

$$\alpha = \delta p/[2(3\mu_R + 2\lambda_R - \gamma_R)], \qquad (2.204b)$$

one eliminates the term linear in u_{kk} and obtains

$$F = \text{constant} + \frac{1}{2} \int d^2r \, \{2\mu_R \phi_{ij}^2 + \lambda_R \phi_{kk}^2 + \gamma_R (\partial_x \phi_y - \partial_y \phi_x)^2$$

$$+ h\{\cos[\mathbf{M}_1 \cdot (\boldsymbol{\phi} + \alpha\mathbf{r})] + \cos[\mathbf{M}_2 \cdot (\boldsymbol{\phi} + \alpha\mathbf{r})]$$

$$+ \cos[\mathbf{M}_3 \cdot (\boldsymbol{\phi} + \alpha\mathbf{r})]\}\}. \tag{2.205}$$

We can now proceed just as in our discussion of $\cos(p\theta)$ perturbations of the two-dimensional XY model (see Section 2.2.5). One finds that the partition function associated with (2.205) may be written as a trace over a Coulomb-gas-like free energy [146]:

$$F_D = -\frac{1}{8\pi} \sum_{\mathbf{r} \neq \mathbf{r}'} \left[K_1' \mathbf{B}(\mathbf{r}) \cdot \mathbf{B}(\mathbf{r}') \ln\left(\frac{|\mathbf{r} - \mathbf{r}'|}{a}\right) \right.$$

$$\left. - K_2' \frac{[\mathbf{B}(\mathbf{r}) \cdot (\mathbf{r} - \mathbf{r}')][\mathbf{B}(\mathbf{r}') \cdot (\mathbf{r} - \mathbf{r}')]}{|\mathbf{r} - \mathbf{r}'|^2} \right]$$

$$+ \ln y_h \sum_{\mathbf{r}} |\mathbf{B}(\mathbf{r})|^2 + iE \sum_{\mathbf{r}} \mathbf{r} \cdot \mathbf{B}(\mathbf{r}), \tag{2.206}$$

with $y_h \approx \frac{1}{2}h$ and

$$E = M_0 \delta p / [2(3\mu_h + 2\lambda_R - \gamma_R)]. \tag{2.207}$$

The allowed values of the $\mathbf{B}(\mathbf{r})$ are reciprocal-lattice vectors of the substrate that are of minimum size M_0, subject to a constraint of vector charge neutrality. In contrast to the case of interacting dislocations, the couplings K_1' and K_2' are *proportional* to the temperature,

$$K_1' = \frac{k_B T (3\mu_R + \lambda_R + \gamma_R)}{4\pi(\mu_R + \gamma_R)(2\mu_R + \lambda_R)}, \tag{2.208a}$$

$$K_2' = \frac{k_B T (\mu_R + \lambda_R - \gamma_R)}{4\pi(\mu_R + \gamma_R)(2\mu_R + \lambda_R)}. \tag{2.208b}$$

The incommensurability parameter δp acts like an "imaginary electric field" on "charges" $\{\mathbf{B}(\mathbf{r})\}$ in the Coulomb-gas representation.

If the periodicity of the substrate is "accidentally commensurate" with that of the adsorbate lattice ($\delta p = E = 0$), the problem is mathematically identical to the statistical mechanics of interacting dislocations with an inverted temperature scale. Upon taking over the results of Young [154], we would expect an instability leading to a commensurate phase (with genuine long-range order) at temperatures low enough that

$$K_1' \lesssim 16\pi. \tag{2.209}$$

Phonon spectra in the low-temperature registered phase will be characterized by a gap, in contrast to spectra of phonons in the incommensurate phase. In order for an accidentally commensurate "floating solid" to exist, one must have both $K_1' \lesssim 16\pi$ and $K_1 \gtrsim 16\pi$. In practice, a window of temperatures within which this is possible will exist provided that

$$\frac{|M_i|}{G_0} \gtrsim \sqrt{12}, \tag{2.210}$$

where G_0 is the minimum reciprocal-lattice vector of the adsorbate. This condition is analogous to the requirement $p > 4$ for an intermediate phase in the $\cos(p\theta)$ XY models.

A portion of a pressure–temperature phase diagram for a physisorbed monolayer is shown in Fig. 2.25. Lines of melting temperatures and orientational Ising transitions are shown, together with the floating–commensurate solid transition (open circle) discussed above. The commensurate–incommensurate transition at pressures other than the accidentally commensurate one $\delta p = 0$ will be discussed further below. The *shape* of the commensurate–incommensurate-transition curve near the open circle follows from the recursion relations (2.198) and the way the incommensurability parameter δp scales under the renormalization transformation. Upon parameterizing the misfit by the magnitude of the imaginary "electric field" appearing in Eq. (2.206), one finds an instability toward increasing E:

$$\frac{dE(\ell)}{d\ell} \approx E(\ell). \tag{2.211}$$

Such a strongly relevant perturbation near the vector Coulomb-gas transition at T_0 gives rise to transition curves $\delta p_c^{\pm}(T)$ of the form

$$\delta p_c^{\pm}(T) = \pm A_{\pm} e^{-B/(|T-T_0|^{\bar{\nu}})}, \tag{2.212}$$

where A_+, A_- and B are positive constants.

Commensurate–incommensurate transitions at pressures $\delta p \neq 0$ have little to do with the statistical mechanics of a vector Coulomb gas. Just on the incommensurate side, one expects that large patches of solid will be pulled into registry. Krypton physisorbed onto graphite, for example, has three equivalent $\sqrt{3} \times \sqrt{3}$ superlattices of substrate sites in which to register. Different registered regions will be connected by soliton-like domain walls across which the phase of the translational order parameter varies rapidly [85, 86]. Figure 2.26 shows a honeycomb arrangement of walls inclined at 120° angles. Striped patterns with all walls parallel are also possible. Although it is locally registered, the

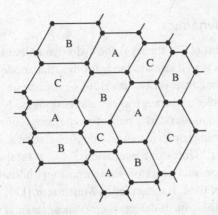

Fig. 2.26. The honeycomb array of domain walls appropriate to krypton physisorbed onto graphite. Walls separate regions of local adsorbate registry into one of three possible graphite superlattices. The substrate lattice and krypton atoms are not shown.

adsorbate lattice will appear incommensurate on scales large relative to the spacing between domain walls. Near a continuous commensurate–incommensurate transition, the wall spacing is very large, becoming infinite in the commensurate phase.

Although one can construct renormalization-group recursion relations perturbatively in $y_h \approx \frac{1}{2}h$ for the Coulomb gas (2.206) in the general case $E \neq 0$, these are of limited use in understanding domain-wall-driven commensurate–incommensurate transitions. In the incommensurate phase, y_h should initially grow with iteration (reflecting regions of strong local registry), but ultimately tend to zero at scales large relative to the wall spacing. Unfortunately, the initial growth takes y_h outside the domain of perturbation theory.

The physics of domain walls has been clarified by Villain [85, 86]. He argues that additional entropy associated with the honeycomb arrangement in Fig. 2.26 causes this configuration to be preferred over the striped phase for large wall spacings. Villain shows further that this entropy drives a first-order commensurate–incommensurate transition. Several authors [174] have suggested that an unbinding of dislocations in the lattice of domain wall nodes in Fig. 2.26 could preempt the incommensurate–commensurate transition. The transition would then be to a highly ordered fluid, interposed between commensurate and incommensurate solid phases.

2.5 Melting dynamics

In this section we review the dynamics of solid, hexatic and liquid phases [175]. Many experiments actually measure the response to both space- and time-dependent perturbations. The behaviors of relaxation times and transport coefficients can be quite spectacular, provided that melting occurs via the dislocation–disclination mechanism discussed in Section 2.4.

2.5.1 Hydrodynamics

The starting point of the theory is the hydrodynamics of liquids, solids and hexatics. As is well known, the hydrodynamic modes in liquids are determined by the conservation laws [176]. Conservation of momentum, number of particles and energy in two dimensions leads to viscous and thermal diffusion modes and a pair of propagating sound frequencies. The quasi-long-range order in a two-dimensional solid leads to two additional long-lived "Goldstone excitations," for a total of six modes. Thermal diffusion and pairs of shear and longitudinal sound waves account for five of these. As observed by Martin *et al.* [177], the sixth mode corresponds to diffusion of defects such as vacancies and interstitials.

The hydrodynamics of hexatics resembles that of nematic liquid crystals. There is a thermal diffusion mode and two longitudinal sound modes, just as in a liquid. The transverse part of the momentum, $g_T(r)$, is coupled to the bond-angle field $\theta(r)$. The linearized equations of motion are [175]

$$\frac{\partial g_T}{\partial t} = K_A(\hat{z} \times \nabla)\nabla^2\theta + \nu\nabla^2 g_T + \xi \tag{2.213a}$$

and

$$\frac{\partial\theta}{\partial t} = \frac{1}{2\rho_0}\hat{z}\cdot(\nabla\times g_T) + \Gamma_6 K_A\nabla^2\theta + Y, \tag{2.213b}$$

where ρ_0 is the equilibrium mass density and K_A is the bond-angle-stiffness constant discussed in Section 2.4. The parameter ν is the kinematic viscosity, while Γ_6 is an orientational relaxation rate. The quantities $\xi(r, t)$ and $Y(r, t)$ are independent Langevin noise sources with autocorrelations in Fourier space,

$$\langle\xi_i(q, t)\xi_j(q', t')\rangle = 2\rho_0\nu k_B T q^2\left(\delta_{ij} - \frac{q_i q_j}{q^2}\right)\delta(q+q')\delta(t-t'), \tag{2.214a}$$

$$\langle Y(q, t)Y(q', t')\rangle = 2\Gamma_6 k_B T\delta(q+q')\delta(t-t'), \tag{2.214b}$$

These relations insure that the Fokker–Planck equations associated with (2.213) leads to an equilibrium probability distribution $\mathcal{P}_0 \propto e^{-F_0/(k_B T)}$, where

$$F_0 = \frac{1}{2}\int d^2r\left(K_A(\nabla\theta)^2 + \frac{1}{\rho_0}|g_T|^2\right). \tag{2.215}$$

The reversible coupling in (2.213b) accounts for the precession of the bond-angle field in the local vorticity. The first term in (2.213a) shows that inhomogeneities in the bond-angle field tend to excite transverse momentum currents. The strength of this effect is controlled by the elastic parameter $K_A(T)$, whose temperature dependence is displayed in Fig. 2.19.

The transverse hexatic eigenfrequencies follow from looking for wave-like solutions of (2.213a), proportional to $e^{i\mathbf{q}\cdot\mathbf{r}-i\omega(q)t}$, in the absence of external noise sources. The transverse eigenfrequencies $\omega_\pm(q)$ are either purely diffusive, or else acquire a real part proportional to q^2. Upon defining

$$\kappa \equiv \Gamma_6 K_A, \tag{2.216}$$

these modes may be written as

$$\omega_\pm(q) = -\tfrac{1}{2} D_\pm q^2 i \tag{2.217}$$

with

$$D_\pm = \nu + \kappa \pm \sqrt{(\nu - \kappa)^2 - K_A/\rho_0}, \tag{2.218}$$

provided that

$$\Delta = K_A/\rho_0 - (\nu - \kappa)^2 < 0. \tag{2.219}$$

For positive Δ, we have

$$\omega_\pm(q) = \tfrac{1}{2}\sqrt{\Delta}\, q^2 - \tfrac{1}{2}(\nu + \kappa) q^2 i. \tag{2.220}$$

For a Lennard-Jones fluid just below T_i, one probably has [175] $\kappa/\nu \gtrsim 7$ and Δ turns out to be negative. We shall see later that ν/κ diverges near T_m and Δ is again negative. Since the quantity $(\nu - \kappa)^2$ must vanish somewhere between T_m and T_i, Δ will be positive over an intermediate range of temperatures. The propagating eigenfrequencies (2.220) will be heavily damped even in this case, however. Including thermal diffusion and the two longitudinal sound modes there are five long-lived hexatic excitations in all.

The dynamics of the bond-angle order parameter

$$\psi(\mathbf{r}, t) = e^{6i\theta(\mathbf{r},t)} \tag{2.221}$$

implied by (2.213) is rather intriguing. Space–time correlations in $\psi(\mathbf{r}, t)$ are easily computed using the result

$$\langle \psi(r, t)\psi^*(\mathbf{0}, 0)\rangle \equiv C(r, t) \approx e^{-18\langle[\theta(\mathbf{r},t)-\theta(0,0)]^2\rangle}, \tag{2.222}$$

which is vaid for Gaussian processes. From (2.213), we find that

$$\langle[\theta(r, t) - \theta(0, 0)]^2\rangle = k_B T \int \frac{d^2 q}{4\pi^2}\left(\frac{A}{q^2}(1 - e^{i\mathbf{q}\cdot\mathbf{r}-i\omega_+(q)t})\right.$$
$$\left. + \frac{B}{q^2}(1 - e^{i\mathbf{q}\cdot\mathbf{r}-i\omega_-(q)t})\right), \tag{2.223}$$

where A and B are complicated functions of the hydrodynamic parameters. For large r at zero time separation one of course recovers the static result (2.183). For very long times, there is algebraic decay with time,

$$\lim_{t \to \infty} \langle \psi(\mathbf{r}, t)\psi^*(\mathbf{0}, 0)\rangle \sim t^{-9k_B T/(\pi K_A)} \sim t^{-\eta_6(T)/2}. \tag{2.224}$$

Thus, correlations decay half as fast in time as they do in space. More generally, we find a scaling form for $C(r, t)$, at large space–time separations, of

$$C(r, t) = \frac{1}{r^{\eta_6}} \Phi(r/t^{1/2}), \tag{2.225}$$

where the scaling function $\Phi(x)$ can be extracted from (2.222) and (2.223).

This discussion has neglected weak singularities [178] connected with the slow decay of time correlation functions in two dimensions. When nonlinearities are added to linearized fluid hydrodynamics, one typically finds weak logarithmic divergences in transport coefficients, at long wavelengths and low frequencies. As shown by Zippelius [179], there are problems of this kind in two-dimensional solids and hexatics, as well as in liquids. It is unlikely, however, that such weak singularities have observable consequences at experimentally accessible wavevectors and frequencies. The hydrodynamics described above is appropriate for standard molecular-dynamics simulations and for free-standing smectic liquid-crystal films. In these systems, the conservation laws appropriate for truly isolated materials remain intact. With some modifications, the theory presented here could be taken over to other interesting experimental situations. Melting of a physisorbed monolayer on glassy substrates, for example, could be treated by discarding conservation of momentum. Conservation of number density would be broken if the adsorbate were in equilibrium with a dense vapor phase. Henceforth, we focus on a particularly simple ideal situation, in which only conservation of energy is discarded. Neglect of conservation of energy is, of course, not a fundamental limitation of the theory.

2.5.2 Dynamics near the dislocation-unbinding transition

Because the shear modulus remains finite at T_m^- (just like the superfluid density at the vortex unbinding transition in a superfluid), the hydrodynamic modes of a solid will propagate even at temperatures above T_m at wavevectors q such that $q\xi_+ \gtrsim 1$ (Fig. 2.27). At long wavelengths just above T_m, we regard the hexatic as a solid with a dilute concentration of free dislocations. A theory analogous to the "Maxwell's equations" treatment of third sound in a superfluid then leads to excitations

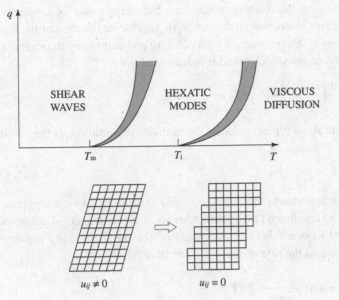

Fig. 2.27. Transverse excitations in two-dimensional matter. Because elastic constants are finite at T_m and T_i, the characteristic excitations of the solid and hexatic spill over to high temperatures at finite wavevector **q**.

Fig. 2.28. An important dynamical process near T_m. Solids can relax a strain imposed by external boundary conditions by first nucleating pairs of dislocations, which then move to the surface. The dislocations effecting this change are not shown.

characteristic of a hexatic, with diverging transport coefficients [175]. Here, we give a simple physical argument for a diverging viscosity and discuss other salient results near T_m.

A basic physical process mediated by dislocations is shown in Fig. 2.28. Pairs of dislocations nucleate in the interior of the sample and relax the strain by unbinding and moving to the edges [153]. An analogous vortex-mediated process for superfluids is shown in Fig. 2.13. Relaxation happens very slowly in a solid, at a rate proportional to a temperature-dependent nonlinear power of the strain. Strains decay much more rapidly when free dislocations are present. Taking over an argument first applied by Shockley [180] to dislocation loops in bulk materials, we show that free dislocations give rise to a fluid-like viscosity.

The rate at which moving dislocations relax a uniform shear strain u_{xy}^0 [153] is

$$\frac{du_{xy}^0}{dt} = \sum_\alpha \frac{a_0}{2\Omega_0} \mathbf{z} \cdot \left(\mathbf{b}_\alpha \times \frac{d\mathbf{r}_\alpha}{dt} \right), \qquad (2.226)$$

where Ω_0 is the area and α indexes the positions $\mathbf{r}_\alpha$ of free dislocations carrying the (dimensionless) Burgers vector $\mathbf{b}_\alpha$. Physically, (2.226) says that u_{xy}^0 slips each time a dislocation goes by. Under conditions of constant applied shear stress σ_{xy}, one expects that free dislocations eventually acquire a velocity proportional to σ_{xy}:

$$\frac{d\mathbf{r}_\alpha}{dt} \sim a_0 b_\alpha \mu(\mathbf{b}_\alpha) \sigma_{xy}. \qquad (2.227)$$

Most vector and tensor indices have been suppressed for simplicity and the dislocation-mobility tensor $\mu_{ij}(\mathbf{b}_\alpha)$ is an even function of the Burgers vector $\mathbf{b}_\alpha$. Upon averaging over (2.226) and expressing the strain rate in terms of the velocity field $\mathbf{v} \equiv d\mathbf{u}/dt$, we have

$$\frac{1}{2}\left(\frac{dv_x}{dy} + \frac{dv_y}{dx}\right) \sim a_0^2 \bar{\mu} n_f \sigma_{xy}, \tag{2.228}$$

where $\bar{\mu}$ is a typical value of the mobility tensor and n_f is the density of free dislocations,

$$n_f \approx \xi_+^{-2}. \tag{2.229}$$

We assume that the mobility remains finite at T_m, like the vortex mobility in superfluids [141]. Remembering that the dynamical shear viscosity $\eta(T)$ in a fluid is the constant of the proportionality between the stress and the rate of the strain, we have finally

$$\eta(T) \sim \frac{1}{a_0^2 \bar{\mu} n_f} \sim \xi_+^2(T). \tag{2.230}$$

As T approaches T_m from above, the shear viscosity mimics the strong divergence (2.168) in the translational correlation length.

The above argument ignores complications associated with the dynamics of the bond-angle field. A more detailed theory [175] also gives (2.230) and shows that $\kappa = \Gamma_6 K_A$ remains finite at T_m^+. Both resulting hexatic frequencies are purely diffusive, with one diffusivity D_+ diverging like ξ_+^2 and the other remaining finite. Longitudinal sound, which has the following characteristic frequency above T_m,

$$\omega_1(q) = \pm c_1 q - \tfrac{1}{2} D_1 q^2 i, \tag{2.231}$$

is also quite singular near T_m. The damping constant D_1 diverges as $T \to T_m^+$ in the hexatic hydrodynamic region [175],

$$D_1 \sim \xi_+^2. \tag{2.232}$$

Compressional waves couple strongly to a relaxational mode that becomes diffusion of defects below T_m and exhibits critical slowing down above T_m, decaying at a rate $\sim \xi_+^{-2}$. Measurements of $D_1(T)$ at *finite* wavevectors should exhibit a maximum above T_m. The physical reason for the maximum is the polarization of weakly bound pairs of dislocations by the sound wave. The velocity of sound $c_1(T)$ will rise smoothly with decreasing temperature to a value appropriate to the solid. Below T_m, there are additional contributions to the damping both of shear and of longitudinal sound due to strain-induced dislocation unbinding. These behave like a temperature-dependent power of q at long wavelengths.

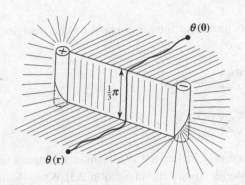

Fig. 2.29. A cut joining two oppositely charged disclinations in the hexatic bond-angle field $\theta(\mathbf{r})$. In order for $\theta(\mathbf{r})$ to have the same value when it is reached by all possible paths from $\theta(0)$, a cut of magnitude $\pi/3$ joining the two disclinations must be introduced.

2.5.3 Dynamics near the disclination-unbinding transition

As shown in Fig. 2.27, hexatic excitations persist into the liquid phase at finite wavevectors, because $K_A(T_i^-)$ is finite. To track the evolution of these modes into liquid-like excitations, we consider effects of a dilute gas of free *disclinations*. A problem immediately arises in generalizing hexatic hydrodynamics to include disclinations, since the bond-angle field $\theta(\mathbf{r}, t)$ becomes multiple-valued unless cuts (Fig. 2.29) joining the pairs of disclinations are introduced. To avoid having to follow the dynamics of the cuts, we introduce a hexatic "superfluid velocity"

$$v_h = \text{``}(\nabla\theta)\text{''}$$
$$\equiv (\psi^*\nabla\psi - \psi\nabla\psi^*)/(12i|\psi|^2), \tag{2.233}$$

where $\psi(\mathbf{r}, t) \propto e^{6i\theta(\mathbf{r},t)}$ is well defined everywhere except at the actual locations of the disclination changes. By "$\nabla\theta$" we mean a gradient with the delta-function contribution arising from the cuts in the θ field subtracted off. This is done automatically by differentiating the ψ field as indicated in (2.233). The field $v_h(\mathbf{r}, t)$ is like a superfluid velocity in that it is nonsingular and single-valued except at the cores of disclinations.

The equations describing hexatics with disclinations are [175]

$$\frac{\partial \mathbf{g}_T}{\partial t} = K_A(\hat{\mathbf{z}}\times\nabla)(\nabla\cdot\mathbf{v}_h) + \nu\nabla^2\mathbf{g}_T, \tag{2.234}$$

$$\frac{\partial \mathbf{v}_h}{\partial t} = \frac{1}{2\rho_0}\nabla[\hat{\mathbf{z}}\cdot(\nabla\times\mathbf{g}_T)] + \kappa\nabla(\nabla\cdot\mathbf{v}_h) + \hat{\mathbf{z}}\times\mathbf{J}_6, \tag{2.235}$$

$$\frac{\partial S_6}{\partial t} + \nabla\cdot\mathbf{J}_6 = 0, \tag{2.236}$$

$$\nabla\times\mathbf{v}_h = S_6, \tag{2.237}$$

where $S_6(\mathbf{r}, t)$ is a disclination charge density and $\mathbf{J}_6(\mathbf{r}, t)$ is the corresponding disclination current density. We have suppressed for simplicity various Langevin noise sources that allow for thermal fluctuations.

In terms of a collection of integer disclination charges s_α located at positions $\mathbf{r}_\alpha$, we have

$$S_6(\mathbf{r}, t) = \frac{\pi}{3} \sum_\alpha s_\alpha \delta[\mathbf{r} - \mathbf{r}_\alpha(t)], \tag{2.238}$$

$$\mathbf{J}_6(\mathbf{r}, t) = \frac{\pi}{3} \sum_\alpha s_\alpha \frac{d\mathbf{r}_\alpha}{dt} \delta[\mathbf{r} - \mathbf{r}_\alpha(t)]. \tag{2.239}$$

These equations may be derived in a fashion analogous to the discussion of "Maxwell's equations" in Section 2.3. When disclinations are bound, $\mathbf{J}_6$ vanishes at long wavelengths and Eq. (2.234) and (2.235) are equivalent to the hydrodynamic results (2.213).

Regarding a liquid just above T_i as a hexatic with a dilute concentration of free disclinations, equations (2.234)–(2.237) can be closed with a constitutive relation for the current,

$$\mathbf{J}_6 = \frac{\pi D_6 K_A(T_m^-)}{3} \frac{1}{k_B T} n_f^{(6)} \hat{\mathbf{z}} \times \mathbf{v}_h + D_6 \nabla S_6, \tag{2.240}$$

where D_6 is a disclination diffusion constant and the density of free disclinations $n_f^{(6)}$ is related to a diverging orientational correlation length

$$n_f^{(6)} \approx \xi_6^{-2} \sim e^{-\text{constant}/|T - T_i|^{1/2}}. \tag{2.241}$$

A dielectric constant describing effects of bound pairs of disclinations has been set to unity for simplicity. One finds two distinct transverse eigenfrequencies above T_i, one describing the relaxation of v_h by disclination currents,

$$\omega_+ \approx -D_6 \xi_6^{-2} i, \tag{2.242}$$

and another which is ordinary viscous diffusion,

$$\omega_- \approx -\nu q^2 i, \tag{2.243}$$

Note that the coefficient of q^2 is just a constant in (2.243), so that the shear viscosity of the liquid remains finite as $T \to T_i^+$.

The effective shear viscosity in the hexatic phase should be *larger* than that in the liquid. If one defines an effective shear viscosity in terms of the transverse momentum autocorrelation function

$$\nu_{\text{eff}}^{-1} \equiv \lim_{q \to \infty} \lim_{\omega \to \infty} \frac{q^2}{2k_B T \rho_0} \langle |\mathbf{g}_T(\mathbf{q}, \omega)|^2 \rangle, \tag{2.244}$$

hexatic hydrodynamics gives

$$\nu_{\text{eff}} = \nu \left(1 + \frac{K_A}{4\kappa\nu\rho_0}\right), \tag{2.245}$$

in contrast to the result $\nu_{\text{eff}} = \nu$ in a liquid. The same effective viscosity (2.245) results from a conventional viscosity measurement in a hexatic fluid, provided that the boundary conditions pin the bond orientations at the walls [175]. The jump discontinuity in ν_{eff} across T_i (which becomes a rounded shoulder at finite frequencies and wavevectors) is an important dynamical signature of the disclination-unbinding transition. A 50% increase in ν_{eff} was in fact observed in a simulation by Frenkel and McTague [166] at what they identified as a disclination-unbinding temperature.

Longitudinal sound does not couple in an important way to vorticity–bond-angle excitations, since it relaxes much more rapidly. Because of mode-coupling nonlinearities neglected here, compressional sound can in fact decay into transverse momentum and bond-orientation modes. Since bond-angle excitations are strongly affected by an unbinding of disclinations, one might expect a small anomaly in the damping of sound waves at T_i.

2.6 Anisotropic melting

2.6.1 Smectics, cholesterics and the Rayleigh–Bénard problem

Additional investigations of defect-driven melting have focused on effects of anisotropy [173, 181, 182], which are crucial to an understanding of, say, smectic liquid-crystal films with tilted molecules and lipid monolayers with tilted hydrocarbon chains. A most extreme type of anisotropic translational order occurs in materials that are liquid-like in one direction and solid-like in another. Examples of such layered systems include smectic and cholesteric films and Rayleigh–Bénard convective rolls.

In principle, it may be possible to study layered, two-dimensional order in nematogens, floating on water, with two hydrophyllic ends [183]. At sufficiently low temperatures, one can envisage the smectic-like in-plane translational order shown in Fig. 2.30(a). Although few nematic-like monolayer films have been made, a closely related ordering occurs in thin films of smectic-C liquid crystals [53, 54]. In freely suspended films containing a few smectic-C monolayers, the projections of the tilted molecules play the role of "nematogens," with a vector symmetry instead of the true nematic symmetry of double-headed arrows. Anisotropic molecules physisorbed onto a periodic substrate may behave in a similar fashion. Although in-plane layered order is always a possibility, one must also allow for crystallization into a genuine two-dimensional solid, with anisotropic translational order in both directions. The connection between melting of this kind of solid and layered order will be discussed in the next subsection.

Cholesteric liquid crystals exhibit layered order, the periodicity of

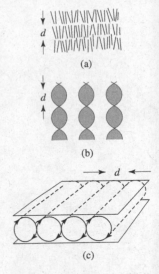

Fig. 2.30. Layered order (a) in two-dimensional smectics, (b) in cholesterics and (c) in the Rayleigh–Bénard problem. In (b), rod-shaped molecules rotate about a vertical axis with pitch d.

which is hundreds of nanometers, rather than the ~ 2–3 nm repeat distance of a smectic. In cholesterics, chiral nematogens rotate slowly in space around a preferred axis (see Fig. 2.30(b)) and the planes of constant phase comprise a kind of layering. One can prepare cholesteric films with thicknesses comparable to the pitch (see Fig. 6.27 of [183]). Unfortunately, the energy scale for fluctuations like dislocations and disclinations is quite high, so it is not clear whether their effects will be visible at experimentally accessible temperatures.

Figure 2.30(c) shows Rayleigh–Bénard convection cells, which are an intriguing example of layered order far from equilibrium. Above a threshold difference in temperature applied across two confining plates, heat is often transferred by a layered pattern of convective rolls. Fluctuation effects have been studied by Graham [184] and by Swift and Hohenberg [185], who use ideas and methods developed for systems in thermal equilibrium. To the limited extent to which a thermal-equilibrium description is appropriate, these systems can be studied in the same way as two-dimensional smectics and cholesterics.

2.6.2 Melting of layered materials

Order in the layered systems described above can be understood in terms of the effective free energy [183]:

$$F = \frac{1}{2}B \int d^2r[(\partial_z u)^2 + \lambda^2(\partial_x^2 u)^2]. \tag{2.246}$$

Here, $u(\mathbf{r}) = u(x, z)$ is the layer displacement in the $\hat{z}$ direction from a preferred arrangement of equidistant layers parallel to the $\hat{x}$ direction. The system is disordered in the $\hat{x}$ direction, but is solid-like perpendicular to the layers. The quantity B is an elastic constant, while λ is a length comparable to the spacing between the layers. The angle the layer normal makes with the $\hat{z}$ axis is, for small displacements,

$$\theta(\mathbf{r}) = -\partial_x u(\mathbf{r}). \tag{2.247}$$

This quantity is analogous to the bond-angle field introduced for triangular lattice melting in Section 2.4. Because the free energy must be invariant under uniform layer rotations, a term proportional to $(\partial_x u)^2 = \theta^2$ is inadmissible in (2.246).

To assess the effect of phonon fluctuations on translational order, we consider correlations in

$$\psi(\mathbf{r}) \equiv e^{iq_0 u(\mathbf{r})}, \tag{2.248}$$

where q_0 is the wavevector associated with the layering. One easily finds from (2.246) that the effect of phonons is quite significant [184]:

(a)   (b)   (c)

Fig. 2.31. Dislocation and disclination defects in a layered medium. Note that the disclinations (b) and (c) can be merged together to make the dislocation (a), shown together with its Burgers circuit.

$$\langle\psi(\mathbf{r})\psi^*(\mathbf{0})\rangle \sim \begin{cases} \exp\left(\dfrac{-q_0^2 k_B T}{B}\sqrt{\dfrac{|z|}{4\pi\lambda}}\right), & x^2 \ll \lambda z, \quad (2.249) \\[3ex] \exp\left(\dfrac{-q_0^2 k_B T}{4B\lambda}|x|\right), & x^2 \gg \lambda z. \quad (2.250) \end{cases}$$

Phonons are considerably less important in decorrelating the layer orientations. Upon defining a nematic-like order parameter

$$\mathbf{N}(\mathbf{r}) \equiv \begin{pmatrix} \cos\theta(\mathbf{r}) \\ \sin\theta(\mathbf{r}) \end{pmatrix}, \tag{2.251}$$

we find that orientational order persists, even in the presence of thermally excited phonons:

$$\lim_{r\to\infty}\langle\mathbf{N}(\mathbf{r})\cdot\mathbf{N}(\mathbf{0})\rangle = \text{constant}. \tag{2.252}$$

It is natural to ask how dislocations and disclinations affect these results (see Fig. 2.31). Because isolated dislocations have a finite energy E_D in layered materials [152, 186], one expects them to occur with density

$$n_D \sim e^{-E_D/(k_B T)} \tag{2.253}$$

in thermal equilibrium, although note from Fig. 2.31 that a dislocation can itself be regarded as a tightly bound pair of oppositely charged disclinations. The combined action of phonon fluctuations and dislocations is to produce a phase at a finite temperature with persistent orientational order in the local layer normal [182]. At scales greater than

$$\xi_D \sim e^{-E_D/(2k_B T)} \tag{2.254}$$

the properties are summarized by a nematic-like effective free energy,

$$F_N = \frac{1}{2}\int d^2r\,\{K_1(T)(\nabla\cdot\mathbf{N})^2 + K_3(T)[\mathbf{N}\times(\nabla\times\mathbf{N})]^2\}. \tag{2.255}$$

Fig. 2.32. (a) The sequence of transitions for melting of triangular lattices. (b) The analogous sequence for layered melting.

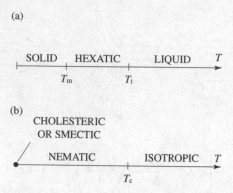

The usual twist Frank constant K_2 is absent in two dimensions. At sufficiently small wavevectors, the Frank constants K_1 and K_3 are driven to a common value $K(T)$ [187] and one finds algebraic decay of orientational order,

$$\langle e^{2i\theta(\mathbf{r})} e^{-2i\theta(0)} \rangle \sim r^{-\eta_2},$$

(2.256a)

where

$$\eta_2 = 2k_B T / [\pi K(T)].$$

(2.256b)

Just as in hexatics, disclinations become important in the presence of free dislocations. The free energy (2.255) produces a weaker logarithmic binding between oppositely charged disclinations. In the vector-like smectic-C liquid-crystal films discussed earlier, the important excitations are logarithmically interacting vortices, rather than nematic disclinations. In either case, the relevant excitations unbind at a finite temperature T_c, producing an isotropic material. Our conclusions for melting of layered structures are summarized and contrasted with ordinary melting in Fig. 2.32. The same sequence of phases is evident for both systems, except that the dislocation-unbinding transition has been suppressed to $T = 0$ in layered materials, due to the peculiar nature of parabolic elasticity theory.

The nematodynamics near the $T = 0$ dislocation-unbinding transition has also been studied [182]. A simple diffusive model of layer motion, coupled to a finite density of dislocations, gives rise to the characteristic relaxation rate $\omega(\mathbf{q})$ of a nematic with no conservation laws,

$$\omega(\mathbf{q}) = -i\Gamma_N (K_1 q_x^2 + K_3 q_z^2).$$

(2.257)

Such a model might be appropriate for layering on a substrate, or for the Rayleigh–Bénard problem. The nematic kinetic coefficient Γ_N vanishes at low temperatures,

$$\Gamma_N \sim \xi_D^{-2}.$$

(2.258)

The rate of decay of uniform strains $\partial_z u$ exhibits critical slowing down, relaxing at a rate proportional to $D\xi_D^{-2}$, where D is a dislocation diffusion constant. This rate is related to the time it takes a dislocation to diffuse one correlation length.

Experiments on tilted liquid-crystal films are often done with an in-plane orienting electric field $\mathbf{E}$. Such a field will stabilize layered order at finite temperatures. Instead of (2.246), the effective free energy is now

$$F = \int d^2r \left(\frac{1}{2} B[(\partial_z u)^2 + \lambda^2(\partial_x^2 u)^2] - E\cos(\partial_x u) \right), \tag{2.259}$$

where we assume that the layers orient themselves perpendicular to $\mathbf{E}$. Since fluctuations in $\partial_x u$ are well behaved even for $E = 0$,

$$\langle (\partial_x u)^2 \rangle < \infty, \tag{2.260}$$

we can expand the argument of the cosine to obtain an anisotropic XY-like free energy to leading order in a gradient expansion,

$$F \approx \text{constant} + \frac{1}{2} \int d^2r \, [B(\partial_z u)^2 + E(\partial_x u)^2]. \tag{2.261}$$

Note that the anisotropy can be controlled experimentally by varying the electric field. Dislocations will now interact logarithmically and unbind into a "nematic" (i.e. a smectic-C liquid crystal) via a Kosterlitz–Thouless transition at a finite temperature. The electric field will wash out the smectic-C–smectic-A vortex-unbinding transition, however. The orienting effect of a periodic substrate on layered order is quite similar [173].

2.6.3 Melting of anisotropic lattices

At low temperatures, anisotropic molecules should eventually crystallize into uniaxial solids. As shown by Ostlund and Halperin [173], melting of anisotropic solids can proceed somewhat differently from the scenario for isotropic triangular lattices. Two anisotropic crystalline structures are shown in Figs. 2.33(a) and (b). In both cases, two dislocations (of type I) have their Burgers vectors aligned along a symmetry direction, while the four remaining ones (of type II) have Burgers vectors at an angle to this reflection axis. A square lattice can be viewed as a special case of such structures. For the situations studied in most detail, the anisotropic molecules are assumed to align along or exactly in between the bonds of the lattice.

The free energy of these solids may be written quite generally [41] as

$$F = \frac{1}{2} \int d^2r \, C_{ijkl} u_{ij} u_{kl}, \tag{2.262}$$

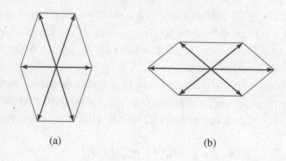

Fig. 2.33. Two different kinds of anisotropic crystalline structures with a plane of reflection symmetry. Elementary Burgers vectors are shown as heavy lines. One pair of these vectors is of a different length from the remaining four. The pair or quartet of Burgers vectors with the shortest length probably unbinds first.

(a)   (b)

where C_{ijkl} is the elastic tensor and u_{ij} is the strain field. Ostlund and Halperin allow for the four independent components of C_{ijkl} consistent with an axis of reflection symmetry. The energies of interaction between oppositely charged pairs of type-I and type-II dislocations of separation r goes like $K_I \ln(r/a)$ and $K_{II} \ln(r/a)$, respectively, where K_I and K_{II} depend on the elastic constants. A naïve application of the Kosterlitz–Thouless "entropy argument" would predict dissociation of the different pairs of dislocations at two distinct temperatures. However, if the type-II dislocations unbind first, they take type-I dislocations with them. The physical reason for this is that two free type-II dislocations of the appropriate orientation are equivalent to one free type-I dislocation (see Fig. 2.33) at wavelengths long relative to their separation. Melting into a nematic-like phase, characterized by quasi-long-range order in both bond angles and molecular orientations, then proceeds.

Type-I melting, which occurs when dislocations along the axis of symmetry unbind first, is more interesting. One then cannot form the remaining dislocations from the unbound ones, so translational order is strongly decorrelated only in the direction perpendicular to the Burgers vectors of the unbound pair. Ostlund and Halperin find that the properties above the dissociation temperature T_m are just those of the two-dimensional smectic discussed in Section 2.6.2. Although the behavior can be smectic-like over a wide range of lengths and temperatures, a small concentration of the remaining type-II dislocations is in fact unbound at the largest separations. Just as in layered materials one passes with increasing temperature or wavelength first through regimes describable in terms of the two-Frank-constant nematic free energy (2.255) and then through a one-Frank-constant ($K_1 = K_3$) nematic regime.

This picture is supported by detailed renormalization-group recursion relations involving the four anisotropic solid elastic constants [173]. In all types of anisotropic melting, there is a translational correlation length that diverges near T_m as in Eq. (2.168a), but with $\bar{\nu} = \frac{1}{2}$. This conclusion also applies as a special case to square lattices, which have only three independent elastic constants.

In addition to anisotropic liquid-crystal films, the above results are applicable to anisotropic incommensurate solids on periodic substrates. The smectic phase in type-I melting is stabilized by the substrate. A commensurate smectic-like physisorbed monolayer has been conjectured from neutron-scattering data by Coulomb *et al.* [188].

2.6.4 Bulk liquid-crystal phases

Bulk smectic liquid crystals may be viewed as stacked layers of two-dimensional matter, so it is not surprising that results obtained in two dimensions might be helpful in classifying smectic phases. Birgeneau and Litster [189] argued that the coupling between smectic layers could lead to a "stacked-hexatic" phase with long-range orientational order, but short-range in-plane translational order. They suggested that single-domain X-ray-diffraction patterns of smectic-B liquid crystals with a hexagonal array of six fuzzy spots constituted evidence for a stacked hexatic phase. More precise scattering measurements [55, 190] have shown that some smectic-B liquid crystals are, in fact, three-dimensional crystals with a large diffuse scattering. Other investigations, however [191, 192], have uncovered an apparently genuine stacked hexatic phase in the compound 650 BC. This discovery is a powerful confirmation of one aspect of the theory of melting discussed in Section 2.4. See [9] for a more up-to-date account of experimental developments.

The Landau-theory discussion of translational and orientational order presented in Section 2.4.1 is also applicable to bulk smectics and is relevant in particular to the smectic-A-to-stacked-hexatic transition. It can be shown [193] that the in-plane X-ray structure factor in the stacked-hexatic phase takes the form

$$S(\mathbf{q}, T) = \sum_{n=0}^{\infty} c_n(q, T)[\cos(6n\theta_{\hat{q}})\,\mathrm{Re}\,\langle\psi^n(\mathbf{r})\rangle + \sin(6n\theta_{\hat{q}})\,\mathrm{Im}\,\langle\psi^n(\mathbf{r})\rangle],$$

$$(2.263)$$

where $\theta_{\hat{q}}$ is the angle the scattering vector **g** makes with the $\hat{x}$ axis and the coefficients $c_n(q, T)$ are weak functions of temperature. The contours of constant scattering in the (q_x, q_y)-plane are as in Fig. 2.24. Thus, single-domain X-ray-diffraction experiments provide a direct measure of the orientational order parameter $\psi(r) \propto e^{6i\theta(r)}$ in bulk hexatics.

There has also been some discussion of smectic-F liquid crystals [194, 195] as possible candidates for a *tilted* stacked hexatic. Possible tilted liquid-crystal phases in a few isolated smectic layers have been studied in some detail [181]. As many as seven distinct phases are possible, including anisotropic (smectic-H) and isotropic (smectic-B) solids, hexatics, liquids (smectic-A) and anisotropic liquids (smectic-C).

Smectic-C liquid crystals have induced 6-fold bond-angle order both in two and in three dimensions. Thus, there is no fundamental distinction between a smectic-C and a tilted hexatic. Induced bond-angle order vanishes quite rapidly near the smectic-C–smectic-A transition in $d = 3$, like $|T - T_c|^{3.59}$ [193].

Although a smectic-F could simply be a smectic-C with an abnormally large amount of orientational order, the theory of anisotropic two-dimensional melting may provide an alternative explanation. As can easily be checked, interlayer couplings stabilize the in-plane layered order found for type-I melting of anisotropic solids. The resulting bulk structure is solid-like in two directions and liquid-like in one! The lower critical dimension for such "bismectics" with solid-like order in $(d-1)$ directions and liquid-like order in one is $d_c = 2.5$. Thus there is genuine long-range translational order in three dimensions. The tilt degree of freedom will superimpose in-plane bond-orientational order [181]. Even if smectic-F crystals are not "bismectics," it would be interesting to search for such bulk liquid-crystal phases experimentally.

One can also speculate on the nature of the bulk cholesteric "blue phases" which occur interposed between isotropic and conventional cholesteric phases [196, 197]. Perhaps one of these blue phases could be a "nematic" of the kind mentioned above as a possibility in two-dimensional cholesteric films. Translational order in the pitch would persist over large chunks or "cybotactic clusters" of material, with the average layer normal playing the role of a nematic order parameter. In equilibrium, there would be a first transition from such a material into an isotropic phase.

2.7 Line singularities in three dimensions

2.7.1 Vortex rings in superfluids and superconductors

Feynman [198] has speculated that the λ transition in superfluid ^{4}He could be understood in terms of a sudden proliferation of unbound vortex rings (see also Popov [199]). The behavior near the λ point is now quite well understood in terms of the $\varepsilon = 4 - d$ expansion [15], which focuses very little on defects like vortex rings. Both the diameter of the core of a vortex and the width of the interface between the core and the surrounding superfluid are believed to scale like the superfluid coherence length $\xi_-(T)$. Since $\xi_-(T)$ diverges as $T \to T_\lambda^-$, a "vortex ring" becomes indistinguishable at long wavelengths from an amorphous blob of normal material. There is no great advantage in describing this phase transition in terms of a set of line singularities like vortex rings. The situation is quite different in two dimensions, in which point vortices have finite cores right up to the transition. The assumption that vortices remain sharp singularities even at long wavelengths is basic to the Kosterlitz–Thouless ideas about superfluid helium films. This assertion amounts to the neglect of fluctuations in the amplitude of the order parameter discussed in Section 2.2.2.

Although amplitude fluctuations are important at long wavelengths in $d = 3$, there is certainly nothing fundamentally wrong in modelling microscopic excitations in superfluid helium in terms of a set of interacting vortex rings. The superfluid velocity is then the solution of

$$\nabla \times \boldsymbol{v}_s = \frac{2\pi\hbar}{m} \mathbf{M}(\mathbf{r}), \tag{2.264}$$

where the integral of the vortex line density $\mathbf{M}(\mathbf{r})$ over a surface gives the total number of lines piercing that surface. The condition of overall "vortex charge neutrality" in two dimensions now becomes the requirement that lines cannot stop or start within the fluid. If we coarse-grain enough that $\mathbf{M}(\mathbf{r})$ can be regarded as a continuous function of $\mathbf{r}$, this restriction means that $\mathbf{M}(\mathbf{r})$ must satisfy

$$\nabla \cdot \mathbf{M}(\mathbf{r}) = 0. \tag{2.265}$$

Given that we have a free-energy density of the form

$$F = \frac{1}{2}\rho_s \int d^3r \, |v_s(\mathbf{r})|^2, \tag{2.266}$$

it is straightforward to evaluate the renormalized superfluid density in the presence of a finite density of unbound vortex rings [200]. The calculation proceeds as in the discussion of the wavevector-dependent superfluid density in Section 2.4.4. Not surprisingly, the renormalized superfluid density vanishes at long wavelengths, as expected in a normal fluid.

It is much more difficult to construct a quantitative theory of the phase transition itself in terms of interacting vortex rings. The energy of the interaction between two rings occupying closed contours C_1 and C_2 is

$$E_{\text{int}} = \frac{1}{2}\frac{\hbar^2}{m^2}\rho_s \oint_{C_1} \oint_{C_2} \frac{d\mathbf{r}_1 \cdot d\mathbf{r}_2}{|\mathbf{r}_1 - \mathbf{r}_2|}, \tag{2.267}$$

where the line elements point in the direction of the local vorticity. There are also "self-energy" contributions from each vortex loop. The statistical mechanics of wiggling vortex rings coupled by a long-range interaction seems quite formidable. Fortunately, the λ point seems rather well described by the more traditional expansions about Landau theory near four dimensions.

The situation appears to be more complicated in bulk superconductors. Halperin *et al.* [201] have argued that, at least near four dimensions, fluctuation-corrected Landau theory always leads to a first-order phase transition. An analysis directly in three dimensions in terms of Abrikosov flux lines leads to different conclusions, however [202]. In contrast to neutral superfluids, the charged currents surrounding vortex lines in superconductors produce a screened, short-range interaction [203]. Helfrich and Müller [204] have observed that low-temperature defect loops with short-range interactions are closely related to the loops appearing in diagrammatic high-temperature series expansions of spin models [15]. Their analysis suggests in particular that a loop-unbinding transition in superconductors is in the same universality class as three-dimensional XY models with an *inverted* temperature scale. These conclusions were reached and confirmed via duality arguments and Monte Carlo simulations for a lattice model of superconductivity by Dasgupta and Halperin [202]. (These authors also argue that loops in neutral superfluids interacting according to Eq. (2.267) exhibit a conventional XY-like phase transition.) The fluctuation-driven first-order transition predicted for bulk superconductors by Halperin *et al.* [201] could be too small to be observed in simulations or experiments. Nevertheless, the conflict with the predictions of the loop-unbinding picture raises important and, at present, unresolved questions about the validity of the ε expansion in $d = 3$.

Interacting polymer rings provide an interesting example of sterically interacting loops. In the polymerization of sulfur solutions in equilibrium with monomers, for example, polymer rings can grow and contract at the expense of the monomers [205]. If closed loops are strongly preferred energetically to open ones, one has a direct physical realization of the kind of problems treated by Helfrich and Müller [204].

2.7.2 Dislocation lines in solids and bond-orientational order

Dislocations have long been proposed as a mechanism for melting of three-dimensional crystalline solids into isotropic liquids. Shockley [180], for example, has argued that bulk melting may be viewed as a sudden proliferation of a tangled array of dislocation loops in the solid. He showed that a finite density of mobile dislocation lines would result in a liquid-like viscosity. Shockley's ideas have been elaborated by a variety of subsequent workers whose efforts are reviewed by Cotterill [206]. Edwards and Warner [207] used methods developed in polymer physics to argue in favor of a dislocation-driven first-order melting transition. One problem with dislocation-mediated bulk melting is that current theories are not sophisticated enough to distinguish between interacting dislocation loops and interacting vortex lines in superfluid helium. The same approximations applied to the λ transition in helium would predict a first-order phase transition, in contradiction with experiment.

The properties of a solid in the presence of a finite concentration of unbound dislocation loops have been studied by Nelson and Toner [200]. Although all elastic constants characteristic of the solid vanish, there remains a residual resistance to torsion that is not present in an isotropic liquid. The quantity

$$\boldsymbol{\theta}(\mathbf{r}) \equiv \tfrac{1}{2}[\nabla \times \mathbf{u}(\mathbf{r})] \tag{2.268}$$

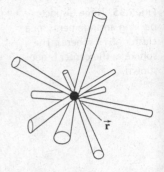

Fig. 2.34. A cluster of bonds surrounding a particle at position **r**. The bonds could be obtained by the Voronoi construction, or, say, by simply searching for the 12 nearest neighbors of a given atom.

is the local rotation from a preferred set of crystallographic axes in a solid characterized by a displacement field $\mathbf{u}(\mathbf{r})$. In the presence of unbound dislocation loops, gradients in $\boldsymbol{\theta}(\mathbf{r})$ are resisted by a term in the free energy,

$$\delta F = \frac{1}{2}\int d^3r[K_a|\nabla \times \boldsymbol{\theta}|^2 + K_b(\nabla \cdot \boldsymbol{\theta})^2], \tag{2.269}$$

where K_a and K_b are given in terms of energies of edge- and screw-dislocation cores,

$$K_a = 2E_c, \qquad K_b = 8E_s - 4E_c. \tag{2.270}$$

Although the above results refer to isotropic solids, only minor changes are introduced when anisotropies are taken into account.

Evidently, unbound dislocation loops in three dimensions produce a bulk analogue of the hexatic phase, instead of the isotropic liquid envisioned by Shockley [180]. An order parameter for such a phase (analogous to $\psi(r) = e^{6i\theta(r)}$ in two dimensions) can be constructed from a cluster of near-neighbor bonds like that shown in Fig. 2.34. Upon expanding the density $\rho(\mathbf{r}, \Omega)$ of points pierced by these bonds on a small sphere inscribed about **r**, we have

$$\rho(\mathbf{r}, \Omega) = \sum_{\ell=0}^{\infty} \sum_{m=-\ell}^{\ell} Q_{\ell m}(\mathbf{r}) Y_{\ell m}(\Omega), \tag{2.271}$$

where the $Y_{\ell m}(\Omega)$ are spherical harmonics. The distribution of bonds as a function of solid angle Ω could be determined by, say, the Voronoi construction [147]. In a liquid, we expect that $\rho(\mathbf{r}, \Omega)$ becomes isotropic upon averaging over particle positions **r**,

$$\langle\rho(\mathbf{r}, \Omega)\rangle = r_0 \equiv \langle Q_{00}\rangle/\sqrt{4\pi}. \tag{2.272}$$

A heavily dislocated solid has persistent bond-orientational order in the sense that Fourier coefficients with $\ell \neq 0$ in Eq. (2.271) fail to vanish

Fig. 2.35. Three distinct packing arrangements for a cluster of 13 spheres. The sphere at the center is not shown.

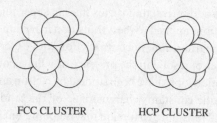

FCC CLUSTER   HCP CLUSTER

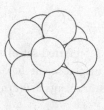

ICOSAHEDRON

upon averaging. For materials with a cubic rotational symmetry, the first coefficients which are nonvanishing correspond to $\ell = 4$,

$$\langle\delta\rho(\mathbf{r}, \Omega)\rangle \equiv \langle\rho(\mathbf{r}, \Omega)\rangle - \rho_0 \approx \sum_{m=-4}^{4} \langle Q_{4m}(\mathbf{r})\rangle Y_{4m}(\Omega). \qquad (2.273)$$

In contrast to bcc, fcc and simple cubic crystals, for which the $\langle Q_{4m}(\mathbf{r})\rangle$ are also nonvanishing, there is no translational order. The collection of $\langle Q_{4m}(\mathbf{r})\rangle$ with $|m| \leqslant 4$ constitutes a nine-component order parameter for such a phase. In a coordinate system chosen such that only Q_{40} and $Q_{4\pm4}$ are non zero, one has [200]

$$\frac{|Q_{4\pm4}|}{|Q_{40}|} = \sqrt{\frac{5}{14}}. \qquad (2.274)$$

Another possibility is bond-orientational order based on an *icosahedral* symmetry. In two dimensions, the natural packing element in dense liquids is a snowflake-like cluster of seven atoms, with one particle at the center. The analogous cluster in three dimensions contains 13 atoms and there are *three* distinct packing arrangements (Fig. 2.35). In addition to crystallographic packings that are seeds for fcc and hcp lattices, there is the icosahedral arrangement shown in Fig. 2.35. For most simple potentials, the icosahedral cluster is energetically preferred over the crystalline packings [208]. The first nonzero spherical harmonic coefficients associated with icosahedral bond-orientational order occur

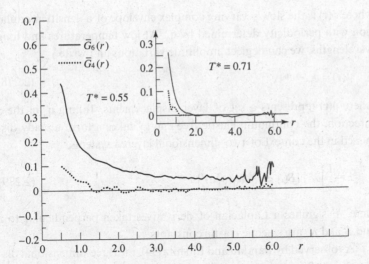

Fig. 2.36. Bond-orientational correlation functions for a supercooled Lennard-Jones liquid for $\ell = 4$ and $\ell = 6$ at two distinct temperatures.

at $\ell = 6$. A coordinate system can always be chosen such that only Q_{60} and $Q_{6\pm5}$ are nonzero; one then has [171]

$$\frac{|Q_{6\pm5}|}{|Q_{60}|} = \sqrt{\frac{11}{7}}. \tag{2.275}$$

Figure 2.36 shows the correlation functions

$$G(r) \propto \frac{1}{2\ell + 1} \sum_{m=-\ell}^{\ell} \langle Q_{\ell m}(\mathbf{r}) Q_{\ell m}^*(\mathbf{0}) \rangle \tag{2.276}$$

in a computer simulation of 864 particles of supercooled argon for $\ell = 4$ and $\ell = 6$ at two distinct temperatures [171]. Evidently, extended orientational correlations suggesting an icosahedral symmetry appear at intermediate length scales upon supercooling from $T^* = 0.71$ to $T^* = 0.55$. Icosahedral bond order would have intriguing consequences for supercooled liquids and possibly for the glass transition. See the preface to Chapter 4 for recent explorations of these ideas.

2.7.3 Melting of smectic-A liquid crystals

The melting of bulk smectic-A liquid crystals, which are solid-like in one direction and liquid-like in two, is an intriguing problem [72]. Unlike other systems with line singularities, bulk smectics are precisely at their "lower critical dimension" and fluctuations are just sufficient to prevent genuine long-range order. The density in such materials may be written [152, 186] as

$$\rho(\mathbf{r}) \approx \rho_0 \mathrm{Re}[1 + \psi(\mathbf{r})e^{i\mathbf{q}_0 \cdot \mathbf{r}}], \tag{2.277}$$

where $\psi(\mathbf{r})$ is the slowly varying complex envelope of a density modulation with periodicity determined by $\mathbf{q}_0$. At low temperatures and long wavelengths, we can neglect amplitude variations and write

$$\psi(\mathbf{r}) = \psi_0 e^{i\mathbf{q}_0 \cdot \mathbf{u}(\mathbf{r})}, \tag{2.278}$$

where $\mathbf{u}(\mathbf{r})$ represents a set of layer displacements. Taking $\mathbf{q}_0$ in the z direction, the continuum elastic free energy takes a form already discussed in the context of two-dimensional layered systems,

$$F = \frac{1}{2} \int d^3 r \, [B(\partial_z u)^2 + K_1(\partial_\perp^2 u)^2]. \tag{2.279}$$

Here, $\partial_\perp^2$ signifies a Laplacian of derivatives taken perpendicular to $\hat{z}$ and B and K_1 are smectic elastic constants.

As observed by Landau and Lifshitz [94], the three-dimensional displacement fluctuations associated with (2.279) are logarithmically divergent. Just as in the two-dimensional systems discussed in Section 2.1, correlations in $\psi(\mathbf{r})$ decay algebraically to zero [40, 209]. For positions separated in the $\hat{z}$ direction one finds

$$\langle \psi^*(\mathbf{r}_\perp, z)\psi(\mathbf{r}_\perp, 0) \rangle \sim 1/z^{\eta(T)}, \tag{2.280a}$$

whereas for separations perpendicular to $\hat{z}$ the result is

$$\langle \psi^*(\mathbf{r}_\perp, z)\psi(0, z) \rangle \sim 1/r_\perp^{2\eta(T)}, \tag{2.280b}$$

where

$$\eta(T) = k_B T q_0^2 / (8\pi \sqrt{K_1 B}). \tag{2.281}$$

This continuously variable power-law decay can be represented by a line of fixed points within renormalization theory. As shown by Grinstein and Pelcovits [210], nonlinear perturbations about this line (which were neglected in (2.279)) renormalize only very slowly to zero at long wavelengths and give rise to a temperature-independent logarithmic correction to the results of Caillé.

Smectics with quasi-long-range translational order can melt into either isotropic liquids or nematic liquid crystals. The transition to an isotropic liquid appears experimentally to be of first order, consistent with the theoretical ideas of Brasovskii [151]. Melting into a nematic phase, however, can be continuous according to Landau theory [152, 186]. The nematic phase plays a role very similar to the bond-oriented hexatic, cubic and icosahedral fluids discussed elsewhere in this review. If $\mathbf{n}$ is a director aligned with the local orientation of the rod-like nematogens, the elastic free energy distinguishing a nematic from an isotropic liquid is [183]

$$F = \frac{1}{2} \int d^3r [K_1(\nabla \cdot \mathbf{n})^2 + K_2[\mathbf{n} \cdot (\nabla \times \mathbf{n})]^2 + K_3 |\mathbf{n} \times (\nabla \times \mathbf{n})|^2], \qquad (2.282)$$

where K_1, K_2 and K_3 are elastic constants for bending, twisting and splaying, respectively. In the smectic phase, $\mathbf{n}$ is locked to the layer normal and K_2 and K_3 are infinite. The elastic constant K_1 also appears in the smectic free energy (2.279).

Halperin and Lubensky [211] and Lubensky and Chen [212] have argued via expansions in $\varepsilon = 4 - d$ that the smectic-A–nematic transition will always be driven to be first order by the coupling to director fluctuations. This conjecture seems to be contradicted, however, by a variety of experiments in which apparently continuous transitions have been observed [72, 213]. The strong fluctuations in bulk smectics suggest that $d = 3$ is in some sense quite far from $d = 4$, so one does have reason to distrust the ε expansion.

Another possibility is a continuous transition mediated by an unbinding of dislocation loops. Dislocation lines in smectics have both an edge and screw character and carry Burgers vectors oriented perpendicular to the smectic planes [214]. A finite transition temperature is suggested by a balance between the energy and entropy of an isolated dislocation loop [215]: The energy per unit layer spacing of smectic dislocation line has a finite value E_c, in contrast to that for vortex lines in superfluid helium. The number of configurations in a large loop of length L (in units of the layer spacing) increases like q^L, where q is a dimensionless constant greater than unity. The free energy of the loop is then approximately

$$F \approx LE_c - k_B T \ln q^L, \qquad (2.283)$$

suggesting a proliferation of loops above a temperature

$$T_c = E_c / (k_B \ln q). \qquad (2.284)$$

The loop mechanism is intriguing, because it has within it the seeds of "anisotropic scaling." A crude physical picture of smectic-like fluctuations in a nematic just above T_c is shown in Fig. 2.37. There are two correlation lengths, parallel and perpendicular to the orientation of the director, which can in principle diverge anisotropically with *different* exponents ($\nu_\parallel \neq \nu_\perp$):

$$\xi_\parallel \sim \frac{1}{|T - T_c|^{\nu_\parallel}}, \quad \xi_\perp \sim \frac{1}{|T - T_c|^{\nu_\perp}}. \qquad (2.285)$$

Fig. 2.37. A schematic illustration of smectic-A fluctuations in the nematic phase.

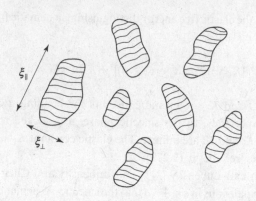

There is some experimental evidence indicating anisotropic scaling [72, 213] and Lubensky and Chen [212] have found that this is at least a theoretical possibility within the ε expansion. Above T_c the Frank constants K_2 and K_3 behave according to

$$K_2 \sim \xi_\perp^2/\xi_\parallel, \qquad K_3 \sim \xi_\parallel, \tag{2.286}$$

while K_1 remains finite. Just below T_c, the elastic constant B characterizing smectic order varies like

$$B \sim \xi_\parallel/\xi_\perp^2. \tag{2.287}$$

Consequences of the dislocation model of smectic melting have been worked out by Nelson and Toner [200]. The correlation lengths $\xi_\parallel$ and $\xi_\perp$ represent the mean spacings between unbound dislocation lines running parallel and perpendicular to the layer normal, respectively, in this picture. This dislocation mechanism reproduces the anisotropic scaling results (2.286) and (2.287) and leads to the specific prediction

$$\nu_\parallel = 2\nu_\perp, \tag{2.288}$$

implying that K_2 and B are finite in the limit $T \to T_c$. A scaling argument below T_c in d dimensions (which does not depend on the dislocation mechanism) gives [200]

$$B \sim \xi_\parallel/\xi_\perp^{d-1}, \qquad K_1 \sim \xi_\perp^{5-d}/\xi_\parallel. \tag{2.289}$$

The Frank constant K_1 is expected to remain finite through the transition, since it does not couple to fluctuations that break up smectic order. This can happen only provided that

$$\nu_\parallel = (5-d)\nu_\perp, \tag{2.290}$$

which agrees with (2.288) in the limit $d \to 3$.

Virtually all experiments to date indicate that B tends to zero near continuous smectic-A–nematic transitions, which would contradict

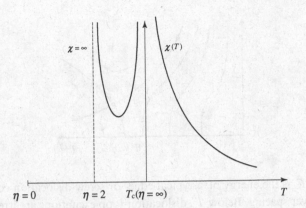

Fig. 2.38. The nonmonotonic susceptibility near the nematic–smectic-A transition implied by a vanishing elastic constant $B(T)$ and the assumption that χ diverges at T_c.

(2.288). Nevertheless, it would be surprising if $B(T)$ actually vanished as $T \to T_c^-$, since this would imply a divergence in the Caillé formula (2.281) for $\eta(T)$. Once $\eta(T)$ exceeds two, the susceptibility $\chi(T)$ to smectic-like perturbations in density is *finite*. Assuming that $\chi(T)$ diverges at T_c [213], one arrives at the rather unphysical nonmonotonic temperature-dependent response function shown in Fig. 2.38. One way out of this implausible scenario is to argue that $\chi(T)$ actually diverges *below* T_c, at a temperature such that $\eta(T) = 2$. The specific heat, however, is insensitive to the value of η and surely diverges precisely at the critical temperature [213]. It seems improbable that the fluctuations in energy embodied in the specific heat could actually be *larger* than the fluctuations in order measured by $\chi(T)$ near T_c.

Another possibility is that $B(T)$ exhibits a small jump discontinuity at T_c, which is obscured by a rather large superimposed cusp singularity. A dimensionless measure of the minimum measured value of B is the Caillé exponent $\eta(T)$, which has never been observed to be larger than about $\frac{1}{3}$ [209]. Recent experiments by Fisch *et al.* [216] on reentrant nematics tend to support this point of view. Their results also suggest that very small reduced temperatures will be required in order to observe anisotropic scaling with $\nu_\parallel = 2\nu_\perp$ directly in the most commonly studied liquid-crystalline compounds.

Note that the limiting value of the Caillé exponent $\eta(T_c^-)$ need not be the same as the corresponding critical exponent measured precisely at T_c. The qualitative renormalization-group flows we expect for the smectic-A–nematic transition are shown in Fig. 2.39. Temperature is measured in terms of the function $\eta(T)$ and we have introduced a dislocation fugacity per unit length y. The dashed locus of initial conditions is given by

$$y = e^{-E_c/(k_B T)}, \tag{2.291}$$

Fig. 2.39. Conjectured renormalization flows for the nematic–smectic-A transition.

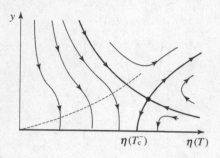

Fig. 2.39. Conjectured renormalization flows for the nematic–smectic-A transition.

where E_c is the energy of a typical edge- or screw-dislocation core per unit layer spacing. Below T_c, dislocation-loop excitations are irrelevant at long wavelengths and y is renormalized to zero. All trajectories then terminate in a line of fixed points parameterized by $\eta(T)$. There is a critical fixed point at *finite* y that controls the actual phase transition. The exponent $\eta(T_c^-)$ is given by the terminus of the heavy trajectory leaving this fixed point and ending in the line $y = 0$. There is a separate critical exponent η^* determined by an eigenvalue of the finite-y fixed point. That such a fixed point exists can be inferred from Helfrich's entropy argument, which suggests that E_c and hence y are finite at T_c. The situation we envision is different from that in, say, the two-dimensional XY model, wherein the fixed point controlling the transition is part of the fixed line (see Fig. 2.6).

Note that the renormalization-group flows of Fig. 2.39 suggest that $\eta(T)$ approaches a system-independent constant as $T \to T_c^-$,

$$\lim_{T \to T_c} \frac{k_B T q_0^2}{8\pi\sqrt{K_1(T)B(T)}} = \text{universal constant} \approx \frac{1}{3},\qquad (2.292)$$

in analogy with the universal jump in the superfluid density of helium films (see Section 2.2). If this is indeed the case, systems with small intrinsic Frank constants K_1 (or large values of $k_B T_c q_0^2$) would have to have correspondingly larger jump discontinuities in $B(T)$.

Acknowledgements

I am indebted to B. I. Halperin for fruitful collaborations on many of the subjects discussed in this chapter. These ideas were recently revisited for melting on a *one*-dimensional periodic substrate: see L. Radzihovsky, E. Frey and D. R. Nelson, *Phys. Rev.* E63, 031503 (2001). This chapter is adapted with permission from *Phase Transitions and Critical Phenomena*, edited by C. Domb and J. L. Lebowitz, vol. 7 (Academic, New York, 1983) pp. 1–99.

References

[1]   J. M. Kosterlitz and D. J. Thouless, *Prog. Low Temp. Phys.*, **78**, 371 (1978).

[2]   V. L. Berezinski, *Zh. Éksp. Teor. Fiz.* **59**, 907 (1970) [*Sov. Phys. JETP* **32**, 493 (1971)]; *Zh. Éksp. Teor. Fiz.* **61**, 1144 (1972) [*Sov. Phys. JETP* **34**, 610 (1972)].

[3]   Feynman's theory is described in R. L. Elgin and D. L. Goodstein, *Monolayer and Submonolayer Films*, edited by J. G. Daunt and E. Lerner (Plenum, New York, 1973), p. 35 and R. L. Elgin and D. L. Goodstein, *Phys. Rev.* A **9**, 2657 (1974). There is a minor error in the coefficient Feynman found for the diverging energy of an isolated dislocation. Feynman eventually took the time to check his calculations and uncovered his error while awaiting surgery for a major illness! (D. L. Goodstein, private communication).

[4]   C. F. Chou, A. J. Jin, S. W. Hui, C. C. Huang and J. T. Ho, *Science* **280**, 1424 (1998) and references therein.

[5]   C. M. Murray, *Bond-Orientational Order in Condensed Matter Systems*, edited by K. Strandburg (Springer, Berlin, 1992).

[6]   K. Zahn, R. Lenke and G. Maret, *Phys. Rev. Lett.* **82**, 2721 (1999); K. Zahn and G. Maret, *Phys. Rev. Lett.* **85**, 3656 (2000).

[7]   R. Seshadri and R. M. Westervelt, *Phys. Rev. Lett.* **66**, 2774 (1991).

[8]   C. Knobler and R. Desai, *Ann Rev. Phys. Chem.* **43**, 207 (1992).

[9]   C. C. Huang, in *Bond-Orientational Order in Condensed Matter Systems*, edited by K. Strandburg (Springer, Berlin, 1992); C. C. Huang, *Adv. Phys.* **42**, 343 (1993).

[10]   H. Strey, R. Podgornik, D. C. Rau, A. Rupprecht and V. A. Parsegian, *PNAS* **93**, 4261 (1996); D. H. Van Winkle, A. Chatterjee, R. Link and R. L. Rill, *Phys. Rev.* E **55**, 4354 (1997); see also R. L. Rill, T. E. Strzelecka, M. W. Davidson and D. H. Van Winkle, *Physica* A **176**, 87 (1991).

[11]   D. J. Bishop, P. L. Gammel and C. A. Murray, in *The Vortex State*, edited by N. Bontemps, Y. Bruynseraede, G. Deutscher and A. Kapitulnik (Kluwer, Dordrecht, 1994), p. 99; see also C. A. Murray, P. L. Gammel, D. J. Bishop, D. B. Mitzi and A. Kapitulnik, *Phys. Rev. Lett.* **64**, 2312 (1990).

[12]   A. Jaster, *Phys. Rev.* E **59**, 2594 (1999).

[13]   K. Bagchi, H. C. Andersen, W. Swope, *Phys. Rev.* E **53**, 3794 (1996).

[14]   F. L. Somer Jr, G. S. Canright and T. Kaplan, *Phys. Rev.* E **58**, 5748 (1998).

[15]   C. Domb and M. S. Green, *Phase Transitions and Critical Phenomena*, Vol. 6 (Academic, London, 1976).

[16]   A. A. Migdal, *Zh. Éksp. Teor. Fiz.* **69**, 1457 (1975) [*Sov. Phys. JETP* **42**, 743 (1976)].

[17]   L. P. Kadanoff, *Ann. Phys. (New York)* **100**, 359 (1976).

[18]   D. J. Wallace, and R. K. P. Zia, *Phys. Rev. Lett.* **43**, 808 (1979).

[19]   N. D. Mermin and H. Wagner, *Phys. Rev. Lett.* **17**, 1133 (1966).

[20]   N. D. Mermin, *J. Math. Phys. (N.Y.)* **8**, 1061 (1967).

[21]   N. D. Mermin, *Phys. Rev.* **176**, 250 (1968).

[22]   P. C. Hohenberg, *Phys. Rev.* **158**, 383 (1967).

[23]   A. M. Polyakov, *Phys. Lett.* B **57**, 79 (1975a).

[24] E. Brézin and J. Zinn-Justin, *Phys. Rev. Lett.* **36**, 691 (1976).

[25] E. Brézin and J. Zinn-Justin, *Phys. Rev.* B **14**, 3110 (1976).

[26] V. L. Berenzinskii, *Zh. Éksp. Teor. Fiz.* **61**, 1144 [*Sov. Phys. JETP* **34**, 610 (1972)].

[27] J. M. Kosterlitz and D. J. Thouless, *J. Phys.* C **5**, L124 (1972).

[28] J. M. Kosterlitz and D. J. Thouless, *J. Phys.* C **6**, 1181 (1973).

[29] F. Bloch, *Z. Phys.* **61**, 206 (1930).

[30] R. E. Peierls, *Ann. Inst. Henri Poincaré* **5**, 177 (1935).

[31] L. D. Landau, *Phys. Z. Sowjetunion* **II**, 26 (1937).

[32] T. M. Rice, *Phys. Rev.* A **140**, 1889 (1965).

[33] J. W. Kane and L. P. Kadanoff, *Phys. Rev.* **155**, 80 (1967).

[34] G. Lasher, *Phys. Rev.* **172**, 224 (1968).

[35] B. Jancovici, *Phys. Rev. Lett.* **19**, 20 (1967).

[36] H. J. Mikeska and H. Schmidt, *J. Low Temp. Phys.* **2**, 371 (1970).

[37] Y. Imry and L. Gunther, *Phys. Rev.* B **3**, 3939 (1971).

[38] F. J. Wegner, *Z. Phys.* **206**, 465 (1967).

[39] V. L. Berenzinskii, *Zh. Éksp. Teor. Fiz.* **59**, 907 (1970) [*Sov. Phys. JETP* **32**, 493 (1971)].

[40] A. Caillé, *C. R. Acad. Sci.* B **274**, 891 (1972).

[41] L. D. Landau and E. M. Lifshitz, *Theory of Elasticity* (Pergamon, New York, 1970).

[42] K. R. Atkins and I. Rudnick, *Prog. Low Temp. Phys.* **6**, 37 (1970).

[43] I. Rudnick, *Phys. Rev. Lett.* **40**, 1454 (1978).

[44] K. R. Atkins, *Phys. Rev.* **113**, 962 (1959).

[45] D. J. Bishop and J. Reppy, *Phys. Rev. Lett.* **40** 1727 (1978).

[46] M. R. Beasley, J. E. Mooij and T. P. Orlando, *Nucl. Phys. Rev. Lett.* **42**, 1165 (1979).

[47] J. Pearl, *Appl. Phys. Lett.* **5**, 65 (1964).

[48] A. F. Hebard and A. T. Fiory, *Phys. Rev. Lett.* **44**, 291 (1981).

[49] D. U. Gubser, T. L. Francavilla, S. A. Wolf and J. R. Liebowitz (eds.), in *Inhomogeneous Superconductors* (American Institute of Physics, New York, 1980).

[50] D. S. Fisher, *Phys. Rev.* B **22**, 1190 (1980).

[51] B. I. Halperin and D. R. Nelson, *J. Low Temp. Phys.* **36**, 599 (1979).

[52] B. A. Huberman and S. Doniach, *Phys. Rev. Lett.* **42**, 1169 (1979).

[52] D. J. Resnick, J. C. Garland, J. T. Boyd, S. Shoemakers and R. S. Newrock, *Phys. Rev. Lett.* **41**, 1542 (1981).

[53] C. Y. Young, R. Pindak, N. A. Clark and R. B. Meyer, *Phys. Rev. Lett.* **40**, 773 (1978).

[54] C. Rosenblatt, R. Pindak, R. A. Clark and R. B. Meyer, *Phys. Rev. Lett.* **42**, 1220 (1979).

[55] D. E. Moncton and R. Pindak, *Phys. Rev. Lett.* **43**, 701 (1979).

[56] R. Pindak, D. J. Bishop and W. O. Springer, *Phys. Rev. Lett.* **44**, 1461 (1980).

[57] C. C. Grimes and G. Adams, *Phys. Rev. Lett.* **42**, 795 (1979).

[58] D. S. Fisher, B. I. Halperin and P. Platzman, *Phys. Rev. Lett.* **42**, 798 (1979) and references therein.

[59] P. Pieranski, *Phys. Rev. Lett.* **45**, 569 (1980).

[60] O. Albrecht, H. Gruler and E. Sackmann, *J. Physique* **39**, 301 (1978).

[61] P. S. Pershan, *J. Physique Coll.* **40**, C3 (1979).

[62] H. Burecki and N. Amer, *J. Physique Coll.* **40**, C3-433 (1979).

[63] J. Doucet, A. Levelut and A. M. Lambert, *Phys. Rev. Lett.* **32**, 301 (1974).

[64] J. Doucet, P. Keller, A. M. Levelut and P. Pourquet, *J. Physique* **39**, 548 (1978).

[65] J. G. Dash, *Films on Solid Surfaces* (Academic Press, New York, 1975).

[66] J. P. McTague, D. Frenkel and M. P. Allen, in *Ordering in Two Dimensions*, edited by S. Sinha (North-Holland, New York, 1980).

[67] J. P. McTague, M. Nielsen and L. Passell, in *Ordering in Strongly Fluctuating Condensed Matter Systems*, edited by T. Riste (Plenum Press, New York, 1980).

[68] M. D. Chin and S. C. Fain Jr, *Phys. Rev. Lett.* **39**, 146 (1977).

[69] D. E. Moncton, P. W. Stephens, R. J. Birgeneau, P. M. Horn and G. S. Brown, *Phys. Rev. Lett.* **46**, 1533 (1981).

[70] M. Nielsen, J. Als-Nielsen, J. Bohr and J. P. McTague, *Phys. Rev. Lett.* **47**, 582 (1981).

[71] P. A. Heiney, R. J. Birgeneau, G. S. Brown, P. M. Horn, D. E. Moncton and P. W. Stephens, *Phys. Rev. Lett.* **48**, 104 (1982).

[72] J. D. Litster, R. J. Birgeneau, M. Kaplan, C. P. Safinya and J. Als-Nielsen, in *Ordering in Strongly Fluctuating Condensed Matter Systems*, edited by T. Riste (Plenum Press, New York, 1980).

[73] M. Pomerantz, *Phase Transitions in Surface Films*, edited by J. G. Dash and J. Ruvalds (Plenum, New York, 1980).

[74] R. Navarro and J. L. de Jongh, *Physica* B **84**, 229 (1976).

[75] K. Hirakawa and K. Ubukoshi, *J. Phys. Soc. Japan* **50**, 1909 (1981).

[76] S. T. Chui and J. D. Weeks, *Phys. Rev. Lett.* **40**, 733 (1978).

[77] J. D. Weeks, in *Ordering in Strongly Fluctuating Condensed Matter Systems*, edited by T. Riste (Plenum Press, New York, 1980) p. 293.

[78] J. M. Kosterlitz and D. J. Thouless, *Prog. Low Temp. Phys.* **78**, 371 (1978).

[79] D. R. Nelson, in *Fundamental Problems in Statistical Mechanics*, Vol. V, edited by E. G. D. Cohen (North-Holland, New York, 1980).

[80] T. Riste, *Ordering in Strongly Fluctuating Condensed Matter Systems* (Plenum, New York, 1980).

[81] J. G. Dash and J. Ruvalds, *Phase Transitions in Surface Films* (Plenum, New York, 1980).

[82] S. Sinha, *Ordering in Two Dimensions* (North-Holland, New York, 1980).

[83] B. I. Halperin, in *Physics of Defects, Proceedings of the Les Houches Summer Institute* (North-Holland, Amsterdam, 1981).

[84] M. Barber, *Phys. Rep.* **59**, 375 (1980).

[85] J. Villain, in *Ordering in Strongly Fluctuating Condensed Matter Systems*, edited by T. Riste (North-Holland, New York, 1980).

[86] J. Villain, in *Ordering in Two Dimensions*, edited by S. Sinha (Plenum Press, New York, 1980).

[87] V. L. Pokrovsky and Talapov, *Zh. Éksp. Teor. Fiz.* **78**, 269 (1980).

[88] H. E. Stanley and T. A. Kaplan, *Phys. Rev. Lett.* **17**, 913 (1966).

[89] H. E. Stanley, *Phys. Rev. Lett.* **20**, 589 (1968).

[90] M. A. Moore, *Phys. Rev. Lett.* **23**, 861 (1969).

[91] P. C. Hohenberg, in *Proceedings of the Enrico Fermi School of Physics*, edited by M. S. Green (Academic Press, New York, 1971).

[92] A. M. Polyakov, *Phys. Lett.* B **59**, 81 (1975).

[93] J. José, L. P. Kadanoff, S. Kirkpatrick and D. R. Nelson, *Phys Rev.* B **16**, 1217 (1977).

[94] L. D. Landau and E. M. Lifshitz, *Statistical Physics* (Addison-Wesley, Reading, MA, 1969) p. 466.

[95] R. A. Pelcovits, PhD Thesis, Harvard University (1978).

[96] J. Toner, PhD Thesis, Harvard University (1981).

[97] D. J. Amit, S.-K. Ma and R. K. P. Zia, *Nucl. Phys.* B **180**, 157 (1981).

[98] D. R. Nelson and J. M. Kosterlitz, *Phys. Rev. Lett.* **39**, 1201 (1977).

[99] J. M. Kosterlitz, *J. Phys.* C **7**, 1046 (1974).

[100] P. C. Hohenberg and P. C. Martin, *Ann. Phys.* **34**, 291 (1965).

[101] D. R. Nelson, *Phys. Rev.* B **18**, 2318 (1978).

[102] P. W. Anderson and G. Yuval, *J. Phys.* C **4**, 607 (1971).

[103] V. Ambegaokar, B. I. Halperin, D. R. Nelson and E. D. Siggia, *Phys. Rev.* B **21**, 1806 (1980).

[104] A. N. Berker and D. R. Nelson, *Phys. Rev.* **19**, 2488 (1979).

[105] S. Solla and E. K. Riedel, *Phys. Rev.* B **23**, 6008 (1981).

[106] J. Tobochnik and G. V. Chester, *Phys. Rev.* B **25**, 6778 (1982).

[107] B. D. Josephson, B. D. *Phys. Rev.* B **21**, 608 (1966).

[108] P. C. Hohenberg, A. Aharony, B. I. Halperin and E. D. Siggia, *Phys. Rev.* B **13**, 2986 (1976).

[109] J. José, L. P. Kadanoff, S. Kirkpatrick and D. R. Nelson, *Phys. Rev.* B **17**, 1477 (1978).

[110] E. Webster, G. Webster and M. Chester, *Phys. Rev. Lett.* **42**, 243 (1979).

[111] J. A. Roth, G. J. Jelatis and J. D. Maynard, *Phys. Rev. Lett.* **44**, 333 (1980).

[112] P. Minnhagen and G. G. Warren, *Phys. Rev.* B **24**, 2526 (1981).

[113] A. P. Young, *J. Phys.* C **11**, L453 (1978).

[114] J. Villain, *J. Physique* **36**, 581 (1975).

[115] L. P. Kadanoff, *J. Phys.* C **11**, 1399 (1978).

[116] V. L. Pokrovsky, and G. V. Uimin, *Phys. Lett.* A **45**, 467 (1973).

[117] E. Lieb and F. Y. Wu, in *Phase Transitions and Critical Phenomena*, Vol. 1, edited by C. Domb and M. S. Green (Academic Press, London, 1972).

[118] E. Lieb, *Phys. Rev. Lett.* **18**, 1046 (1967).

[119] L. P. Kadanoff, *Phys. Rev. Lett.* **39**, 903 (1977).

[120] H. J. F. Knops, in *Fundamental Problems in Statistical Mechanics*, Vol. V, edited by E. G. D. Cohen (North-Holland, New York, 1980).

[121] L. P. Kadanoff and A. C. Brown, *Ann. Phys. (New York)* **121**, 318 (1979).

[122] M. P. M. den Nijs, *J. Phys.* A **12**, 1857 (1979).

[123] J. L. Black and V. J. Emery, *Phys. Rev.* B **23**, 429 (1981).

[124] S. Elitzur, R. Pearson and J. Shigemitzu, *Phys. Rev.* D **19**, 3698 (1979).

[125] D. Horn, M. Weinstein and S. Yankidowicz, *Phys. Rev.* D **19**, 3715 (1979).

[126] A. Houghton, R. D. Kenway and S. C. Ying, *Phys. Rev.* B **23**, 298 (1981).

[127] J. Cardy and S. Ostlund, *Phys. Rev.* B **25**, 6899 (1981).

[128] J. José, *Phys. Rev.* B **20**, 2167 (1979).

[129] P. C. Hohenberg, B. I. Halperin and D. R. Nelson, *Phys. Rev.* B **22**, 2373 (1980).

[130] L. D. Landau and E. M. Lifshitz, *Fluid Mechanics* (Pergamon, New York, 1959) Chap. 1.

[131] D. J. Bergman, *Phys. Rev.* **188**, 370 (1969).

[132] D. J. Bergman, *Phys. Rev.* A **3**, 2058 (1971).

[133] R. A. Ferrell, N. Menyhard, H. Schmidt, F. Schwabl and P. Szepfalusy, *Phys. Rev. Lett.* **18**, 891 (1967).

[134] B. I. Halperin and P. C. Hohenberg, *Phys. Rev.* **117**, 952 (1969).

[135] G. Ahlers, *Phys Rev. Lett.* **21**, 1159 (1968).

[136] T. Tyson, *Phys. Rev. Lett.* **21**, 1255 (1968).

[137] J. D. Jackson, *Classical Electrodynamics* (Wiley, New York, 1967).

[138] P. W. Anderson, *Rev. Mod. Phys.* **38**, 298 (1966).

[139] J. S. Langer and M. E. Fisher, *Phys. Rev. Lett.* **19**, 560 (1967).

[140] J. S. Langer and J. D. Reppy, *Prog. Low Temp. Phys.* **6**, 1 (1970).

[141] R. G. Petschek and A. Zippelius, *Phys. Rev.* B **23**, 3483 (1981).

[142] B. Ratnam and J. Mochel, *Phys. Rev. Lett.* **25**, 711 (1970).

[143] J. Maps and R. B. Hallock, *Physica* B + C **107**, 399 (1981).

[144] G. Agnolet and J. D. Reppy, *Physica* B + C **107**, 4, 5 (1981).

[145] B. I. Halperin and D. R. Nelson, *Phys. Rev. Lett.* **41**, 121; **41**, 519 (1978).

[146] D. R. Nelson and B. I. Halperin, *Phys. Rev.* B **19**, 2457 (1979).

[147] R. Collins, in *Phase Transitions and Critical Phenomena*, edited by C. Domb and M. S. Green (Academic, London, 1972) Vol. 2 p. 271.

[148] D. R. Nelson, M. Rubenstein and F. Spaepen, *Phil. Mag.* A **46**, 105 (1982).

[149] G. Baym, H. A. Bethe and C. Pethich, *Nucl. Phys.* A **175**, 1165 (1971).

[150] S. Alexander and J. P. McTague, *Phys. Rev.Lett.* **40**, 702 (1978)

[151] S. A. Brasovskii, *Zh. Éksp. Teor. Fiz.* **68**, 42 [*Sov. Phys. JETP* **41**, 85 (1975)].

[152] P. G. de Gennes, *Mol. Cryst. Liq. Cryst.* **21**, 49 (1973).

[153] F. R. N. Nabarro, *Theory of Dislocations* (Clarendon Press, Oxford, 1967).

[154] A. P. Young, *Phys. Rev.* B **19**, 1855 (1979).

[155] B. J. Alder and T. E. Wainwright, *Phys. Rev. Lett.* **127**, 359 (1962).

[156] S. Toxvaerd, *J. Chem. Phys.* **69**, 4750 (1978).

[157] R. Bruinsma, B. I. Halperin and A. Zippelius, *Phys. Rev.* B **25**, 579 (1982).

[158] J. P. Hansen and I. R. McDonald, in *Theory of Simple Liquids* (Academic Press, London, 1976).

[159] R. Morf, *Phys. Rev. Lett.* **43**, 931 (1979).

[160] R. G. Gann, S. Chakravarty and G. V. Chester, *Phys. Rev.* B **20**, 326 (1979).

[161] D. S. Fisher, B. I. Halperin and R. Morf, *Phys. Rev.* B **20**, 4692 (1979).

[162] S. T. Chui, *Phys. Rev. Lett.* **48**, 933 (1982).

[163] J. Q. Broughton, G. H. Gilmer and J. D. Weeks, *Phys. Rev.* B **25**, 4651 (1982).

[164] R. K. Kalia, P. Vashishta and S. D. Leeuw, *Phys. Rev.* B **23**, 4794 (1981).

[165] R. M. J. Cotterill and L. B. Peterson, *Solid State Commun.* **10**, 439 (1972).

[166] D. Frenkel and J. P. McTague, *Phys. Rev. Lett.* **42**, 1632 (1979).

[167] F. F. Abraham, *Phys. Rev. Lett.* **44**, 463 (1980).

[168] S. Toxvaerd, *Phys. Rev. Lett.* **44**, 1002 (1980).

[169] J. A. Barker, D. Henderson and F. F. Abraham, in *STATPHYS 14*, edited by J. Stephensen (North-Holland, Amsterdam, 1981), p. 226.

[170] R. Morf, in *Proceedings of the International Conference on the Physics of Intercalation Compounds, Trieste, Italy*, edited by L. Pietronero and E. Tosatti (Springer, Berlin, 1981).

[171] P. Steinhardt, D. R. Nelson and M. Ronchetti, *Phys. Rev. Lett.* **47**, 1297; *Phys. Rev.* B **28**, 784 (1983).

[172] A. D. Novaco and J. P. McTague, *Phys. Rev. Lett.* **38**, 1286 (1977).

[173] S. Ostlund and B. I. Halperin, *Phys. Rev.* B **23**, 335 (1981).

[174] S. Coppersmith, D. S. Fisher, B. I. Halperin, P. A. Lee and W. F. Brinkman, *Phys. Rev. Lett.* **46**, 549 (1981); **46**, 869; see also *Phys. Rev.* B **25**, 349 (1982).

[175] A. Zippelius, B. I. Halperin and D. R. Nelson, *Phys. Rev.* B **22**, 2514 (1980).

[176] D. Forster, *Hydrodynamic Fluctuations, Broken Symmetry, and Correlation Functions* (Benjamin, Reading, MA, 1975).

[177] P. C. Martin, O. Parodi and P. S. Pershan, *Phys. Rev.* A **6**, 2401 (1972).

[178] Y. Pomeau and P. Résibois, *Phys. Rev.* **196**, 641 (1975).

[179] A. Zippelius, *Phys. Rev.* A **22**, 732 (1980).

[180] W. Shockley, *L'Etat Solid, Proceedings of the Neuvième Conseil de Physique*, edited by R. Stoops (Institute International de Physique Solvay, Brussels, 1952).

[181] D. R. Nelson and B. I. Halperin, *Phys. Rev.* B **21**, 5312 (1980).

[182] J. Toner and D. R. Nelson, *Phys. Rev.* B **23**, 316 (1981).

[183] P. G. de Gennes, *The Physics of Liquids Crystals* (Oxford University Press, London, 1974).

[184] R. Graham, in *Fluctuations, Instabilities, and Phase Transition* edited by T. Riste, (Plenum, New York, 1975).

[185] J. Swift and P. D. Hohenberg, *Phys. Rev.* A **15**, 319 (1977).

[186] P. G. de Gennes, *Solid State Commun.* **10**, 753 (1972).

[187] D. R. Nelson and R. A. Pelcovits, *Phys. Rev.* B **16**, 2191 (1977).

[188] J. P. Coulomb, J. P. Biberian, J. Suzanne, A. Therny, G. J. Trott, H. Taub, H. R. Danner and F. Y. Hanson, *Phys. Rev. Lett.* **43**, 1878 (1979).

[189] R. J. Birgeneau and J. D. Litster, *J. Physique Lett.* **39**, L399 (1978).

[190] P. S. Pershan, G. Aeppli, J. D. Litster and R. J. Birgeneau, *Mol. Cryst. Liq. Cryst.* **67**, 205 (1981).

[191] R. Pindak, D. E. Moncton, S. C. Davey and J. W. Goodby, *Phys. Rev. Lett.* **46**, 1135 (1981).

[192] C. C. Huang, J. M. Viner, R. Pindak and J. W. Goodby, *Phys. Rev. Lett.* **46**, 1289 (1981).

[193] R. Bruinsma and D. R. Nelson, *Phys. Rev.* B **23**, 402 (1981).

[194] A. J. Leadbetter, J. P. Gaughan, B. Kelley, G. W. Gray and J. Goodby, *J. Physique* **40**, C3-178 (1979).

[195] J. J. Benattar, J. Doucet, M. Lambert and A. M. Levelut, *Phys. Rev.* A **20**, 2505 (1979).

[196] P. Pollman and G. Scherer, *Z. Naturforsch*, **34a**, 225 (1979).

[197] S. Meiboom and M. Sammon, *Phys. Rev. Lett.* **44**, 882 (1980).

[198] R. P. Feynman, *Prog. Temp. Phys.* **1**, 67 (1955).

[199] V. N. Popov, *Zh. Éksp. Teor. Fiz.* **64**, 672, [*Sov. Phys. JETP* **37**, 341 (1973)].

[200] D. R. Nelson and J. Toner, *Phys. Rev.* B **24**, 363 (1981).

[201] B. I. Halperin, T. C. Lubensky and S.-K. Ma, *Phys. Rev. Lett.* **32**, 292 (1974).

[202] C. Dasgupta and B. I. Halperin, *Phys. Rev. Lett.* **47**, 1556 (1981).

[203] M. Tinkham, *Introduction to Superconductivity* (McGraw-Hill, New York, 1975).

[204] H. Helfrich and W. Müller, in *Continuum Models of Discrete Systems* (University of Waterloo Press, Waterloo, 1980) p. 753.

[205] J. C. Wheeler, S. J. Kennedy and P. Pfeuty, *Phys. Rev. Lett.* **45**, 1748 (1980).

[206] R. M. J. Cotterill, in *Order in Strongly Fluctuating Condensed Matter Systems*, edited by T. Riste (Plenum, New York, 1980).

[207] S. F. Edwards and M. Warner, *Phil. Mag.* **40**, 257 (1979).

[208] F. C. Frank, *Proc. Roy. Soc. London* A **215**, 43 (1952).

[209] J. Als-Nielsen, J. D. Litster, R. J. Birgeneau, M. Kaplan, C. R. Safinyia, A. Lindegaard-Anderson and S. Mathiesen, *Phys. Rev.* B **22**, 312 (1980).

[210] G. Grinstein and R. A. Pelcovits, *Phys. Rev. Lett.* **47**, 856 (1981).

[211] B. I. Halperin and T. C. Lubensky, *Solid State Commun.* **14**, 997 (1974).

[212] T. C. Lubensky and J.-H. Chen, *Phys. Rev.* B **17**, 366 (1978).

[213] R. J. Birgeneau, C. W. Garland, G. B. Kasting and B. M. Ocko, *Phys. Rev.* A **24**, 2624 (1981).

[214] P. S. Pershan, *J. Appl. Phys.* **45**, 1590 (1975).

[215] H. Helfrich, *J. Physique* **39**, 1199 (1978).

[216] M. Fisch, L. B. Sorensen and P. S. Pershan, *Phys. Rev. Lett.* **48**, 943 (1982).

Chapter 3
Order, frustration and two-dimensional glass

Preface

Understanding the glass transition is a formidable challenge [1]. One might hope that this problem would simplify in two dimensions. Although ubiquitous in three-dimensional covalent bonded materials (SiO_2, As_2Se_3), rapidly cooled binary alloys (metallic glasses), polymers (polystyrene, Se) and molecular glass formers (toluene, isopentane), good glasses are hard to find in two dimensions. One reason is the lack of topological frustration in $d = 2$ for identical particles interacting with approximately isotropic, short-range potentials. A crystal can easily nucleate from a liquid around an equilateral triangle composed of three identical particles in close contact – such an arrangement is easily extended to tile the plane in a triangular close-packed lattice. The barrier to forming a crystalline ground state at low temperatures is low, especially if freezing proceeds via continuous instead of first-order phase transitions. As discussed in Chapter 4, the situation in three dimensions is very different, because there is no natural packing of identical tetrahedra and freezing is a strongly first-order transition with large nucleation barriers.

To study glasses in two dimensions, it is helpful introduce a source of frustration in an effort to stabilize an amorphous phase at low temperatures. One possibility is to inject particles with a noncrystalline 5-fold symmetry. The figure (opposite) shows the result of "annealing" a high-density "liquid" of small aluminum pentagons in a vibrating shake-table apparatus [2]. This particle configuration was obtained by gradually increasing the density of pentagons. Twelve of these pentagons would form an unfrustrated regular dodecahedron if they were used to tile the surface of a sphere. In the plane, however, the figure shows that essentially every particle is six-coordinated and that the centers of the pentagons form a regular triangular lattice! Evidently, the 6-fold coordination favored by planar particle configurations is very important. This conclusion that the geometrical environment is more important than particle shape is supported by the structure of the polyomavirus, SV40 [3]. In this unusual virus, 72 identical pentagonal

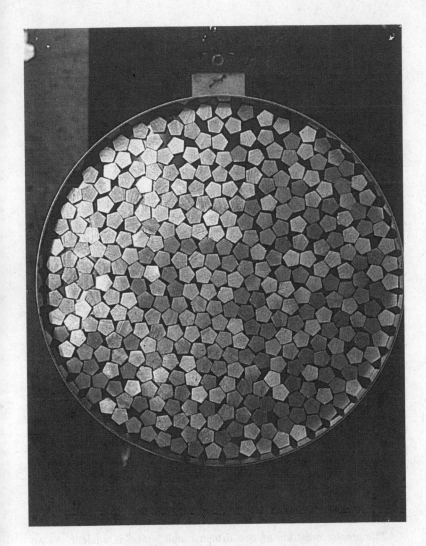

"morphological units" (protein pentamers) are arrayed on the surface of a sphere. Twelve of these units sit in a 5-fold environment, similar to the 12 pentagonal panels on the surface of a soccer ball. However, the remaining 60 units are all surrounded by *six* nearest neighbors, oblivious to the fact of their pentagonal shape, just as in the figure.

A more effective way of introducing geometrical frustration into particle arrays is to undercool liquids with particles of two different sizes, or to anneal identical particles on surfaces of constant positive or negative Gaussian curvature. The extension of ideas about defect-mediated phase transitions to two-dimensional glasses formed in this way is the subject of this chapter. One conclusion is that it should be possible to form a metastable hexatic glass phase for binary mixtures of particles cooled in the plane rapidly enough to prevent phase separation. For an important modification of the theoretical phase diagram for quench random impurity disorder proposed in Fig. 3.4, see [4]. Although initial efforts to find hexatic glasses in Monte Carlo simulations of two-dimensional Lennard-Jones systems were unsuccessful [5], results consistent with the presence of an intermediate hexatic phase were found recently in a simulation with polydispersed particle sizes [6].

3.1 Introduction

Compared with the explosion of theoretical work on randomness in spin systems [7], relatively few analytical results are available on disordered solids. Of course, one might hope that an understanding of, say, spin glasses would lead to insights into technologically more important problems such as the glass transition. Nevertheless, information on statistical mechanical models of disorder in liquids and solids, without appealing to spin-glass "analogies," would clearly be quite valuable. In this paper, we discuss models of quenched randomness in two-dimensional fluids and crystals. We shall restrict our attention to two-dimensional analogues of metallic glass formers, which do not exhibit directional bonding. Some of what we say probably applies to directionally bonded systems as well. Unlike the situation in three dimensions, analytical theories of melting and crystallization in pure systems are available [8–11]; they are based on the dislocation mechanism proposed by Kosterlitz and Thouless [12] and reviewed in Chapter 1. Because the predicted melting transitions are continuous, one can bring to bear renormalization-group methods and ideas developed for understanding critical phenomena [13]. It appears that the distinction between translational and orientational order, which plays a role in two-dimensional melting [9], may also be relevant in parameterizing structure in glass.

The *precise* meaning of two-dimensional "glass" is unclear. As we shall discuss later, there are important topological differences between the ways in which particles pack together in two and three dimensions. Nevertheless, it is probably possible to find thin films with particles of two different sizes that exhibit something like a glass transition with decreasing temperature. Computer simulations could be quite illuminating, even though the limited time scales available even on the fastest machines obscure a genuine glass transition in three dimensions [14]. Results of real experiments, with vastly longer equilibration times, may soon become available. With the advent of high-resolution synchrotron X-ray spectroscopy [15] and elegant techniques for measuring shear moduli and viscosities in thin films [16, 17], one can hope for detailed structural and dynamical information on the effects of disorder in lipid films on water [18], freely suspended liquid-crystal films [19], colloidal crystals at air–water interfaces [20] and physisorbed rare-gas monolayers [21]. In particular, it would be interesting to know whether the shear viscosity $\eta_s(T)$ (or related quantities such as the particle diffusivity) exhibits a Vogel–Fulcher-law behavior near an intrinsic glass-transition temperature T_0,

$$\eta_s(T) \sim \exp[\text{constant}/(T - T_0)]. \tag{3.1a}$$

As discussed in Chapter 2, the dynamical theory of two-dimensional melting predicts a somewhat similar temperature dependence for the shear viscosity just above the equilibrium hexatic-to-solid transition temperature T_m[11]:

$$\eta_s(T) \sim \exp[\text{constant}/(T - T_m)^{0.37}]. \tag{3.1b}$$

There are some obvious advantages in trying to understand disordered solids and liquids in two, rather than three, dimensions. As mentioned above, one can build on analytical results for pure systems. In three dimensions, there are numerous speculations that defects such as dislocation lines, disclination lines, grain boundaries and small elements of "free volume" are important in understanding supercooled liquids and glasses [22]. In $d = 2$, most of these defects are point singularities, which are much easier to treat theoretically than are lines and surfaces. There are clever ways to visualize two-dimensional defects in thin films and in computer experiments, when they are viewed from the third dimension [23, 24]. One can hope to extract quantitative predictions from various hypotheses and compare them with results of the experiments mentioned above. Ultimately, insights into disordered materials in $d = 3$ may be possible.

There is at least one glaring problem with the above program. An important topological feature of dense random-packing models of metallic glasses [25, 26] in three dimensions is the tendency to maximize the local density by forming tetrahedral clusters. Because tetrahedra cannot be closely packed to fill space, highly frustrated or "jammed" particle configurations result. See Chapter 4 for a more detailed discussion. The situation in two dimensions is quite different, for triangular packing units not only maximize the local density but also combine neatly to form a space-filling lattice [23]. It is difficult to imagine disordered arrays of identical particles that would be stable at low temperatures or high densities, rather than racking up like billiard balls into a triangular lattice. In $d = 3$, tetrahedral clusters combine naturally into an icosahedron [27]. The 12 particles arranged around a central sphere in an icosahedral cluster are not packed perfectly, because the distance between pairs of these particles is about 5% greater than the distance to the center. In a crude sense this mismatch is the source of the excess free volume or entropy in simple theories of glass formation [28, 29]. In two dimensions, the figure most analogous to the icosahedron is the hexagon, for which this frustration effect does not exist. Here, we use the word "frustration" to mean a source of disorder that leads to there being many competing ground states at low temperatures.

There are two ways around this problem. One is to consider films with particles of two different sizes, on time scales such that phase separation is impossible. Frustration is present because one must now pack an assortment of equilateral and isosceles triangles. Two different particle sizes are in fact usually required in order to obtain stable metallic glasses at cooling rates currently available in the laboratory even in three dimensions. In two dimensions, one can introduce a controllable amount of randomness by varying the concentration of particles of the "wrong" size. There is such simple theoretical limit in $d = 3$.

Another, more radical, approach for making glass physics in two dimensions more like the situation in $d = 3$ is to consider the statistical mechanics of particles cooled on a manifold of constant negative curvature. As we shall discuss later, the packing problems of triangles in such a space are very similar to frustration effects involved in packing tetrahedra in ordinary flat three-dimensional space. A related point of view has been taken in three dimensions by Kléman and Sadoc [30], who start with the observation that tetrahedra *do* pack into a space-filling structure on three-dimensional manifolds with appropriately chosen curvatures. They then map this structure into flat space by introducing a minimal number of defects.

In Section 3.2, we pursue the first approach and describe experimental [31] and theoretical [32] work on order in crystalline and fluid films with a quenched distribution of impurities. One finds that crystals with quenched impurities are stable only at low temperatures and weak disorder strength. Otherwise, thermally excited pairs of dislocations are broken apart by the random impurity potential. The hexatic fluid which results when the disorder is strong can persist down to $T = 0$. These conclusions are consistent with results of experimental studies on vibrating ball-bearing arrays, which show that dislocations are often quite effectively trapped by impurities. One finds that many nominally "amorphous" arrays exhibit orientational order on a scale much larger than the relatively short translational correlation length. In Section 3.3, we discuss some ideas about what happens to arrays of uniformly sized particles cooled in a space of constant negative curvature. By choosing the curvature appropriately, one may be able to obtain insights into three-dimensional dense random packing.

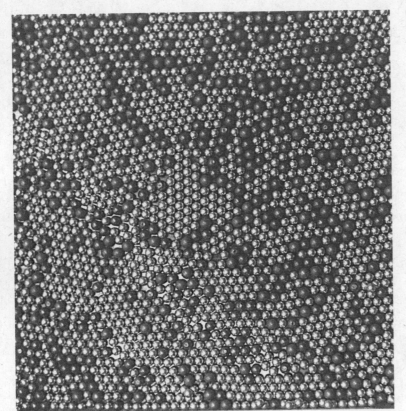

Fig. 3.1. A disordered ball-bearing array.

3.2 Order and frustration in quenched binary arrays

Figure 3.1 shows a ball-bearing array containing spheres of two different sizes "annealed" in a vibrating-dish apparatus [31] originally developed by Turnbull and Cormia [33]. The vibrations are a crude approximation to the fluctuations present in thermal equilibrium and the array is "frustrated" by the inhomogeneities in sphere size. Crystallization is prevented by this frustration, unless there is sufficient time for the ball bearings to undergo phase separation and form two coexisting crystals. Phase separation is out of the question on the time scale of these experiments and on the time scales of many of the thin-film systems mentioned in the introduction at low temperatures. The disorder is quenched in the sense that neighboring spheres do not have time to exchange places by diffusion. Although long-wavelength collective excitations are possible, the coordinates of a particle relative to the surrounding matrix are fixed.

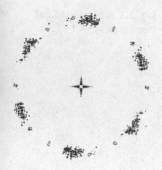

Fig. 3.2. Contours of constant scattering intensity for the array in Fig. 3.1.

Figure 3.2 shows the contours of constant intensity in the structure function [31]

$$S(\mathbf{q}) = \langle |\rho(\mathbf{q})|^2 \rangle, \tag{3.2}$$

where $\rho(\mathbf{q})$ is the Fourier transform of the particle density. Despite the rather disordered appearance of the particle configuration in Fig. 3.1, there is a striking 6-fold asymmetry in $S(\mathbf{q})$. Note that all information about this modulation would be lost in the usual "powder-average" X-ray diffraction pattern. The array is certainly not crystalline, because of the finite radial width of the peaks. Orientational order, as opposed to the translational order measured in terms of $S(\mathbf{q})$, can be defined locally via the bond orientational order parameter discussed in Chapter 2 [9],

$$\psi_6(\mathbf{r}) = \exp[6i\theta(\mathbf{r})], \tag{3.3}$$

where $\theta(\mathbf{r})$ is the angle the line joining the centers of two neighboring particles makes with respect to some reference axis. The factor of six appears in the exponent of Eq. (3.3) because of the predominance of hexagonal clusters in two-dimensional liquids and solids. A direct measure of correlations in the local "crystallographic axes" is the function

$$G_6(r) = \langle \psi(\mathbf{r})\psi^*(\mathbf{0}) \rangle, \tag{3.4}$$

where the angle brackets indicate an average over the area of the sample. In isotropic liquids or glasses one would expect that $G_6(r)$ ultimately decays exponentially to zero,

$$G_6(r) \sim \exp(-r/\xi_6), \tag{3.5}$$

which defines an orientational correlation length ξ_6. The modulation evident in Fig. 3.2 can occur only provided that ξ_6 is comparable to the size of the approximately 1400-particle sample. The translational correlation length ξ_T, defined by the inverse radial width of the peaks in $S(\mathbf{q})$, is only about four mean particle diameters. The orientational and translational correlation lengths associated with Fig. 3.1 clearly satisfy the inequality

$$\xi_6 \gg \xi_T. \tag{3.6}$$

These disparate orientational and translational length scales suggest that supercooled liquids can expel entropy with decreasing temperature, by developing extended orientational correlations, despite the topological frustration and without an appreciable increase in the amount of crystalline, translational order. The amount of "orientational entropy" available to be expelled should be comparable to the difference in entropy between an isotropic liquid and a well-ordered hexatic phase [9]. This expulsion of entropy need not occur via a genuine phase transition. In the presence of quenched disorder, it is possible that $\xi_6(T)$ grows slowly but steadily with decreasing temperature, rather than diverging to infinity as it would at a second-order phase transition. It is perhaps worth emphasizing that Fig. 3.1 would be rather poorly modelled by set of *randomly* oriented micro-crystallites. The orientational and translational correlation lengths in such systems are comparable.

Correlations in dense binary arrays are relatively easy to understand in the limit of a dilute concentration of particles of the "wrong" size embedded in an otherwise uniform matrix [31]. We assume that motion of dislocations via glide diffusion (movement parallel to the Burgers vector) is still possible even at very high densities. The relatively mobile dislocations interact with an impurity pinned at the origin in a solid-like matrix via the long-range potential [34]

$$V(\mathbf{r}) = \frac{(\mu + \lambda)}{\pi(2\mu + \lambda)} \Omega_0 \frac{\hat{\mathbf{z}} \cdot (\mathbf{b} \times \mathbf{r})}{r^2}, \tag{3.7}$$

where μ and λ are elastic constants and $\mathbf{b}$ is the Burgers vector of an impurity at position $\mathbf{r}$. The quantity Ω_0 is the change in area, relative to a pure system, associated with the impurity. Taking $\mathbf{b}$ in the x direction, we see that Eq. (3.7) predicts a repulsive or attractive interaction, depending on whether the dislocation is above or below the x axis. If Ω_0 is, say, positive, the dislocation would clearly like to position itself so that the impurity partially compensates for its missing row of atoms. Because of this impurity–dislocation interaction, one might expect dislocations to become trapped on the impurities when liquid-like particle arrays are quenched by increasing the density or lowering the temperature.

Assuming that dislocations are indeed trapped on a finite fraction of impurities, it is straightforward to determine the range of translational and orientational order. Translational order will clearly be dephased on a scale comparable to the spacing of trapped dislocations, just as in the equilibrium melting scenario discussed in Chapter 2. Orientational order, on the other hand, is more robust. The twist $\theta(\mathbf{r})$ in the local crystallographic axes due to a set of dislocations at positions $\mathbf{r}'$ and with Burgers vectors $\mathbf{b}(\mathbf{r}')$ is [9]

$$\theta(\mathbf{r}) = -\frac{1}{2\pi} \sum_{\mathbf{r}'} \frac{\mathbf{b}(\mathbf{r}') \times (\mathbf{r} - \mathbf{r}')}{|\mathbf{r} - \mathbf{r}'|^2}. \tag{3.8}$$

If the positions and Burgers vectors of the dislocations are uncorrelated, we have

$$\langle b_i(r)b_j(r') \rangle = a_0^2 c \delta_{ij} \delta_{\mathbf{r}, \mathbf{r}'}, \tag{3.9}$$

where a_0 is the length of the smallest Burgers vector in the defect-free lattice and c is proportional to the concentration of dislocation-bearing impurities. With this assumption, it is straightforward to show that the bond orientational correlation function decays algebraically [31],

$$G_6(r) \sim 1/r^{\eta_6}, \tag{3.10}$$

with

$$\eta_6 \approx 9c/\pi. \tag{3.11}$$

Because of the slow power-law decay defined in Eq. (3.10), the orientational correlation length ξ_6 defined by Eq. (3.5) is infinite. The translational correlation length, however, is of order

$$\xi_T = a_0/c^{1/2}. \tag{3.12}$$

The inequality $\xi_6 \gg \xi_T$ characteristic of a hexatic glass is certainly satisfied in this case!

Figure 3.3 shows the centers of large and small spheres in a ball-bearing array with a dilute concentration of large particles like that shown in Fig. 2.16(a) [31]. Also shown are the coordination numbers of those particles surrounded by a number of near neighbors different from six. On a flat two-dimensional surface, the average coordination number must be exactly six [35]. Displaying the deviant coordination numbers provides a graphic way of exhibiting various kinds of defects [23, 24]. Evidently, a certain amount of topological debris surrounds all the large-sphere inhomogeneities. As indicated explicitly by the Burgers construction, several impurities have, in fact, managed to trap dislocations. Isolated dislocations show up a coordination-number plot as a pair of 5- and 7-coordinated particles. An example is contained within the Burgers contour in the upper-left-hand-portion of Fig. 3.3. Although this array appears locally crystalline, translational order is in fact broken up on a scale comparable to that of the trapped dislocations. The orientations of the shaded hexagons containing seven particles in the lower-left- and upper-right-hand corners of Fig. 3.3 are almost identical. Evidently, Fig. 3.3 provides a nice illustration of the kind of order apparent in the more disordered mixture displayed in Figs. 3.1 and 3.2. For other ratios of sphere sizes, networks of grain boundaries, which are absent from highly disordered mixtures, were generated [31].

It is interesting to examine the stability of crystalline solids at finite temperatures in the presence of quenched impurity disorder [32]. One can model such crystals by the continuum elastic free energy

$$F = \tfrac{1}{2} \int d^2x [2\mu u_{ij}^2 + \lambda u_{kk}^2 - 2w\,\delta c(\mathbf{r})u_{kk}], \tag{3.13}$$

where $u_{ij}(r)$ is the symmetrized strain tensor associated with a fluctuating phonon displacement field $\mathbf{u}(\mathbf{r})$. The quantity $\delta c(\mathbf{r})$ is a frozen-in fluctuation in the local concentration of impurities. The parameter w describes the coupling between the local elastic dilation $\nabla \cdot \mathbf{u} = u_{kk}$ and the quenched volume fluctuations due to impurities, while μ and λ are the same elastic constants (characterizing a triangular solid) as those appearing in Eq. (3.7). Ball-bearing arrays at finite temperature would be described by a free energy like Eq. (3.13) composed of purely entropic contributions. This free energy is, of course, also relevant to the experimental systems described in the introduction, which are characterized by softer potentials. We assume that a given configuration $\delta c(\mathbf{r})$ occurs with probability

$$P[\delta c(\mathbf{r})] \propto \exp\left(-\frac{1}{2\sigma} \int d^2 r [\delta c(r)]^2 \right). \tag{3.14}$$

Small deviations from this Gaussian distribution turn out to be unimportant at long wavelengths. Because the positions of the impurities within the host matrix are quenched, it is the macroscopic free energy,

Fig. 3.3. A coordination-number plot for an array with a dilute concentration of large spheres, indicated by diamonds. Small particles are indicated by crosses. Note that the shaded hexagons in the lower left and upper right corners have almost identical orientations.

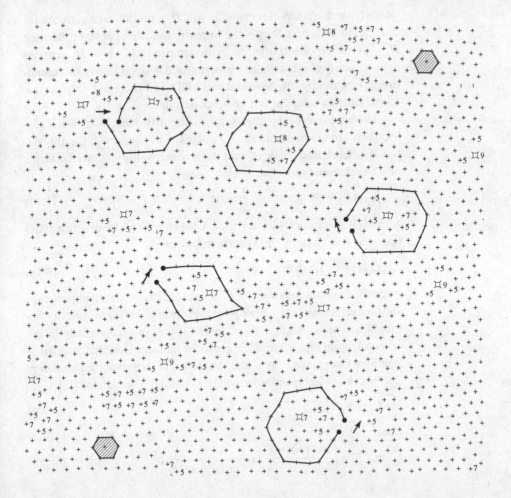

$$F = -k_B T \ln \left[\int \mathcal{D}u(r) \exp\left(-\frac{F}{k_B T} \right) \right],$$ (3.15)

rather than the partition function, which should be averaged over this probability distribution.

We shall assume that dislocations have time to reach a thermal equilibrium consistent with the fixed distribution of impurities. Since movement perpendicular to the Burgers vector of the dislocation (dislocation climb) is activated by diffusion of vacancies and interstitials [34], this sort of motion will be impossible for the same reasons that impurity diffusion is forbidden. Dislocations can still reach thermal equilibrium via glide diffusion (which does not require vacancies or interstitials), provided that there is time to diffuse across the entire sample. To insure impurity quenching and equilibration of dislocations, the experimental time scale t_{exp} must be such that

$$L^2/D_g \ll t_{exp} \ll a_0^2/D_{imp},$$ (3.16)

where L is a characteristic crystallite size, D_g is the diffusion constant for dislocation glide and D_{imp} is the impurity diffusion constant. Because D_{imp} drops exponentially fast with temperature while D_g remains approximately constant, many experimental systems satisfy this condition at sufficiently low temperatures [32], even though L is a macroscopic length whereas a_0 is a microscopic one.

There are interesting effects due to impurity quenching even if dislocations are ignored entirely. As discussed in Chapter 2, the translational order characteristic of a crystal is conveniently measured in terms of a translational order parameter,

$$\rho_G(\mathbf{r}) = e^{i\mathbf{G} \cdot \mathbf{u}(\mathbf{r})},$$ (3.17)

where $\mathbf{G}$ is a reciprocal lattice vector. The behavior of the structure function near $\mathbf{G}$ is given by the Fourier transform of

$$C_G(\mathbf{r}) = \overline{\langle \rho_G(\mathbf{r})\rho_G^*(0) \rangle},$$ (3.18)

where the angle brackets indicate a conventional thermal ensemble average and the overbar signifies a subsequent average over the frozen impurity distribution. The correlation function (3.18) decays exponentially in the arrays shown in Figs. 3.1 and 3.3, due to the finite translational correlation lengths in these systems. It is straightforward to show that the free energy (3.13) leads to algebraically decaying translational correlations [32],

$$C_G(\mathbf{r}) = 1/r^{\eta_G},$$ (3.19a)

where

$$\eta_G(T) = \left(\frac{k_B T(3\mu + \lambda)}{4\pi\mu(2\mu + \lambda)} + \frac{\sigma w^2}{4\pi(2\mu + \lambda)^2} \right) |\mathbf{G}|^2.$$ (3.19b)

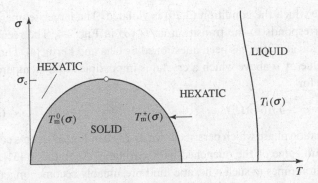

Fig. 3.4. Solid, liquid and hexatic phases as a function of temperature and the degree of disorder σ. The translational correlation length diverges on the path marked by the arrow.

The long-range order we would expect in a three-dimensional solid at finite temperatures even with quenched impurities has been destroyed. Note that $\eta_G(T)$ remains finite at $T=0$, due to impurity-induced volume fluctuations.

One can apply the theory of dislocation-mediated melting in pure systems [8–12] to this problem. The conclusions as a function of temperature T and the strength of the impurity randomness σ are summarized in Fig. 3.4 [32]. The randomness parameter σ should be an increasing function of the concentration of impurity. For sufficiently small σ, there is the usual dislocation-mediated continuous melting transition at a temperature $T_m^+(\sigma)$. This transition is into a hexatic phase with residual 6-fold bond orientational order. A second transition $T_i(\sigma)$ is necessary in order to destroy orientational order and complete the process of melting. Ref. [32] argued for a reentrant melting transition at a low temperature $T_m^0(\sigma)$. At these low temperatures, pairs of dislocations can be ripped apart by a random impurity potential, which can be represented by a superposition of terms like Eq. (3.7). Cha and Fertig have recently argued that the line $T_m^0(\sigma)$ should be replaced by a horizontal line connecting $T_m^+(\sigma_c)$ to the point σ_c at $T=0$ [4]. For σ greater than a critical value σ_c, the crystalline phase is destroyed entirely.

A useful criterion for assessing the stability of the solid is the condition

$$K - \sigma(\Omega_0 K/a_0)^2 > 16\pi, \tag{3.20}$$

where we have used the estimate [32] $w = \Omega_0(\mu + \lambda)$ and the dependence on temperature is contained in

$$K = \frac{4\mu(\mu + \lambda)}{(2\mu + \lambda)k_B T}. \tag{3.21}$$

When $\sigma = 0$, Eq. (3.20) reduces to the usual Kosterlitz–Thouless criterion for the stability of a two-dimensional solid, which is obtained by balancing the energy and entropy of an isolated dislocation [12]. For nonzero randomness, there is a temperature interval, $T_- < T < T_+$, outside which the condition (3.20) is violated. The larger temperature T_+ corresponds to the transition at $T_m^+(\sigma)$ in Fig. 3.4. The reentrant transition at $T_m^0(\sigma)$ has been questioned by Cha and Fertig [4]. The critical value of σ above which a crystal is impossible at any temperature occurs for

$$\sigma = \sigma_c = a_0^2/(64\pi\Omega_0^2). \tag{3.22}$$

The hexatic phase which persists down to $T = 0$ in Fig. 3.4 seems consistent with some of the quenched ball-bearing arrays found in [31]. The relaxation times in such a hexatic fluid presumably become longer and longer with decreasing temperature, with a diverging Arrhénius temperature dependence in the hexatic shear viscosity. As discussed in Chapter 2, this viscosity also diverges as one approaches the hexatic–solid phase boundary [11]:

$$\eta_s \sim \xi_T^2, \tag{3.23}$$

where ξ_T is the translational correlation length. This length diverges near the hexatic–solid phase boundary at $T_m^+(\sigma)$,

$$\xi_T \sim \exp(\text{constant}/|T - T_m|^{\bar{\nu}(\sigma)}), \tag{3.24}$$

where $\bar{\nu}(\sigma = 0) \approx 0.37$. This exponent decreases to zero as σ approaches σ_c [32].

3.3 Order and frustration in spaces of incommensurate curvature

X-ray diffraction experiments on dense liquids reveal something like 12 particles in the first coordination shell of every atom. As observed many years ago by Sir Charles Frank [36], it is energetically preferable to arrange these particles in the icosahedral configuration shown in Fig. 3.5(a), rather than into nuclei that can form the seeds of fcc or hcp crystals. (See section 2.7.2.)

Frank invoked the icosahedron to explain the remarkable degree to which simple liquid metals can be supercooled in the laboratory [37]. Computer simulations of supercooled liquid argon suggest that short-range icosahedral order begins to develop at about 10% below the equilibrium melting temperature [38], leading to an inequality analogous to Eq. (3.6). Icosahedra also play an important role in dense random-packing models of metallic glasses, since they can be viewed as slightly distorted arrangements of 20 tetrahedra. Like triangles in two dimensions, tetrahedra of particles maximize the local density. The 5-fold symmetry axis evident in Fig. 3.5(a) precludes a space-filling structure composed of tetrahedra, however. As discussed in the introduction,

(a)

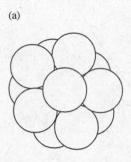

(b)

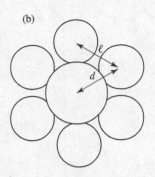

Fig. 3.5. (a) An icosahedron in flat space. (b) A hexagon in curved space.

there is leftover space when 12 hard spheres are packed in an icosahedral arrangement around a central sphere of the same size. This extra volume is an important source of degeneracy and frustration in dense random-packing models of the glassy state.

The obvious analogue of the icosahedron in two dimensions occurs when six triangles combine to form a hexagon. Unlike the icosahedron, however, the resulting cluster of seven identical disks can be packed neatly to form a space-filling triangular lattice. There is, moreover, no leftover space between the disks on the perimeter. Although these topological features favor formation of equilibrium hexatic phases (intermediate between a solid and a liquid), they also preclude a sharp analogy with dense random-packing problems in $d = 3$. One way of correcting for this deficiency is to consider the statistical mechanics of dense liquids on a surface of constant negative curvature. For an appropriately chosen curvature, seven-atom hexagonal clusters are quite similar to 13-atom icosahedral clusters in three-dimensional flat space. Although understanding liquids and glasses on a curved surface is more complicated than it is in flat space, this problem is in many ways considerably easier than that of understanding liquids and glasses in *three-dimensional* flat space!

To specify positions and distances on a curved surface, it is convenient to use polar coordinates r and ϕ. The metric in a space with negative Gaussian curvature K is [39]

$$d^2s = d^2r + \frac{\sinh^2(\kappa r)}{\kappa^2} d^2\phi, \tag{3.25a}$$

where

$$\kappa = \sqrt{-K} \tag{3.25b}$$

is a curvature parameter that tends to zero as space becomes flat. As we shall see, κ can be related to the mismatch in the way six identical disks fit around a central one. A similar curvature parameter can be defined

Fig. 3.6. The distance ℓ
between perimeter disks in a
hexagon as a function of the
distance to the center d. The
dashed $\ell = d$ characteristic of
flat space is shown for
comparison.

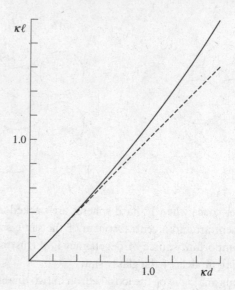

Fig. 3.6. The distance ℓ between perimeter disks in a hexagon as a function of the distance to the center d. The dashed $\ell = d$ characteristic of flat space is shown for comparison.

in three-dimensional flat space in terms of the mismatch in an icosahedral cluster. Physically, κ^{-1} is the length scale on which frustration effects become important. It is straightforward to integrate up the metric in Eq. (3.25a), to find the distance ℓ_{ab} between two particles with polar coordinates (r_a, ϕ_a) and (r_b, ϕ_b), given implicitly by

$$\cosh(\kappa\ell_{ab}) = \cosh(\kappa r_a)\cosh(\kappa r_b) - \sinh(\kappa r_a)\sinh(\kappa r_b)\cos(\phi_a - \phi_b).$$
(3.26)

This formula is closely related to the distance between two points on a sphere with imaginary radius $R_0 = i/\kappa$. Let us apply Eq. (3.26) to the arrangement of seven disks shown in Fig. 3.5(b), taking the origin of our polar coordinate system to be the center of the middle disk. The apparent difference in size of the disks is due to projecting a small portion of curved space onto a flat piece of paper; the disks all have an identical diameter d. Assuming that the disks on the perimeter are symmetrically arranged so as just to touch the disk at the center, we find that that the geodesic distance joining their centers is

$$\ell = \frac{1}{\kappa}\cosh^{-1}\left(\frac{1}{2}\cosh^2(\kappa d) + \frac{1}{2}\right).$$
(3.27)

As shown in Fig. 3.6, ℓ exceeds d for $\kappa > 0$, so there is indeed extra space between the perimeter disks, just as in the case of spheres on the surface of an icosahedron.

Evidently, it should be possible to tune the curvature so that hexagonal packings mimic as closely as possible frustration effects associated

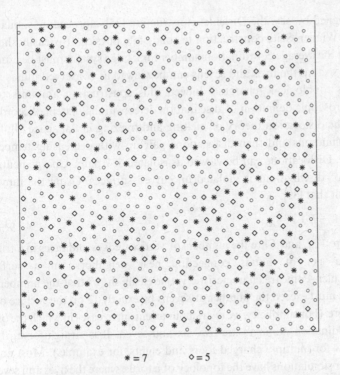

Fig. 3.7. A coordination-number plot for an isotropic liquid of identical particles. Asterisks denote particles with seven near neighbors while diamonds indicate 5-coordinated particles. The remaining particles (circles) have six near neighbors.

$$* = 7 \qquad \diamond = 5$$

with three-dimensional icosahedra. One possibility is to choose κ so that

$$\ell/d = 4/\sqrt{10 + 2\sqrt{5}}$$
$$\simeq 1.051462, \tag{3.28}$$

which is the ratio of ℓ to d associated with an icosahedron [40]. Probably a better choice would be to adjust κ so that the "free area" associated with Fig. 3.5(b) matches the free volume associated with an icosahedron. This free volume is such that a 13th sphere can be brought fairly close to the central one if the other spheres on the surface are pushed aside. Interesting behavior should also occur for "commensurate" curvatures such that an integral number of disks (larger than six) pack perfectly around the central one. It can easily be shown that the curvatures κ_7 and κ_8 for packing seven and eight disks, respectively, are given by

$$\kappa_7 d = 1.09054966, \qquad \kappa_8 d = 1.52857092. \tag{3.29}$$

Here, we shall be primarily concerned with what happens when the curvature is slightly incommensurate, $\kappa d \ll 1$.

Figure 3.7 shows the particle positions in a computer simulation of a hot liquid in flat space [41]. Particles with anomalous 5- and 7-fold coordination numbers are highlighted [23, 24]. Near neighbors are assigned by, say, the Dirichlet construction [35], a procedure similar to the Wigner–Seitz construction in solid-state physics. It is easily shown that the average coordination number for a large system must be exactly six, so that the fives and sevens can be viewed as conserved minus and plus topological "charges." Indeed, this construction amounts to a microscopic definition of disclination defects in the system. According to the dislocation theory of melting and freezing in flat space [9], first disclinations and then dislocations pair up with decreasing temperature. Dislocations can be viewed as isolated 5–7 pairs. Charge neutrality of disclinations is violated on curved surfaces. Using the famous result that [39]

$$V - E + F = \chi, \tag{3.30}$$

where V, E and F are the numbers of vertices, edges and faces in a triangular net of bonds connecting near neighbors on a surface of Euler–Poincaré characteristic χ, one easily shows that the number of fives must exceed the number of sevens by exactly 12 on the surface of a sphere. (For simplicity, we restrict our attention to cases in which the only coordination numbers are five, six and seven. It is straightforward to allow for multiply charged fours and eights, for example.). Most computer simulations have the topology of a torus, where the fives and sevens must occur in equal numbers.

It is easy to see that a finite net disclination charge density must characterize a space of constant negative curvature. We imagine triangulating an array of particles on such a manifold with a set of geodesic lines joining neighboring atoms. The integral curvature of one of the geodesic triangles obtained in this way is [39]

$$\iint_{\triangle ABC} -\kappa^2 dS = A + B + C - \pi, \tag{3.31}$$

where A, B and C are the three angles characterizing the triangle. Summing over all triangles i in a large region and ignoring edge effects, we obtain

$$-\kappa^2 S = \sum_i (A_i + B_i + C_i - \pi)$$
$$= 2\pi V - \pi F, \tag{3.32}$$

where V is the number of vertices (particles) in the net, F is the number of triangular faces and S is the total surface area of the region. Since the number of edges (bonds) in the net is $E = 3F/2$, we also have $E = \bar{q}V/2$, where $\bar{q}$ is the average coordination number. Equation (3.32) can be reexpressed in the form [42]

$$\bar{q} = 6 + 3s\kappa^2/\pi, \tag{3.33}$$

where s is the surface area per particle. Evidently, the effect of a small nonzero curvature is to break "charge neutrality" by creating an excess of 7-fold disclinations.

The situation described above is just a two-dimensional version of one of the proposals of Kléman and Sadoc [30] for describing glasses in three-dimensional flat space (see also Chapter 4). Because disclinations are simple point defects rather than lines in two dimensions, there is some hope of determining analytically what happens when particles are cooled on such a curved manifold. Some insights are possible on the basis of the theory of equilibrium melting in flat space [8–12]. We assume that freezing in flat space proceeds via a pairing of disclination charges at a high temperature T_i, followed by a pairing of dislocations (5–7 disclination pairs) at a lower temperature T_m. If we now impose a small negative curvature, the transition at T_i will be smeared, due to the excess of 7-fold disclination charges. In a hot liquid like that shown in Fig. 3.7, the curvature-induced charge imbalance should be hardly noticeable. The distance between unbound disclinations in flat space defines a diverging orientational length scale as $T \to T_i$ [9],

$$\xi_6(T) \sim \exp[\text{constant}/(T - T_i)^{1/2}]. \tag{3.34}$$

Curvature effects will become important when ξ_6 becomes comparable to the curvature scale, i.e., for

$$\kappa \xi_6(T) \lesssim 1. \tag{3.35}$$

Six-fold orientational order can clearly never extend over distances greater than the spacing between the curvature-induced net density $n_7 = \kappa^2$ of 7-fold disclination charges.

The effect of curvature on the disclination unbinding transition is rather like the smearing of the Kosterlitz–Thouless transition in a dirty superconducting film when one imposes a uniform magnetic field [43]. Here, the magnetic field induces an imbalance of Abrikosov flux vortices of a particular sign. Another analogy with superconductivity becomes apparent when we consider the energy of a disclination in curved space near the temperature T_i. Generalizing the hydrodynamic treatment of hexatic liquids in flat space reviewed in Chapter 2, we expect that the free energy associated with slow variations in the bond-angle field $\theta(r, \phi)$ is approximately

$$F \approx \frac{1}{2} K_A \int_0^R \frac{\sinh(\kappa r)}{\kappa} dr \int_0^{2\pi} d\phi |\nabla \theta(r, \phi)|^2, \tag{3.36}$$

where K_A is an elastic constant and we have integrated in polar coordinates over a circular portion of curved space with a large radius R. The gradient operator in a space of constant negative curvature is

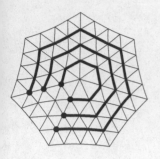

Fig. 3.8. Burgers circuits around a disclination.

$$\nabla = \hat{r}\frac{\partial}{\partial r} + \hat{\phi}\frac{\kappa}{\sinh(\kappa r)}\frac{\partial}{\partial\phi}, \tag{3.37}$$

where $\hat{r}$ and $\hat{\phi}$ are unit vectors in the r and ϕ directions. For a disclination at the origin, we require that the line integral of $\nabla\theta$ around the origin be an integral multiple of $\pi/3$,

$$\oint \nabla\theta \cdot d\ell = \frac{2\pi m}{6}, \tag{3.38}$$

where m is an integer which is $+1$ for a 7-coordinated particle and -1 for a 5-coordinated particle. It is easily checked that the solution

$$\theta = \frac{m}{6}\phi, \qquad \nabla\theta = \frac{\kappa}{\sinh(\kappa r)}\hat{\phi}, \tag{3.39}$$

satisfies Eq. (3.38) and also obeys the Euler–Lagrange equation association with Eq. (3.36), namely

$$\frac{1}{\sinh(\kappa r)}\frac{\partial}{\partial r}\left(\sinh(\kappa r)\frac{\partial\theta}{\partial r}\right) + \frac{\kappa^2}{\sinh^2(\kappa r)}\frac{\partial^2\theta}{\partial\phi^2} = 0. \tag{3.40}$$

The energy of this disclination is readily found to be

$$F = \frac{\pi m^2}{36}K_A\ell n\left(\frac{\tanh(\frac{1}{2}\kappa R)}{\tanh(\frac{1}{2}\kappa a)}\right), \tag{3.41}$$

where a is the diameter of the disclination core. This expression reduces to the usual logarithmically diverging flat-space result when the system's size R is much less than the curvature scale κ^{-1}. For $\kappa R \gg 1$, however, the energy tends to a finite value for large R. A more complete calculation would include a "vector potential" term in (3.36), which represents screening of the disclination by the nonzero Gaussian curvature of the curved surface. Very similar results hold for a type-II superconductor [44], in which the role of the curvature scale is played by the London penetration depth.

Below the smeared liquid-to-hexatic transition at T_i, one presumably has a gas of unpaired 7-fold disclinations together with a finite concentration of dislocations. The existence of unpaired disclination charges implies an irreducible density of "unbound" dislocations, even after dislocations with equal and opposite Burgers vectors are paired up. As shown in Fig. 3.8, the Burgers construction suggests that an amount of dislocation charge proportional to r be associated with a region of size r^2 surrounding an isolated disclination in a solid. Here, we assume that r is less than or comparable to the curvature scale, so that flat-space estimates of distance and surface area can be used. Taking r

to be the curvature scale κ^{-1}, we see that each disclination has a number of dislocations of order $1/(\kappa d)$ associated with it. If these dislocations are spread out uniformly on the curved manifold, their spacing defines a translational correlation length, which is easily seen to be

$$\xi_T = \sqrt{\xi_6 d} = \sqrt{d/\kappa}. \tag{3.42}$$

Note that, if the curvature scale is large, one has

$$\xi_6 \gg \xi_T, \tag{3.43}$$

which is the condition found for glassy binary mixtures in flat space in Section 3.2.

What happens at lower temperatures is less clear. One scenario is that the dislocations condense into grain boundaries attached to a superlattice of isolated 7-fold disclinations. If the curvature is slight, one would expect regions of hexagonal-close-packed crystal between the grain boundary arms. As the curvature increases until it reaches the special value κ_7 in Eq. (3.29), one might expect these crystalline regions to shrink until we are left with a lattice of 7-coordinated particles. It is not clear, however, whether such a configuration is really accessible to the system. First, there is the problem of packing an integral number of particles into a fixed grain size. There are probably commensurate–incommensurate packing effects as a function of curvature. Second, there are important topological constraints on the motion of disclinations in a quasi-crystalline medium. Disclinations can move only by laying down a string of dislocations [46]. It is not clear whether the unpaired disclinations will be able to move to the desired superlattice positions on the time scale of, say, a computer simulation. One may simply become trapped in a region of phase space characterized by the inequality in Eq. (3.43).

The above discussion assumes that there are no pathologies associated with statistical mechanics on a manifold of constant negative curvature. Certainly, one does not expect problems in more understandable spaces of positive curvature, such as the surface of a sphere sometimes introduced as an alternative to periodic boundary conditions [47]. If the scenario described above is also appropriate to supercooled liquids in three-dimensional flat space, one would expect an intrinsic upper limit on the icosahedral correlation length ξ_6 measured in [38]. Strictly speaking, long-range orientational order would be impossible.

It is interesting to note the excitations predicted by continuum elastic theory appear to be rather different for solids packed in curved as opposed to flat space. Phonon-like eigenfunctions of the wave equation become damped out on scales large relative to the curvature scale.

It would be interesting to try to explore these questions further via computer simulations, which seem relatively straightforward to carry out on manifolds of constant curvature.

Acknowledgements

Much of the work described in Section 3.2 was carried out in collaboration with Michael Rubinstein and Frans Spaepen. This chapter is adapted with permission from *Topological Disorder in Condensed Matter*, edited by F. Yonezawa and T. Ninomiya (Springer, Berlin, 1983) pp. 164–180.

References

[1] R. Zallen, *The Physics of Amorphous Solids* (Wiley, New York, 1983).

[2] D. R. Nelson and F. Spaepen, unpublished.

[3] R. L. Garcea and R. C. Liddington, in *Structural Biology of Viruses*, edited by W. Chiu, R. M. Burnett and R. L. Garcea, Chapter 7 (Oxford University Press, Oxford, 1997).

[4] M.-C. Cha and H. A. Fertig, *Phys. Rev. Lett.* **74**, 4867 (1995).

[5] Y. J. Wong and G. V. Chester, *Phys. Rev.* B **35**, 3506 (1987).

[6] M. R. Sadr-Lahijany, P. Ray and H. E. Stanley, *Phys. Rev. Lett.* **79**, 3206 (1997).

[7] See, e.g., *Proceedings of the 1978 Les Houches Summer School: Ill-Condensed Matter*, edited by R. Balian, R. Maynard and G. Toulouse (North-Holland, New York, 1979).

[8] D. R. Nelson, *Phys. Rev.* B **18**, 2318 (1978).

[9] B. I. Halperin and D. R. Nelson, *Phys. Rev. Lett.* **41**, 121; *Phys. Rev.* E **41**, 519 (1978); D. R. Nelson and B. I. Halperin, *Phys. Rev.* B **19**, 2437 (1979).

[10] A. P. Young, *Phys. Rev.* B **19**, 1855 (1979).

[11] A. Zippelius, B. I. Halperin and D. R. Nelson, *Phys. Rev.* B **22**, 2514 (1980).

[12] J. M. Kosterlitz and D. J. Thouless, *J. Phys.* C **6**, 1181 (1973).

[13] C. Domb and M. S. Green, *Phase Transitions and Critical Phenomena*, Vol. 6 (Academic, New York, 1976).

[14] C. A. Angell, J. H. R. Clarke and L. V. Woodcock, in *Advances in Chemical Physics*, Vol. 48, edited by I. Prigogine and S. Rice (Wiley, New York, 1981).

[15] See, e.g., P. A. Heiney, R. J. Birgeneau, G. S. Brown, P. M. Horn, D. E. Moncton and P. W. Stephens, *Phys. Rev. Lett.* **48**, 104 (1982).

[16] R. Pindak, D. J. Bishop and W. O. Springer, *Phys. Rev. Lett.* **44**, 1461 (1980).

[17] R. N. Abraham, K. Miyano and J. B. Ketterson, in *Ordering in Two Dimensions*, edited by S. K. Sinha (North-Holland, Amsterdam, 1980).

[18] P. S. Pershan, *J. Physique Coll.* **40**, C3-423 (1979).

[19] C. Y. Young, R. Pindak, N. A. Clark and R. B. Meyer, *Phys. Rev. Lett.* **40**, 773 (1978).

[20] P. Pieranski, *Phys. Rev. Lett.* **45**, 569 (1980).

[21] J. P. McTague, M. Nielson and L. Passell, in *Ordering in Strongly Fluctuating Condensed Matter Systems*, edited by T. Riste (Plenum, New York, 1980).

[22] See, e.g., F. Spaepen, in *Proceedings of the Les Houches Summer School: Physics of Defects*, edited by J. P. Poirier and M. Kléman (North-Holland, Amsterdam, 1981).

[23] F. Spaepen, *J. Non-Cryst. Solids* **31**, 207 (1978).

[24] J. P. McTague, D. Frenkel and M. Allen, *Proceedings of the Conference on Ordering in Two Dimensions*, edited by S. Sinha (North-Holland, Amsterdam, 1980).

[25] C. S. Cargill, *Ann. N. Y. Acad. Sci.*, **279**, 208 (1976).

[26] M. R. Hoare, *J. Non-Cryst. Solids* **31**, 157 (1978).

[27] See, e.g., F. Spaepen in [23].

[28] M. H. Cohen and D. Turnbull, *J. Chem. Phys.* **31**, 1164 (1959); D. Turnbull and M. H. Cohen, *J. Chem. Phys.* **34**, 120 (1961).

[29] J. H. Gibbs and E. A. DiMarzio, *J. Chem. Phys.* **28**, 373 (1958); G. Adam and J. H. Gibbs, *J. Chem. Phys.* **43**, 139 (1965).

[30] M. Kléman and J. F. Sadoc, *J. Phsyique Lett.* **40**, L569 (1979).

[31] D. R. Nelson, M. Rubinstein and F. Spaepen, *Phil. Mag.* A **46**, 105 (1982).

[32] D. R. Nelson, *Phys. Rev.* B **27**, 2902 (1983).

[33] D. Turnbull and R. L. Cormia, *J. Appl. Phys.* **31**, 674 (1960).

[34] A. M. Kosevich, in *Dislocations in Solids*, edited by F. R. N. Nabarro (North-Holland, Amsterdam, 1979).

[35] See, e.g., R. Collins, in *Phase Transitions and Critical Phenomena*, edited by C. Domb and M. S. Green (Academic, New York, 1972), Vol. II.

[36] F. C. Frank, *Proc. R. Soc. London*, A **215**, 43 (1952).

[37] D. Turnbull, *J. Chem. Phys.* **20**, 411 (1952).

[38] P. Steinhardt, D. R. Nelson and M. Ronchetti, *Phys. Rev. Lett.* **47**, 1297 (1981); *Phys. Rev.* B **28**, 784 (1983).

[39] For a lucid introduction to the differential geometry of surfaces, see H. S. M. Coxeter, *Introduction to Geometry* (Wiley and Sons, Inc., New York, 1969), Chapters 19–21.

[40] *CRC Standard Mathematical Tables*, edited by S. M. Selby (The Chemical Rubber Co., Cleveland, 1970), pp. 15–16.

[41] This picture is taken from a computer simulation of particles interacting with a repulsive $1/r$ potential by R. Morf.

[42] For a discussion of boundary effects see, e.g., J. F. Sadoc and R. Mosseri, *Geometrical Frustration* (Cambridge University Press, Cambridge, 1999).

[43] D. S. Fisher, *Phys. Rev.* B **22**, 1190 (1980) and references therein.

[44] M. Tinkham, *Introduction to Superconductivity* (McGraw-Hill, New York, 1975).

[45] D. R. Nelson, *Phys. Rev.* B **28**, 5515 (1983).

[46] M. Kléman, in *Dislocations in Solids*, edited by F. R. N. Nabarro (North-Holland, New York, 1980) Vol. 5. p. 243.

[47] J. P. Hansen, D. Levesque and J. J. Weise, *Phys. Rev. Lett.* **43**, 979 (1979).

Chapter 4
The structure and statistical mechanics of three-dimensional glass

Preface

This chapter describes an order-parameter theory of the structure and statistical mechanics of metallic glasses in three dimensions, inspired by old ideas about simple undercooled liquids and dense random packing by Turnbull [1], Frank [2] and Bernal [3]. The theory exploits a three-dimensional generalization of the Dirichlet construction to describe polytetrahedral particle packings in terms of disclination lines embedded in a medium with short-range icosahedral order. This theory of geometrical frustration has the attractive feature that it does *not* (incorrectly) predict easy glass formation in one- and two-dimensional arrays of identical particles, where frustration is absent. To obtain an equivalent statistical mechanical problem in two dimensions, one would have to consider particles cooled on a surface of constant negative curvature (see Chapter 3). Good agreement with experimentally measured glassy structure functions in three dimensions can be obtained with this approach [4]. In this picture, the slow dynamics of the glass transition is attributed to topological entanglements of disclination lines, which have a large barrier to line crossings. Sachdev has used the order-parameter theory to study the *dynamics* of metallic glasses [5].

Various icosahedral order parameters have been studied in computer simulations of undercooled liquids by Nosé and Yonezawa [6]. The extended icosahedral order which appears in simulations of binary mixtures near the glass transition has been analyzed by Jonsson and Anderson [7]. Straley has cooled 120 identical particles on the curved surface of a three-dimensional sphere, so that geometrical frustration is absent [8]. An icosahedral crystal nucleates even on the limited time scales available in computer simulations, in striking contrast to the behavior of a similar number of particles cooled in flat space. For a more extensive review of glass physics, including the physics of icosahedral quasicrystals, see [9].

Our understanding of simple glass formers suffers because conventional diffraction patterns average over orientational and positional information. Fortunately, the situation is more favorable in glasses and dense liquids formed in solutions containing *colloidal* particles, which are large enough to be imaged directly in real space using confocal microscopy [10, 11]. Considerable short-range icosahedral order does appear in these experiments, usually in the form of the pentagonal bond spindles discussed in this chapter. For a beautiful indication of local 5-fold symmetry in liquid lead, obtained using evanescent X-rays at a solid–liquid interface, see the recent work by Reichert *et al.* [12].

4.1 A physical picture

In contrast to the explosion of theoretical work on spin glasses during the 1970s [13], there have been relatively few attempts to understand the statistical mechanics of "real" glasses. At first glance, this imbalance is surprising, in view of the technological importance of materials such as amorphous SiO_2 (window glass!) and the existence of particularly simple "hard-sphere" disordered particle configurations modelled by metallic glasses [14]. The theory of "real" glasses suffers from the lack of generally accepted simplifications of reality analogous to the Edwards–Anderson model of spin glasses [15]. Such models should capture the essential physics, but should also be sufficiently simple to allow analytical calculations. In this chapter, we summarize attempts [16, 17] to develop models [18, 19] for the structure and statistical mechanics of glass. We shall argue that "real" glasses are, in many ways, like an intricate, non-Abelian, "uniformly frustrated" spin glass.

We shall focus on the statistical mechanics of particles that have been cooled fast enough to avoid nucleating, say, a stable face-centered cubic (fcc) crystal. We shall be primarily interested in atoms interacting via isotropic pair potentials, such as metallic glass-formers. The more common covalently bonded glasses, such as SiO_2, have direction-dependent forces and are complicated by a strong bias toward angles of $\sim 109.5°$ between neighboring chemical bonds. It may be possible, however, to apply ideas developed for metallic glasses to covalently bonded materials using a decoration procedure; the method is similar to the way in which one obtains a covalently bonded diamond lattice with two atoms per unit cell from a close-packed fcc structure in conventional solid-state physics [20]. Formation of metallic glasses with cooling rates now obtainable in the laboratory usually requires particles of two differ- ent sizes. We shall assume that identical particles would form similar structures if they were cooled sufficiently rapidly and indicate briefly how the complication of two different particle sizes can be incorporated into the theory.

The ground state of four identical particles interacting via isotropic pair potentials is clearly a perfect tetrahedron, with particles at the

Fig. 4.1. Frustration in a tetrahedral particle packing and around a plaquette P in a spin glass.

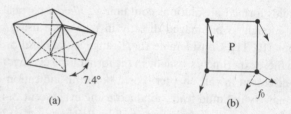

(a)   (b)

vertices. Frustration appears, however, when we try to fill space with such tetrahedra. As shown in Fig. 4.1(a), five perfect tetrahedra wrapped around a common bond leave a gap with a dihedral angle of about 7.4°. The atom near the gap is "frustrated" because it cannot simultaneously sit in the minima provided by the pair potentials of its near neighbors. Because of this frustration, there is no regular lattice of perfect tetrahedra that fills ordinary three-dimensional space. Familiar close-packed regular lattices, like the fcc structure, contain octahedra as well as tetrahedra. The octahedra are necessary in order to obtain a global tessellation of space, even though they do not minimize the energy locally.

Twenty tetrahedra combine with only slight distortions to form a regular 13-particle icosahedron. Sir Charles Frank [18] appealed to an abundance of icosahedra to explain the remarkable degree of supercooling possible in simple liquid metals. Extended correlations in the orientations of neighboring icosahedra have been observed in computer simulations on supercooled liquids [6, 7, 19], even though a global icosahedral lattice is impossible.

The frustration shown in Fig. 4.1(a) is reminiscent of *uniformly* frustrated spin glasses [16]. If the atoms of a glass composed of identical particles condense in a structureless vacuum, the frustration must clearly be the same at every point in space – unlike spin models with quenched-in, spatially varying randomness. One of the simplest uniformly frustrated spin glasses is exemplified by XY spins on a square lattice described by the nearest-neighbor Hamiltonian [21]

$$H = -J \sum_{\langle i,j \rangle} \cos(\theta_i - \theta_j - f_{ij}) \tag{4.1}$$

Here, θ_i is the angle each spin makes with some reference axis and frustration is embodied in the link parameters f_{ij}. As shown in Fig. 4.1(b), the twisting of successive spin angles induced by a fixed distribution of f_{ij} terms cannot in general be accommodated around a plaquette. The frustration is "uniform" if the sum of the twists f_{ij}^P around every plaquette P is the same

$$f_{12}^P + f_{23}^P + f_{34}^P + f_{41}^P \equiv f_0. \tag{4.2}$$

Provided that f_0 is not an integral multiple of 2π, it will be impossible to find a spin configuration such that H assumes the obvious lower bound

Fig. 4.2. A projection of the vertices and nearest-neighbor bonds of polytope {3, 3, 5}.

associated with Eq. (4.1), that is, $H_{min} = -N_b J$, where N_b is the total number of nearest-neighbor bonds.

As pointed out by Teitel and Jayaprakash [22], the Hamiltonian (4.1) is an excellent model for type-II superconducting films subjected to a perpendicular magnetic field of strength f_0. Provided that $f_0 \ll 1$, so that the discreteness of the lattice can be neglected, the ground state is well known [23] to be an Abrikosov flux lattice of identical vortices with, say, positive circulation. Although vortices of both signs will be present at finite temperatures, there will always be a bias toward vortices of a particular sign to accommodate the frustration produced by the magnetic field. As we shall see, one can construct a similar, non-Abelian gauge-field-theory model for particles with short-range tetrahedral order in three dimensions.

In order to understand a spin model like Eq. (4.1), it is helpful to first turn off the frustration by lowering the magnetic field strength to zero. The frustrated ground state is, in fact, simply a lattice of defects in the unfrustrated minimum-energy configuration of aligned spins. To turn off the frustration in a tetrahedral particle array, we must curve space! As described in detail in the books by Coxeter [24, 25], 600 perfect tetrahedra can be embedded without frustration on the surface of a four-dimensional sphere to form a regular lattice of 120 particles. Every one of the vertices in this four-dimensional Platonic solid (or "polytope") sits in an identical, 12-particle icosahedral coordination shell. A two-dimensional projection [25] is shown in Fig. 4.2. This icosahedral crystal is called polytope {3, 3, 5} because it is composed of tetrahedra (denoted by the symbol {3, 3} because *three* equilateral *triangles* meet at every vertex) with *five* tetrahedra wrapped around every bond. The role of {3, 3, 5} in describing dense random packing of hard spheres in flat space was recognized by Coxeter in 1958 [26]. It was suggested as a model for metallic glasses by Sadoc in 1980 [27]. Kléman and Sadoc [28] considered a variety of tetrahedral particle arrays in spherical and hyperbolic spaces and suggested that defects like cut surfaces and disclinations would be necessary in order to map such configurations onto flat space.

An interesting "mean-field" or "effective-medium" approach for dealing with frustration in flat-space tetrahedral particle packings has been discussed by Coxeter [24, 26]. Coxeter argued that some of the properties of dense-packed hard spheres could be modelled by use of a fictitious space-filling "statistical honeycomb" polytope $\{3, 3, q\}$, where, in the simplest of several models discussed by him,

$$q = 2\pi/\cos^{-1}\tfrac{1}{3}$$
$$\doteq 5.104\,299.$$

(4.3)

By definition, this polytope is composed of perfect tetrahedra, with a fractional number, q, wrapped around every bond. Since the dihedral angle of a tetrahedron is $\cos^{-1}\tfrac{1}{3}$, one clearly needs $2\pi/\cos^{-1}\tfrac{1}{3}$ of them to fill the 2π radians of angle around a bond. The fractional part, $q - 5 \approx 0.104\,299$, is a measure of the size of the crack in Fig. 4.1(a).

It is, of course, impossible to have a fractional number of perfect tetrahedra associated with the bonds of a real physical system. It is more realistic to allow for some dispersion in the near-neighbor bond lengths; an arrangement of variable-length bonds can be uniquely associated with any particle configuration using the Voronoi construction [29]. The bonds combine to form a space-filling array of distorted tetrahedra. Figure 4.3 shows a construction, generalizing early work by Frank and Kasper [30], which allows us to view any such configuration as regions of short-range "{3, 3, 5} order" permeated with disclinations [16]. Most bonds in a dense liquid will be surrounded by five tetrahedra, just as in polytope {3, 3, 5}. Links of plus and minus 72° wedge disclination lines in the {3, 3, 5} crystal are associated with 4- and 6-fold tetrahedral bi-pyramids, respectively. These anomalous links combine to form a network of disclination lines that cannot stop or start inside the medium [16]. Because of the crack in Fig. 4.1(a), we would expect to have, on average, more 6-fold lines than 4-fold ones.

This is precisely what one finds in the Frank–Kasper phases of transition metal alloys [30]. These are crystalline phases, with very large unit cells, in which most atoms have 12-particle icosahedral coordination shells. This local icosahedral order is interrupted by particles with "anomalous" coordination numbers of 14, 15 and 16. The geometrical construction described above shows that these exceptional sites are all

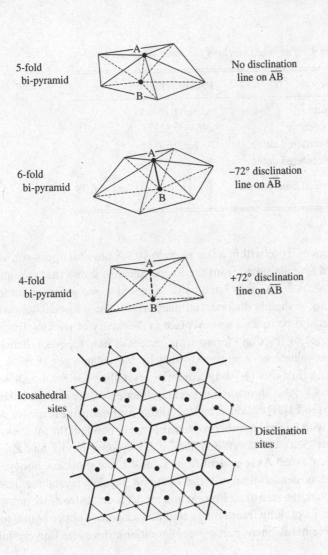

Fig. 4.3. Microscopic constructions for disclination lines.

5-fold bi-pyramid — No disclination line on $\overline{AB}$

6-fold bi-pyramid — $-72°$ disclination line on $\overline{AB}$

4-fold bi-pyramid — $+72°$ disclination line on $\overline{AB}$

Fig. 4.4. Disclination lines in the σ-phase of Co–Cr.

Icosahedral sites

Disclination sites

connected by 6-fold links, which form a contiguous disclination network threading through an otherwise icosahedral medium. A Frank–Kasper phase with a 30-atom unit cell (the σ-phase of Co–Cr) is shown in Fig. 4.4. The cells in a stack of planar disclination networks are threaded by perpendicular quartets of disclination lines. All the defect lines are $-72°$ wedge disclinations; $+72°$ disclinations are absent. This ground state is much like a type-II superconductor in a magnetic field, except that the defect lines run in all directions. The curvature mismatch of flat space with the positive Gaussive curvature required to make polytope {3, 3, 5} plays the role of an applied magnetic field.

The Frank–Kasper phases are closely related to the statistical honeycomb model. The *average* number of tetrahedra per bond $\bar{q}$ in these structures agrees with the statistical honeycomb prediction

Table 4.1 *Packing fractions for some common lattices*

Lattice	Packing fraction
Diamond	$\sqrt{3}\pi/16 \doteq 0.34$
Simple cubic	$\pi/6 \doteq 0.52$
Base-centered cubic	$\sqrt{3}\pi/8 \doteq 0.68$
Face-centered cubic	$\sqrt{2}\pi/6 \doteq 0.74$
Statistical honeycomb	$\frac{1}{\sqrt{2}}\left(6\cos^{-1}\frac{1}{3} - 2\pi\right) \doteq 0.78$

$\bar{q} = 2\pi/\cos^{-1}\frac{1}{3}$ to within a few parts in 10^4! A physical argument, which uses a formula taken from the Regge calculus, shows that the number $\bar{q} = 2\pi/\cos^{-1}\frac{1}{3}$ is associated with a special, energetically preferred packing of slightly distorted tetrahedra [16]. The statistical honeycomb model also occupies a special place in the theory of packing fractions. The packing fraction f for an arrangement of hard spheres is defined to be the volume of a sphere divided by the volume per particle. The packing fractions for some common crystalline lattices are shown in Table 4.1. Also shown is the packing fraction of the statistical honeycomb model [31] obtained by analytically continuing the result for structures with integral numbers of tetrahedra per bond off the integers – the Wigner–Seitz cell for the statistical honeycomb model has $Z \doteq 13.4$ identical faces! As is evident from Table 4.1, the statistical honeycomb model is denser than all the common regular crystalline lattices. Although the statistical honeycomb model is an abstraction, not a real lattice, its packing fraction f_{ideal} is in fact a rigorous upper bound for all possible hard-sphere particle configurations, disordered or crystalline [32].

A hot liquid can be viewed as a dense, tangled mass of icosahedral disclinations. At high temperatures, the frustration-induced bias between oppositely charged defect lines should be hardly noticeable. As one cools the melt, lines with opposite charges pair, until an excess of $-72°$ disclinations becomes evident. As we shall see, there are topological reasons why icosahedral disclination lines cannot cross at low temperatures. Entanglement of disclination lines can lead to severe kinetic constraints, which cause the system to drop out of equilibrium at a glass transition. A disordered network of disclination lines (as opposed to the ordered networks one finds in the Frank–Kasper phases) is an appealing model for structure in metallic glasses [16].

It is instructive to consider similar particle-packing problems in two dimensions. The natural packing element is an equilateral triangle

(a)

$\ell \approx 1.05d$

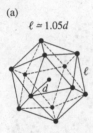

(b) $\ell > d$

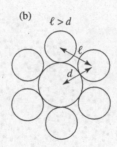

Fig. 4.5. Frustration in three and two dimensions: (a) an icosahedron in flat space and (b) a hexagon in (negatively) curved space.

composed of three identical particles. There is no frustration in flat space, because six triangles combine to form a hexagon, leading to a space-filling hexagonal close-packed lattice. Frustration, similar to that in three-dimensional flat space, can be introduced by embedding the particles in the surface of a sphere. A familiar example is the 32-panel surface of a soccer ball. If we think of each panel as the Wigner–Seitz cell of a particle, we find a "Frank–Kasper phase" of 12 anomalous 5-coordinated particles immersed in a sea of 20 hexagonal sites. The 5-coordinated defect sites form an icosahedral superlattice.

As discussed in Chapter 3, an attractive and theoretically interesting alternative is to pack the particles on an infinite surface of constant negative Gaussian curvature [16, 33, 34]. The properties are defined by a metric that reads, in polar coordinates,

$$d^2s = d^2r + \frac{\sinh^2(\kappa r)}{\kappa^2} d^2\phi. \tag{4.4}$$

The quantity κ^{-1} is a tunable frustration length scale that tends to infinity as space becomes flat ($\kappa \to 0$). When κ is nonzero, cracks open up between six particles symmetrically arranged around a central atom to form a hexagon, in analogy with a similar property of the icosahedron (See Fig. 4.5). One then has the same difficulty in filling space with equilateral triangles as for tetrahedra in three-dimensional flat space. The ratio ℓ/d is a measure of the frustration in both cases.

Figure 4.6 shows a hard-disk particle configuration obtained via a deterministic packing algorithm for hard disks with $\kappa d = 0.5$, where d is the diameter of the disks [34]. Figure 4.6 is a two-dimensional analogue of dense random packing. The coordination-number histogram in the inset shows that most particles have coordination number six, with an asymmetry between the remaining 5- and 7-coordinated particles (circles and squares, respectively). If the fives and sevens are regarded as microscopic point-disclination charges, we see that introducing frustration produces a bias toward 7-fold disclinations. Although the ground state could be a Frank–Kasper-like superlattice of sevens embedded in an otherwise

Fig. 4.6. Disordered packing of hard disks projected out of a space of constant negative curvature. The inset shows the relative proportions of 5-, 6- and 7-coordinated particles.

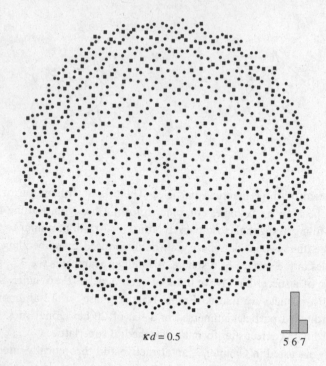

$\kappa d = 0.5$

5 6 7

6-coordinated medium [16, 34], one might expect a glassy configuration like that in Fig. 4.6 when particles are cooled rapidly on a frustrated two-dimensional surface with a constant negative Gaussian curvature.

A theory based on these ideas for systems in three-dimensional flat space [17] will be sketched in Section 4.2.

4.2 The model free energy

In this section, we show how to describe a liquid with short-range icosahedral order using a set of uniformly frustrated order parameters [17]. Our approach is similar to Landau's theory [35] of a liquid just above a conventional freezing transition, wherein there is an order parameter for every reciprocal-lattice vector in the incipient crystalline solid. As we shall see, non-Abelian matrices play the role of reciprocal lattice vectors in materials whose short-range order can be modelled by polytope {3, 3, 5}. It is their noncommutivity that leads to frustration and makes the physics of glass a challenging problem.

We first review how to describe order in a conventional flat-space crystalline solid. Translational periodicity shows up when we perform Fourier expansion of the particle density,

$$\rho(\mathbf{r}) = \sum_{\mathbf{q}} \rho_{\mathbf{q}} e^{i\mathbf{q} \cdot \mathbf{r}}. \tag{4.5}$$

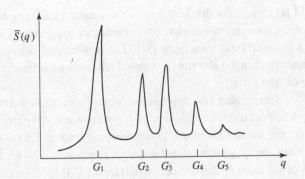

Fig. 4.7. A powder-averaged structure function in flat space. Note the peaks at reciprocal-lattice vectors.

In a gas, or high-temperature liquid, one would expect all Fourier coefficients to occur with roughly equal weights. In a crystalline solid, however, the components ρ_q such that q equals a special set of reciprocal-lattice vectors $\{G\}$ become very large. For a given crystal structure there is a well-known algorithm for determining the allowed G's, which assigns one reciprocal-lattice vector to every set of Bragg planes [36]. In an X-ray scattering experiment, one measures the structure function $S(\mathbf{q})$, which is a thermal average,

$$S(\mathbf{q}) = \langle |\rho_q|^2 \rangle. \tag{4.6}$$

In practice, one usually works with a sample composed of randomly oriented micro-crystallites and effectively averages over the various directions of $\mathbf{q}$. As shown schematically in Fig. 4.7, the angularly averaged function $\bar{S}(q)$ has sharp peaks at the reciprocal-lattice vectors which characterize the crystalline solid.

Following Landau [35], we can develop an order-parameter theory of the process of freezing by defining *local* Fourier components at the reciprocal-lattice positions via

$$\rho_G(\mathbf{r}) = \frac{1}{\Delta V} \int_{\Delta V} d^3r' \, e^{-i\mathbf{G}\cdot\mathbf{r}'} \rho(\mathbf{r}') \tag{4.7}$$

where ΔV is a hydrodynamic averaging volume centered at position $\mathbf{r}$. In practice, one usually works with the Fourier modes corresponding to the smallest nonzero reciprocal-lattice vectors. The free energy which describes the freezing transition is a rotationally and translationally invariant expansion in the ρ_G (compare with Section 2.4.1),

$$F = \frac{1}{2} K \sum_G |(\nabla - i\mathbf{G} \times \boldsymbol{\theta})\rho_G|^2$$

$$+ \frac{1}{2} r \sum_G |\rho_G|^2 + w \sum_{G_1 + G_2 + G_3} \rho_{G_1}\rho_{G_2}\rho_{G_3} + O(\rho_G^4). \tag{4.8}$$

The field $\theta(\mathbf{r})$ measures the deviation of the local crystallographic axes from some reference orientation; its presence in the gradient term is required by rotational invariance [37]. The presence of a third-order term shows that, in the absence of strong fluctuations, the freezing transition is of first order.

Before generalizing this approach to describe order in a flat-space glass, it is helpful to first consider what happens when 120 particles are cooled on the surface S3 of a four-dimensional sphere. This system is unfrustrated, so the ground state for simple pair potentials presumably consists of particles at the vertices of polytope $\{3, 3, 5\}$. Condensation into a $\{3, 3, 5\}$ "crystal" upon cooling reflects a broken $SO(4)$ symmetry in the particle density $\rho(\hat{\mathbf{u}})$, where $\hat{\mathbf{u}}$ is a four-dimensional unit vector describing various positions on the surface of the sphere. The radius of the sphere necessary to accommodate a polytope with geodesic distance d between the particles is [16]

$$R \doteq 1.591\,549\,d$$
$$\equiv \kappa^{-1}. \tag{4.9}$$

For carrying out manipulations on the vertices of $\{3, 3, 5\}$, we can exploit the isomorphism between the group $SU(2)$ of 2×2 complex unitary matrices with unit determinant and the points on the surface of a four-dimensional sphere. This isomorphism is similar to the familiar identification of a circle with the set of complex numbers of unit modulus. Every point $\hat{\mathbf{u}} = (u_0, u_x, u_y, u_z)$ is associated with an $SU(2)$ matrix,

$$\hat{\mathbf{u}} \rightarrow \begin{pmatrix} u_0 + iu_z & iu_x + u_y \\ iu_x - u_y & u_0 - iu_z \end{pmatrix}. \tag{4.10}$$

Points on S3 can now be multiplied together, using the multiplication rules of $SU(2)$ matrices. Many computations simplify because the 120 vertices of polytope $\{3, 3, 5\}$ can be oriented so that they form a group when combined in this way [38]. This group, which we shall call Y', is the lift of the 60-element icosahedral point group into $SU(2)$.

The density $\rho(\hat{\mathbf{u}})$ for an arbitrary configuration of 120 particles on S3 can be expanded in hyperspherical harmonics $Y_{n,m_1 m_2}(\hat{\mathbf{u}})$,

$$\rho(\hat{\mathbf{u}}) = \sum_{n=0}^{\infty} \sum_{m_1, m_2} Q_{n,m_1 m_2} Y_{n,m_1 m_2}(\hat{\mathbf{u}}). \tag{4.11}$$

The coefficients $Q_{n,m_1 m_2}$ in this expansion are like the Fourier coefficients in Eq. (4.5). The non-negative integer subscript n indexes $(n+1)^2$-dimensional irreducible representations of the four-dimensional rotation group $SO(4)$ [39]. Unlike the usual three-dimensional spherical harmonics $Y_{\ell m}(\theta, \phi)$, there are *two* azimuthal quantum numbers, m_1 and m_2, which assume values with integer steps in the range

$$-n/2 \leqslant m_1, m_2 \leqslant n/2. \tag{4.12}$$

Although they are unfamiliar to many physicists, these hyperspherical harmonics turn out to be proportional to the standard $SU(2)$ Wigner matrices, so their properties can readily be determined [39].

The flat-space structure function displayed in Eq. (4.6) is both rotationally and translationally invariant when averaged over the various directions of $\mathbf{q}$. The analogous $SO(4)$-invariant correlation function for particles in thermal equilibrium in S3 is [17]

$$S_n = \frac{1}{(n+1)^2} \sum_{m_1, m_2} \langle |Q_{n,m_1 m_2}|^2 \rangle. \tag{4.13}$$

The quantity S_n becomes proportional to $\bar{S}(q)$ in the limit of a large number of particles embedded in a sphere of infinite radius [17]. In analogy with the peaks in $\bar{S}(q)$ at the reciprocal-lattice positions for crystalline solids (Fig. 4.7), we would expect that S_n can be nonzero only for certain special values of n for particles in the configuration $\{3, 3, 5\}$. The 7200-element group G of symmetry operations of polytope $\{3, 3, 5\}$ which can be continuously connected to the identity has the formal decomposition [38]

$$G = (Y' \times Y')/Z_2. \tag{4.14}$$

In [17], these symmetries are used to show that the only allowed spherical harmonics for polytope $\{3, 3, 5\}$ occur for

$$n = 0, 12, 20, 24, 30, 32, 36, 40, 42, 44, 48, 50, 52, 54, 56, 60, \tag{4.15}$$

and all even $n > 60$. This set of allowed values of n is the curved-space analogue of the allowed magnitudes of reciprocal-lattice vectors in a flat-space crystal. Straley [40] has evaluated the function S_n for 120 particles interacting via a repulsive $1/r^{12}$ potential and cooled rapidly via a Monte Carlo computer simulation. Although the particle configuration which results is amorphous (all S_n are nonzero), there is a pronounced peak at $n = 12$, the smallest allowed nonzero value for $\{3, 3, 5\}$.

We now want to adapt these ideas in order to obtain an order parameter for liquids and glasses in *flat* space. The Fourier components for particles in S3 are given by integrating the density over the surface of the sphere,

$$Q_{n,m_1 m_2} = \int d\Omega_{\hat{u}}\, \rho(\hat{u})\, Y^*_{n,m_1 m_2}(\hat{u}). \tag{4.16a}$$

Figure 4.8 illustrates how to define a local $n = 12$ order parameter via stereographic projection of a small flat-space particle configuration

Fig. 4.8. Particles projected onto S^3 from three-dimensional flat space.

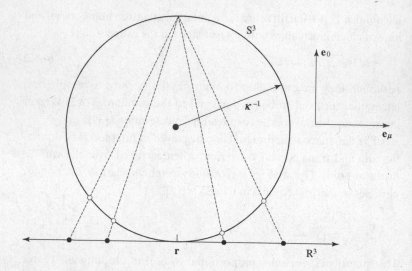

onto a tangent sphere of radius κ^{-1}. If $\Delta V'$ is the projection of the averaging volume ΔV onto S3 and $\rho'(\hat{\mathbf{u}})$ is the projected particle density, the Fourier coefficients associated with the point $\mathbf{r}$ are

$$Q_{12,m_1 m_2}(\mathbf{r}) = \frac{1}{\Delta V'} \int_{\Delta V'} d\Omega \, \hat{u} \rho'(\hat{\mathbf{u}}) Y^*_{12,m_1 m_2}(\hat{\mathbf{u}}). \qquad (4.16b)$$

Equation (4.16b) replaces Eq. (4.7) in the usual Landau approach to flat-space crystalline solids; it differs from Eq. (4.16a) only in the restricted range of the angular integration. We have focused on $n=12$ because it corresponds to the smallest nonzero reciprocal-lattice vector in a medium with local icosahedral order.

The symmetry group of this order parameter is $SO(4)$, modulo the symmetry group G of polytope $\{3, 3, 5\}$. An $SO(4)$-invariant Landau expansion for the free-energy density reads

$$F_{12} = \frac{1}{2}\kappa|D_\mu \mathbf{Q}_{12}|^2 + \frac{1}{2}r|\mathbf{Q}|^2$$

$$+ w \sum_{\substack{m_1 m_2 m_3 \\ m_1' m_2' m_3'}} \begin{pmatrix} 6 & 6 & 6 \\ m_1 & m_2 & m_3 \end{pmatrix}\begin{pmatrix} 6 & 6 & 6 \\ m_1' & m_2' & m_3' \end{pmatrix} Q_{12,m_1,m_1'} Q_{12,m_2,m_2'} Q_{12,m_3,m_3'}$$

$$+ O(Q_{12}^4). \qquad (4.17)$$

For notational convenience, we have replaced the 13×13 matrix $Q_{12,m_1 m_2}$ by a 169-component vector $\mathbf{Q}_{12}$ in the first two terms. Ordinary $SO(3)$ Wigner $3j$ symbols are used to make a rotationally invariant third-order term [17]. The frustration is embedded (see below) in the gradient operator D_μ, which is defined by

$$D_{\mu}Q_{12} = (\partial_{\mu} - i\kappa L^{(12)}_{0\mu})Q_{12} \tag{4.18}$$

Following a low-temperature continuum elastic treatment of glasses proposed by Sethna [41], we require that neighboring particle configurations be related by rolling a {3, 3, 5} template between them; the matrices $L^{(12)}_{0\mu}$ are the generators for the $n = 12$ representation of $SO(4)$ for rolling in the $(0, \mu)$ plane [39]. The quantity κ is the inverse radius of the tangent sphere in Fig. 4.8.

In [17], it is argued that (for large, negative r) the polynomial part of Eq. (4.17) is minimized when the 169 numbers $\{Q_{12,m_1,m_2}\}$ occur in the same proportions as the polytope {3, 3, 5}. Minimizing the polynomial part, however, determines the order parameter only up to a local $SO(4)$ rotation. Minimizing the *gradient* term means that the order parameters at r and at $r + \delta$, where δ is a small separation vector, are related,

$$Q_{12}(r + \delta) = \exp(i\kappa L^{(12)}_{0\mu}\delta^{\mu})Q_{12}(r). \tag{4.19}$$

Equation (4.19) is just a restatement of the requirement that neighboring textures be related by rolling. The quantity $\exp(i\kappa L^{(12)}_{0\mu}\delta^{\mu})$ is an $SO(4)$ Wigner matrix appropriate to the $n = 12$ representation. In a flat-space crystal, the deviation of the density from its average value due to the reciprocal-lattice vector G varies like

$$\Delta\rho(r + \delta) = e^{iG \cdot \delta} \Delta\rho(r). \tag{4.20}$$

Comparison of Eqs. (4.19) and (4.20) makes it clear that the non-Abelian matrices

$$\{G_{\mu}\} = \{\kappa L^{(12)}_{0\mu}, \kappa L^{(26)}_{0\mu}, \kappa L^{(24)}_{0\mu}, \ldots\} \tag{4.21}$$

play the role of reciprocal-lattice vectors in this theory.

Equation (4.19) means that the gradient term in the free energy (4.18) vanishes along the path connecting r and $r + \delta$. The gradient term is frustrated, however, in the sense that it cannot be made to vanish everywhere. To see this, one need only multiply together the matrices corresponding to the path shown in Fig. 4.9. One finds that the net change in the order parameter upon completing a circuit around this small plaquette is [17]

$$Q_{12}(r) \rightarrow (1 - \kappa^2 a^2 [L^{(12)}_{0\mu}, L^{(12)}_{0\mu}])Q_{12}(r)$$

$$\approx \exp(-i\kappa^2 a^2 L^{(12)}_{\mu\nu})Q_{12}(r), \tag{4.22}$$

where $L^{(12)}_{\mu\nu}$ is the generator of rotations in the (μ, ν) plane. This change can be accommodated only by a density of negatively charged wedge-disclination lines threading the plaquette, in agreement with the physical picture developed in Section 4.1.

It is interesting to compare the Landau theory described above with

Fig. 4.9. Frustration around a plaquette in the (μ, ν) plane.

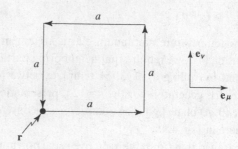

Fig. 4.10. Equilibrium phases associated with the free energy (4.17).

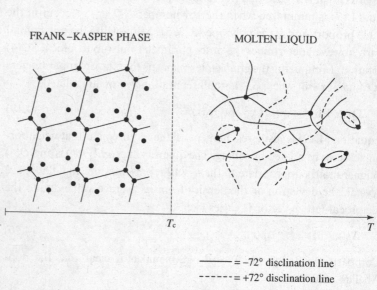

the continuum elastic approach suggested in [41]. Sethna worked with a "fixed-length", $n = 1$ representation of $SO(4)$. Upon applying a similar fixed-length approximation to the Landau expansion in Eq. (4.18), one recovers Sethna's elastic free energy, except that his 4×4 $SO(4)$ matrices are replaced by 169×169 ($n = 12$) representation matrices. (Both theories depend only on the six $SO(4)$ Euler angles in this limit.) Landau's approach is more general, however, because it allows for fluctuations in the *magnitude* of the order parameter and should be more useful at high temperatures, or in the presence of many defects. Since the magnitude of the order parameter vanishes at disclination cores, amplitude fluctuations will appear when one performs coarse graining over a volume containing several disclinations. If the physical picture developed in Section 4.1 is correct, this will be an appropriate description at all temperatures down to T_g.

Our expectation for the equilibrium statistical mechanics associated with the Landau theory is summarized in Fig. 4.10. As usual, we assume that the mass parameter r changes sign with decreasing temperature and

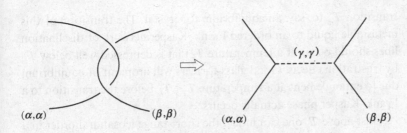

Fig. 4.11. Entanglement of disclination lines.

that the other couplings are temperature-independent. At high temperatures, we have a molten liquid, which can be regarded as a tangled mass of disclination lines. The low-temperature phase is a Frank–Kasper crystal consisting of a regular lattice of $-72°$ disclinations. The transition temperature T_c separating these two phases is like the transition at $H_{c2}(T)$ in a type-II superconductor. We expect that the effect of (annealed) inhomogeneities in particle size can be modelled by a renormalization of r.

We shall now argue that many physical systems will drop out of equilibrium upon cooling, instead of condensing into a Frank–Kasper phase, due to entanglement of defect lines [16]. The order parameter discussed above has the symmetry $SO(4)/G$. The algebra of line defects in such a medium is given by the homotopy group $\pi_1(SO(4)/G)$ [42]. As shown in [17], this group is just the direct product of Y' with itself,

$$\pi_1(SO(4)/G) = Y' \times Y'. \tag{4.23}$$

This result means that every defect is characterized by a pair of $SU(2)$ matrix charges, each chosen from the 120-element group Y'. The diagonal subgroup of elements of the form $(\alpha, \alpha) \equiv \alpha$ consists of the disclinations forced into the medium by the curvature incommensurability. The defect-combination laws are given by $SU(2)$ matrix multiplication. The allowed nodes in networks of $-72°$ disclination lines predicted by this formalism are just those found in the Frank–Kasper phases [16]. As shown in Fig. 4.11, it is very difficult for defect lines characterized by non-Abelian charges to cross. The lines cannot simply break and reform, but must instead leave behind an "umbilical defect" with charge [42]

$$\gamma = \alpha\beta\alpha^{-1}\beta^{-1}. \tag{4.24}$$

This umbilical string ties the two defects together and impedes their crossing. As shown in [16], two $\pm72°$ disclination lines corresponding to rotations about different symmetry axes always produce another $\pm72°$ umbilical defect when they try to cross.

One would expect the kinetic constraints associated with entanglement to scale with the amount of short-range icosahedral order, becoming more and more severe as a liquid is supercooled below the freezing transition T_m to, say, an equilibrium fcc crystal. The transition of this metastable liquid to an ordered Frank–Kasper network of disclination lines should occur at a temperature T_c that is depressed well below T_m by frustration effects. Good glass-formers will drop out of equilibrium due to entanglement at a temperature $T_g < T_m$ before the transition to a Frank–Kasper phase actually occurs.

Just above T_g one can model the short-range icosahdral order in a dense liquid by a sum of free energies like Eq. (4.17), one for each of the allowed values of $n = 0, 12, 20, 24, \ldots$. We truncate this expansion at quadratic order in the $\mathbf{Q}_n$, since the large intrinsic density of defects forces these coarse-grained order parameters to be small. The particle density is given in terms of the $\mathbf{Q}_n$ by [17]

$$\rho(\mathbf{r}) = \sum_{n=0,\,12,\,20,\,24,\,\ldots} \sum_{m_1 m_2} Q_{n,m_1 m_2}(\mathbf{r}) Y^*_{n,m_1 m_2}(-1), \tag{4.25}$$

where the spherical harmonics are evaluated at the south pole of the tangent sphere in Fig. 4.8. It is straightforward to calculate the density–density correlations in the liquid using Eq. (4.25) and this set of free energies. One obtains a remarkably good description [43] of the X-ray structure functions observed experimentally in a variety of metallic glasses, which exhibit peaks corresponding to $n = 12, 20$ and 24.

Once glassy X-ray structure functions are understood, one can try to understand other problems in solid-state physics, such as band structure and transport properties. The kinetics of entanglement presumably controls the viscosity and diffusion constants near T_g. The transition to the Frank–Kasper state may have something to do with an "intrinsic" but experimentally inaccessible glass transition at a temperature $T_0 < T_g$. It is encouraging that the density of defects in a medium of icosahedra turns out to be so high. In the Frank–Kasper phases, for example, approximately 40%–60% of the atoms occupy defect sites. Specifying a disordered array of disclination lines that cannot cross essentially specifies the positions of *all* atoms up to small vibrations. Thus, one can begin to understand the experimental observation that glasses have a very low entropy, even though they are disordered.

Acknowledgements

It is a pleasure to acknowledge many stimulating interactions with my collaborators Michael Rubinstein, Subir Sachdev and Michael Widom. This chapter is adapted with permission from *Applications of Field Theory to Statistical Mechanics*, edited by L. Garrido (Springer, Berlin, 1985) pp. 13–30.

References

[1] D. Turnbull, *J. Chem. Phys.* **20**, 411 (1952).

[2] F. C. Frank, *Proc. R. Soc. London* A **215**, 43 (1952).

[3] J. D. Bernal, *Proc. R. Soc. London* A **280**, 299 (1964).

[4] S. Sachdev and D. R. Nelson, *Phys. Rev.* B **32**, 4592 (1985); see also D. R. Nelson and M. Widom, *Nucl. Phys.* B **240**, 113 (1984).

[5] S. Sachdev. *Phys. Rev.* B **33**, 6395 (1986).

[6] S. Nosé and F. Yonezawa, *J. Chem. Phys.* **84**, 1803 (1986).

[7] H. Jonsson and H. C. Anderson, *Phys. Rev. Lett.* **60**, 2295 (1988).

[8] J. P. Straley, *Phys. Rev.* B **34**, 405 (1986).

[9] D. R. Nelson and F. Spaepen, *Solid State Phys.* **42**, 1 (1989).

[10] A. Van Blaaderen and P. Wiltzius, *Science* **270**, 1177 (1995).

[11] E. R. Weeks, J. C. Crocker, A. C. Levitt, A. Schofield and D. A. Weitz, *Science* **287**, 627 (2000).

[12] H. Reichert, O. Klein, H. Dosch, M. Denk, V. Honkimaki, T. Lippmann and G. Reiter, *Nature* **408**, 839 (2000); see also F. Spaepen, *Nature* **408**, 781 (2000).

[13] See, e.g., the reviews in *Ill-Condensed Matter*, edited by R. Balian, R. Maynard and G. Toulouse (North-Holland, New York, 1979).

[14] For a readable introduction to the physics of glasses, see R. Zallen, *The Physics of Amorphous Solids* (Wiley, New York, 1983).

[15] S. F. Edwards and P. W. Anderson, *J. Phys.* F **5**, 905 (1975).

[16] D. R. Nelson, *Phys. Rev. Lett.* **50**, 982 (1983); *Phys. Rev.* B **28**, 5515 (1983).

[17] D. R. Nelson and M. Widom, *Nucl. Phys.* B FS **12**, 113 (1984).

[18] F. C. Frank, *Proc. R. Soc. London* A **215** 43 (1952).

[19] P. J. Steinhardt, D. R. Nelson and M. Ronchetti, *Phys. Rev. Lett.* **47**, 1297 (1981); *Phys. Rev.* B **28**, 784 (1983).

[20] J. F. Sadoc and R. Mosseri, *Phil. Mag.* B **45**, 467 (1982).

[21] E. Fradkin, B. A. Huberman, and S. H. Shenker, *Phys. Rev.* B **18**, 4789 (1978) and references therein.

[22] S. Teitel and C. Jayaprakash, *Phys. Rev.* B **27**, 598 (1983).

[23] M. Tinkham, *Introduction to Superconductivity* (McGraw-Hill, New York, 1975).

[24] H. S. M. Coxeter, *Introduction to Geometry* (Wiley, New York, 1969).

[25] H. S. M. Coxeter, *Regular Polytopes* (Dover, New York, 1973).

[26] There are several of these "statistical honeycomb" models. See H. S. M. Coxeter, *Ill. J. Math.* **2**, 746 (1958). The version used here is described on p. 411 of [24].

[27] J. F. Sadoc, *J. Physique Coll.* **41**, C8–326 (1980).

[28] M. Kléman and J. F. Sadoc, *J. Physique Lett.* **40**, L569 (1979).

[29] See, e.g., R. Collins, in *Phase Transitions and Critical Phenomena*, edited by C. Domb and M. S. Green (Academic, New York, 1972), Vol. II.

[30] F. C. Frank and J. Kasper, *Acta Crystallogr.* **11**, 184 (1958); *Acta Crystallogr.* **12**, 483 (1959).

[31] See, e.g., D. R. Nelson, *J. Non-Cryst. Solids* **61 & 62**, 475 (1984).

[32] C. A. Rogers, *Proc. London Math. Soc.* **8**, 609 (1958); see also N. J. A. Sloane, *Scient. Am.*, January, 1984, p. 116.

[33] D. R. Nelson, in *Topological Disorder in Condensed Matter*, edited by T. Ninomiya and F. Yonezawa (Springer, Berlin, 1983).

[34] M. Rubinstein and D. R. Nelson, *Phys. Rev.* B **28**, 6377 (1983).

[35] *The Collected Papers of L. D. Landau*, edited by D. ter Haar (Gordon and Breach–Pergamon, New York, 1965); p. 193; see also G. Baym, H. A. Bethe and C. Pethick, *Nucl. Phys.* A **175**, 1165 (1971).

[36] N. W. Ashcroft and N. D. Mermin, *Solid State Physics* (Holt, Rinehart, and Winston, New York, 1976).

[37] See, e.g., D. R. Nelson and J. Toner, *Phys. Rev.* B **24**, 363 (1981), Section III-D.

[38] P. Du Val, *Homographies, Quaternions and Rotations* (Oxford University Press, Oxford, 1964).

[39] L. C. Biedenharn, *J. Math. Phys.* **2**, 433 (1961); M. Bander and C. Itzykson, *Rev. Mod. Phys.* **38**, 330 (1966).

[40] J. P. Straley, *Phys. Rev.* B **34**, 405 (1986).

[41] J. Sethna, *Phys. Rev. Lett.* **51**, 2198 (1983).

[42] N. D. Mermin, *Rev. Mod. Phys.* **51**, 591 (1979).

[43] S. Sachdev and D. R. Nelson, *Phys. Rev.* **32**, 1480 (1985).

Chapter 5
The statistical mechanics of crumpled membranes

Preface

This chapter provides an introduction to how the physics of crumpled polymer chains changes when linear objects are replaced by crumpled *membranes*. In contrast to covalently bonded polymer chains, membranes can exist in liquid, hexatic or crystalline phases. As a result, this subject is extremely rich and complex!

Depending on temperature and salt concentrations, both liquid and hexatic order are probably realized when lipid bilayers close to form spherical vesicles. The large thermal fluctuations which lead to particularly interesting physics can be enhanced by adding double-headed "bola lipids" to, say, dimyristoyl phosphatidyl choline (DMPC) to reduce the rigidity against bending, or "bending rigidity" [1].

Because of the instability of crystalline order with respect to low-energy buckled dislocations in membranes without permanent covalent bonds (see Chapters 1 and 6), polymerization is required in order to produce objects with two-dimensional shear moduli, which can then strongly influence bending and crumpling into a third dimension. The best experimental realization of this problem to date is probably the spectrin skeleton of red blood cells discussed in Chapter 1, which exhibits a pronounced flat phase with properties in good agreement with theoretical predictions [2]. Sheets of graphite oxide have a low bending rigidity and appear to display a crumpled phase [3], but the experimental situation remains unclear [4]. The "rag" form of molybdenum disulfide [5] is another candidate for the crumpled phase. Cross polymerization of lipid bilayers, monolayers or other objects provides another way to obtain well-defined planar sheets, which can then be allowed to crumple and fold in solution [6–8].

Membranes can be polymerized in various phases and the resulting "tethered surfaces" will often exhibit quenched random disorder relative to a perfect crystalline topology in the form of frozen-in impurities, vacancies, interstitials, dislocations, disclinations and grain boundaries. Several theoretical investigations have dealt with quenched random disorder in polymerized membranes. Although the flat phase is stable against random copolymerization of two different monomer species [9, 10], there is an instability leading to the possibility of spin-glass-like behavior at zero temperature. Random spontaneous curvature leads to a new type of flat phase at $T = 0$, which can be studied via the renormalization group and computer simulations [10, 11]. Morse *et al.* have shown that the flat phase remains stable at finite temperature after polymerization of a liquid of monomers, which necessarily has frozen-in disclinations [12]. Some of this work was motivated by the discovery by Mutz, Bensimon and Brienne of a remarkable wrinkling transition in polymerized vesicles [13]. This wrinkling appears as a (reversible!) first-order transition at a temperature T_w when partially polymerized vesicles are cooled well below the chain-melting temperature T_m of the pristine lipids. For studies of effects due to the (buckled) grain boundaries which may arise in this case, see [14, 15].

Computer simulations have played an important role in our understanding of polymerized membranes. The structure function of a dilute solution of randomly oriented self-avoiding polymerized membranes in the flat phase can be studied quite easily [16]. By adding attractive interactions, one can also produce a compact, collapsed phase [17] analogous to a linear polymer chain in a bad solvent. The flat and collapsed phases have fractal dimensions 2 and 3, respectively. It has proved more difficult to uncover an intermediate crumpled phase, with fractal dimension $d_F \approx 2.5$, using computer simulations of flexible, self-avoiding membranes. There may be some fundamental reason why any self-avoidance rules out a crumpled phase entirely. On the other hand, it may just be difficult to find the appropriate parameter ranges and produce sufficiently long computer equilibration times, as was the case with computer modelling for some years after an intermediate hexatic phase had been suggested for two-dimensional melting.

Computer simulations *have* revealed a remarkable intermediate *tubule* phase [18], which was predicted [19] and explored in some detail via analytical methods [20]. Computer simulations have also found exciting evidence for easy proliferation of low-energy dislocation defects (which is expected to lead to a hexatic phase at low temperatures) in flexible particle arrays with a small rigidity against bending [21].

For excellent accounts of recent developments in this active field, see the reviews by Weise [22] and by Bowick and Travesset [23].

5.1 Flat surfaces

5.1.1 The roughening transition

Interesting problems in statistical mechanics arise even for surfaces constrained by surface tension to be fairly flat. A particularly well-studied

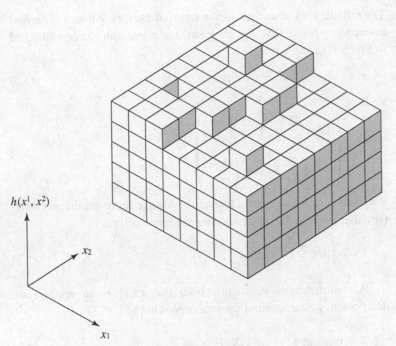

Fig. 5.1. The height function $h(x^1, x^2)$ used to describe the configuration of a crystal–vapor interface.

$h(x^1, x^2)$

x_2

x_1

example is the roughening transition of crystalline interfaces [24]. As shown in Fig. 5.1, we imagine a crystal in equilibrium, with, say, its own vapor. The position of the interface is described by a height function $h(x^1, x^2)$. Such a description implicitly ignores "overhangs" (which cannot be described by a single-valued $h(x^1, x^2)$), islands of crystal in the vapor phase and islands of vapor in the crystal. These complications are believed to be irrelevant variables in the long-wavelength limit [24]. Microscopically, the interface height is quantized in units of the spacing between the Bragg planes normal to the h axis.

At high temperatures, this discreteness is washed out by thermal fluctuations and we can describe the free energy of the interface in terms of a surface tension σ. It is a useful pedagogic exercise to describe this free energy using differential geometry, which, although inessential here, is a very useful language for crumpled membranes. For an arbitrary parameterization of the surface $\mathbf{r}(\zeta^1, \zeta^2)$, the free energy is the surface tension times the surface area,

$$F = \sigma \int \sqrt{g}\, d^2\zeta, \tag{5.1}$$

where g is the determinant of the metric tensor, $g = \det g_{ij}$,

$$g_{ij} = \frac{\partial \mathbf{r}}{\partial \zeta^i} \cdot \frac{\partial \mathbf{r}}{\partial \zeta^j}. \tag{5.2}$$

The formula for the surface area in terms of the metric tensor is derived in many textbooks [25]. For the particular parameterization embodied in Fig. 5.1, i.e.,

$$\mathbf{r}(x^1, x^2) = (x^1, x^2, h(x^1, x^2)),\tag{5.3}$$

we have

$$g_{ij} = \begin{pmatrix} 1 + \left(\dfrac{\partial h}{\partial x^1}\right)^2 & \left(\dfrac{\partial h}{\partial x^1}\right)\left(\dfrac{\partial h}{\partial x^2}\right) \\ \left(\dfrac{\partial h}{\partial x^1}\right)\left(\dfrac{\partial h}{\partial x^2}\right) & 1 + \left(\dfrac{\partial h}{\partial x^2}\right)^2 \end{pmatrix}.\tag{5.4}$$

With this coordinate system (called the Monge representation in differential geometry), Eq. (5.1) assumes the familiar form

$$F = \sigma \int d^2x \sqrt{1 + |\nabla h|^2}.\tag{5.5}$$

At temperatures sufficiently high that (5.1) is an appropriate description, we can expand the square root in (5.5),

$$F \approx \text{constant} + \frac{1}{2}\sigma \int d^2x |\nabla h|^2\tag{5.6}$$

and calculate, for example, the height–height correlation function

$$\langle (h(\mathbf{y}) - h(\mathbf{0}))^2 \rangle = \frac{\int \mathcal{D}h(\mathbf{x}) |h(\mathbf{y}) - h(\mathbf{0})|^2 e^{-F/k_B T}}{\int \mathcal{D}h(\mathbf{x}) e^{-F/k_B T}}.\tag{5.7}$$

The effects of higher-order gradients in Eq. (5.6) can be absorbed into a renormalized surface tension. The Gaussian functional integral is easily carried out in Fourier space, with the result

$$\langle (h(\mathbf{y}) - h(\mathbf{0}))\rangle^2 = \frac{2k_B T}{\sigma} \int \frac{d^2q}{(2\pi)^2} \frac{1}{q^2}(1 - e^{i\mathbf{q}\cdot\mathbf{y}})$$

$$\approx \frac{k_B T}{\pi\sigma} \ln(y/a), \qquad \text{as } y \to \infty,\tag{5.8}$$

where a is a microscopic length. The large-y behavior is the signature of a high-temperature rough phase.

At low temperatures, on the other hand, one might expect a "smooth" interface, i.e., one that has become localized at an integral multiple of a, the spacing between Bragg planes. To see how quantization of the interface height affects the prediction (5.8), we add a periodic perturbation to Eq. (5.6), which tends to localize the interface at $h = 0, \pm a, \pm 2a, \ldots$, and consider the free energy

$$F = \text{constant} + \frac{1}{2}\int d^2x \left\{\sigma |\nabla h|^2 + 2y\left[1 - \cos\left(\frac{2\pi h}{a}\right)\right]\right\}.\tag{5.9}$$

This sine-Gordon model can be solved directly by renormalization-group methods, or by first mapping the problem via a duality transformation onto an X–Y model or the two-dimensional Coulomb gas [24, 26] like that considered in Chapter 2. There is a finite-temperature roughening transition, which is in the universality class of the Kosterlitz–Thouless vortex-unbinding transitions. At sufficiently high temperatures ($T > T_R \simeq \pi \sigma a^2/k_B$), the periodicity is irrelevant and the interface behaves according to Eq. (5.8). For $T < T_R$, however, the interface localizes in one of the minima of the periodic potential and the effective free energy at long wavelengths can be approximated by expanding the cosine

$$F = \text{constant} + \frac{1}{2} \int d^2x \left[\sigma |\nabla h|^2 + \left(\frac{4\pi^2 y}{a^2} \right) h^2 \right]. \tag{5.10}$$

It is easily shown from Eq. (5.10) that the height–height correlation function (5.7) now tends to a constant,

$$\langle (h(\mathbf{y}) - h(\mathbf{0}))^2 \rangle \approx \text{constant}, \qquad \text{as } y \to \infty, \tag{5.11}$$

in contrast to Eq. (5.8).

The analogy with vortex-unbinding transitions leads to many detailed predictions about the roughening transition [24]. This analogy is only approximate, however, so it is important to have rigorous proofs of phase transitions in this and related models [27]. Although Eq. (5.9) is a plausible model of roughening, a more faithful representation of the microscopic physics is the solid-on-solid model, in which interface heights $\{h_i\}$ sit on a lattice of sites $\{i\}$ and are themselves quantized at all temperatures, $h_i = 0, \pm a, \pm 2a, \ldots, \forall\, i$. The Hamiltonian is

$$H = J \sum_{\langle ij \rangle} |h_i - h_j|, \tag{5.12}$$

where the sum is over nearest-neighbor lattice sites and $J > 0$ is a microscopic surface energy. Equation (5.12) measures directly the increase in interfacial area associated with discrete steps in the interface. We call (5.12) a "Hamiltonian" because it is a microscopic energy, in contrast to "free energies" like (5.9), which are supposed to be coarse-grained descriptions, embodying both energy and entropy.

5.1.2 Wetting transitions

Interesting transitions in interfacial surfaces also occur in wetting layers [28]. Consider, in particular, the approach to a liquid–gas phase boundary in the presence of a wall that microscopically prefers to be wet by the liquid, as opposed to the gas. The interfacial profile is shown in Fig. 5.2(a).

Because the wall prefers the denser liquid, there is a thin layer of

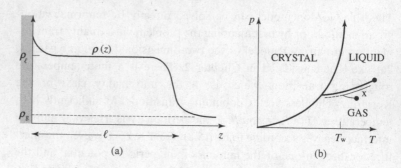

Fig. 5.2. (a) The density profile near a wall in the bulk gas phase close to liquid–gas coexistence. The density starts at a large value ρ_ℓ appropriate to the nearby liquid phase and drops to a smaller value ρ_g appropriate to the gas a distance ℓ from the wall. (b) A pressure–temperature phase diagram with regions of first-order and continuous wetting transitions along the liquid–gas-coexistence curve indicated by dashed and solid lines, respectively. A first-order "prewetting" transition terminating in a critical point extends into the gas phase. The density profile in (a) corresponds to the situation near a wall at the point x.

liquid present, even though the chemical potential of the bulk gas is slightly slower than that of the bulk liquid. This wetting layer extends a distance $\ell(T, p)$ into the gas phase, terminating at a liquid–gas interface whose width is comparable to the correlation length $\xi(T, p)$.

Two distinct behaviors are possible as the liquid–gas-coexistence curve is approached from the gas phase (see Fig. 5.2(b)). Far from the critical point, $\ell(p, T)$ usually remains finite along the liquid–gas-coexistence curve (i.e., along the dashed line in Fig. 5.2(b)). Closer to the critical point, however, $\ell(p, T)$ diverges (logarithmically, in simple model calculations) as the coexistence curve is approached (along the solid line in Fig. 5.2(b)). This divergence may be preceded by a first order "prewetting" transition in the bulk liquid signalled by an upward jump in the liquid density at the wall. The point at which $\ell(p, T)$ diverges to infinity along the coexistence curve, at $T = T_w$, locates a wetting transition, which has been the subject of considerable theoretical interest. This transition can be of first order, or it can proceed via a rather exotic second-order transition [29].

5.2 Crumpled membranes

5.2.1 Experimental realizations

Membranes can be regarded as two-dimensional generalizations of linear polymer chains, for which there is a vigorous theoretical and experimental literature [30, 31]. Flexible membranes should exhibit even more richness and complexity, for two basic reasons. The first is that important geometrical concepts such as intrinsic curvature, orientability and genus, which have no direct analogues in linear polymers, appear naturally in discussions of membranes: Our understanding of the interplay between these concepts and the statistical mechanics of membranes is just beginning. The second reason is that surfaces can exist in a variety of different phases. The possibility of a two-dimensional shear modulus

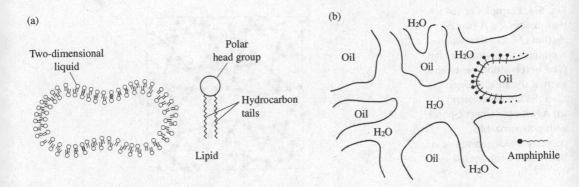

(a)

Two-dimensional liquid

Polar head group

Hydrocarbon tails

Lipid

(b)

H_2O

Oil

Oil

H_2O

Oil

H_2O

H_2O

Oil

Oil

H_2O

Amphiphile

Fig. 5.3. Examples of liquid-like membranes: (a) a red blood cell and (b) a microemulsion.

in membrances shows that we must distinguish between solids and liquids when these objects are allowed to crumple into three dimensions. We shall argue later that hexatic membranes, with extended 6-fold bond orientational order, are another important possibility. There are no such sharp distinctions for linear polymer chains.

Figure 5.3 shows two examples of *liquid* membranes. Figure 5.3(a) is a caricature of an erythrocyte, or red blood cell. The cell wall is a membrane, composed of a bilayer of amphiphillic molecules, each with one or more hydrophobic hydrocarbon tails and a polar head group. The membrane has a spherical topology, as do artificial vesicles formed from bilayers. Although these membranes could, in principle, crystallize upon cooling, they exhibit an almost negligible shear modulus at biologically relevant temperatures. The small shear modulus that is observed for erythrocytes may be due to an additional protein skeleton like spectrin [32].

Figure 5.3(b) illustrates the topology of a microemulsion, which is a transparent solution in which an oil (e.g., dodecane) and water mix in essentially all proportions [33]. This remarkable mixing of two "antagonistic" liquids is only possible because of the addition of significant amounts of an amphiphile such as sodium dodecyl sulfate, which sits at the interface between oil and water and reduces the surface tension almost to zero.

The size of the oil-rich and water-rich regions, which are constantly shifting as the interface fluctuates, is of order 100 Å. Usually, the presence of a cosurfactant like pentanol is necessary in order to stabilize the microemulsion. For more about liquid membranes, see [34, 35].

Although careful experimental investigations are only just beginning, there are also many examples of *solid* membranes. One can, for example, investigate the properties of flexible sheet polymers, "tethered surfaces." Tethered surfaces can be synthesized by polymerizing Langmuir–Blodgett films or amphiphillic bilayers [36]. Although lipid

Fig. 5.4. Examples of solid-like membranes. (a) A planar section of B_2O_3, that when it is crumpled, describes a glass. (The boron atoms centered in each of the small oxygen triangles are not shown.) (b) A lyotropic smectic phase with polymerizable polyacrylamide monomer in the watery interstices.

(a)

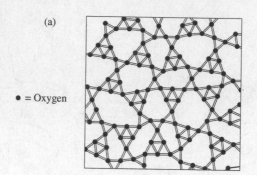

● = Oxygen

(b)

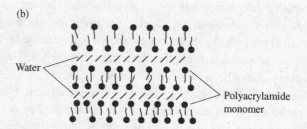

Water

Polyacrylamide monomer

monolayers polymerized at an air–water interface would initially be flat, they could be inserted into a neutral solvent like alcohol and their fluctuations made visible by attaching a fluorescent dye. There are also fascinating accounts of cross-linked methyl-methacrylate polymer assembled on and then extracted from the surface of sodium montmorillonite clays [37].

Two less familiar examples of solid membranes are illustrated in Fig. 5.4. Figure 5.4(a) shows a model of a large sheet molecule believed to be an ingredient of glassy B_2O_3 [38]. Similar structures, also in crumpled form, may exist in chalcogenide glasses such as As_2S_3. Although it may be difficult to obtain dilute solutions in a good solvent, we might hope to produce a dense melt of such surfaces, in analogy with polymer melts or models of amorphous selenium [39].

Figure 5.4(b) illustrates an idea for synthesizing a large number of surfaces of two-dimensional polyacrylamide gel, which I have pursued in collaboration with R. B. Meyer at Brandeis University. We first form a lyotropic smectic liquid crystal of amphiphillic bilayers, similar to those discussed above. The bilayers are separated by water and, if necessary, can be pushed further apart by the addition of oil or water [40, 41]. If the lipids have multiple double bonds, one could of course polymerize the bilayers as discussed above. An attractive alternative for producing flexible surfaces is to introduce polyacrylamide gel into the watery interstices between the bilayers. Meyer and I have succeeded in stabilizing a smectic phase in which each roughly 20-Å-thick water-rich region contains about 15% by weight of acrylamide and *bis*-acrylamide monomers. By shining ultraviolet light on this mixture, it may be possible to produce many slabs of two-dimensional cross-linked polyacrylamide gel. The lipid bilayers, which are used simply as spacers in this experiment, would then be washed away.

A third class of membrane surfaces is possible if we replace fixed covalent cross links like those in Fig. 5.4(a) by weaker van der Waals forces. Van der Waals interactions will tend to crystallize the lipid bilayers discussed above at sufficiently low temperatures. Although these surfaces will have a nonzero shear modulus when they are confined to a plane, they are unstable with respect to the formation of free dislocations when they are allowed to buckle into the third dimension [42]. Dislocations necessitate broken bonds and thus would require prohibitively large energies in covalently bonded systems. As discussed in Chapter 1, the presence of a finite concentration of unbound dislocations at any temperature means that unpolymerized lipid bilayers will in fact be hexatic liquids with residual bond-orientational order at low temperatures [42, 43]. The properties of hexatic membranes are intermediate between those of liquid and solid surfaces and will be discussed at the end of this chapter.

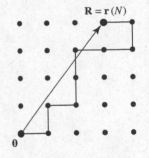

Fig. 5.5. A polymer configuration extending from the origin to **R** on a square lattice.

5.2.2 Results from polymer physics

One route toward understanding crumpled membranes is to generalize various results from polymer physics. As illustrated in Fig. 5.5, we can catalogue polymer configurations on a lattice by first counting the number of self-avoiding walks starting at the origin and terminating at position **R**. The function $\mathbf{r}(s)$ gives the position of the walk after the sth step. If $\mathcal{N}_N(\mathbf{R})$ is the number of walks of length N starting at the origin and terminating at R, the *total* number of walks of length N is given by

$$\mathcal{N}_N^{\text{tot}} = \sum_{\mathbf{R}} \mathcal{N}_N(\mathbf{R}). \tag{5.13}$$

A typical polymer size is given by the radius of gyration R_G,

$$R_G = \left(\frac{1}{N^2} \sum_{s=1}^{N} \sum_{s'=1}^{N} \langle |\mathbf{r}(s) - \mathbf{r}(s')|^2 \rangle \right)^{1/2}, \tag{5.14}$$

where the average is over all polymer configurations. Polymer critical exponents are defined by the asymptotic large-N behavior of R_G and $\mathcal{N}_N^{\text{tot}}$,

$$R_G \sim N^{\nu}, \tag{5.15}$$

$$\mathcal{N}_N^{\text{tot}} \sim (\bar{z})^N N^{\gamma - 1}. \tag{5.16}$$

Here, $\bar{z}$ is a nonuniversal effective "coordination number" reduced from the actual coordination number by self-avoidance constraints. The radius-of-gyration exponent ν is increased by self-avoidance from the random walk result $\nu = \frac{1}{2}$ to the universal result $\nu \approx 0.59 \approx \frac{3}{5}$ in three dimensions. The exponent $\gamma \approx 1.18$ is also universal for polymers with free ends, although it changes for ring polymers [31]. The effect of self-avoidance on the exponents vanishes for $d > d_c = 4$, which is the upper critical dimension for linear polymers, and forms the basis for a $d = 4 - \varepsilon$ expansion for the exponents ν and γ [30, 31].

A simple, but remarkably accurate, estimate of the exponent ν can be obtained by working with a *continuum* model of self-avoiding polymers in a good solvent. The free energy F associated with a coarse-grained polymer configuration $\mathbf{r}(s)$ in d dimensions is assumed to be [30]

$$\frac{F}{k_B T} = \frac{1}{2} K \int_0^N ds \left(\frac{d\mathbf{r}}{ds} \right)^2 + \frac{1}{2} b \int_0^N ds \int_0^N ds' \, \delta^{(d)}[\mathbf{r}(s) - \mathbf{r}(s')]. \tag{5.17}$$

The first term represents a nearest-neighbor elastic energy, possibly entropic in origin, while the second counts the number of self-intersections and assigns them an excluded volume penalty b. Following a famous approximation scheme due to Flory [30, 31], we estimate the first term for a polymer of size R_G by dimensional analysis and note that the second term should be proportional to the probability of self-intersection $(N/R_G^d)^2$, times the volume R_G^d over which a self-intersection is likely to occur. In this way, we find that

$$\frac{F}{k_B T} \approx \frac{1}{2} \frac{K R_G^2}{N} + \frac{1}{2} b \left(\frac{N}{R_G^d} \right) R_G^d. \tag{5.18}$$

Upon minimizing with respect to R_G, we obtain the classic Flory result $R_G \sim N^\nu$, with

$$\nu = 3/(d+2), \tag{5.19}$$

i.e., $\nu \approx 0.60$ in three dimensions.

5.2.3 Generalization to tethered surfaces

Membrane generalizations of the theory of polymer chains are conveniently presented in the language of differential geometry. The partition of the resulting tethered surfaces without self-avoidance is a special case of a more general partition function that first arose in the study of bosonic strings, namely [44]

$$Z = \int \mathcal{D}g_{0,ab} \int \mathcal{D}\mathbf{r}(\zeta^1, \zeta^2) e^{-\frac{1}{2}K \int d^2\zeta \sqrt{g_0} g_0^{ab} \partial_a \mathbf{r} \cdot \partial_b \mathbf{r}}. \tag{5.20}$$

The "action" is composed of surface gradients $\partial_a \mathbf{r}$ contracted with a metric tensor g_0^{ab}. The integrations are over all possible metrics $g_{0,ab}$, as well as over all possible surface configurations $\mathbf{r}(\zeta^1, \zeta^2)$. Although the underlying metric and the surface are independent variables, Polyakov [44] has shown that a relation analogous to Eq. (5.2), i.e.,

$$g_{0,ab} = \frac{\partial \mathbf{r}}{\partial \zeta^a} \cdot \frac{\partial \mathbf{r}}{\partial \zeta^b} \tag{5.21}$$

is recovered in the low-temperature, strong-coupling ($K \rightarrow \infty$) limit.

A microscopic physical interpretation of Eq. (5.20) is illustrated in Fig. 5.6. For a fixed metric $g_{0,ab}$, the surface is represented by a fixed triangulation of particles connected by harmonic springs. The action in Eq. (5.20) is the continuum limit of the energy associated with these springs. The particle positions can be stretched to approximate *any* particular simply connected surface with free boundaries $\mathbf{r}(\zeta^1, \zeta^2)$; there is, however, a significant energetic cost associated with large deviations from the surfaces preferred by the underlying connectivity or "background metric." To carry out the functional integral (5.20) on a computer, one would first integrate over all particle positions for a fixed triangulation and then sum over different triangulations.

Fig. 5.6. A lattice of Gaussian springs with a fixed connectivity.

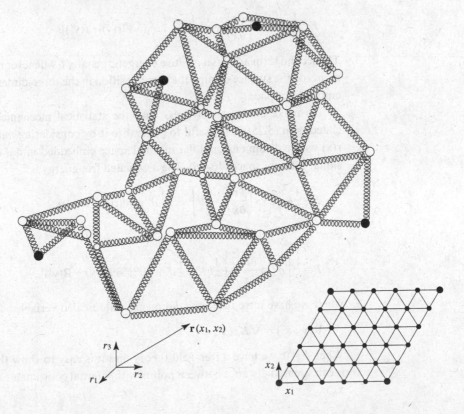

Tethered surfaces, discussed more completely in [45, 46], are an example of the string partition function (5.20), specialized to a single "flat" triangulation, in which every particle is connected to exactly six nearest neighbors. Although the springs are usually replaced by a square-well tethering potential, the particles behave as if they were connected by springs at long wavelengths for entropic reasons [45, 46]. The "background metric" is

$$g_{0,ab} = \delta_{ab} \tag{5.22}$$

and the partition function is

$$Z_0 = \int \mathcal{D}\mathbf{r}(x_1, x_2) e^{-F_0} \tag{5.23}$$

where

$$F_0 = \frac{1}{2} K \int d^2 x \left[\left(\frac{\partial \mathbf{r}}{\partial x_1} \right)^2 + \left(\frac{\partial \mathbf{r}}{\partial x_2} \right)^2 \right]. \tag{5.24}$$

As it stands, we now have a model for "phantom" polymerized membranes, without self-avoiding interactions between distant particles. To obtain a model for a real self-avoiding membrane, we replace the free energy F_0 by

$$F = \frac{1}{2} K \int d^2 x \left(\frac{\partial \mathbf{r}}{\partial x} \right)^2 + \frac{1}{2} b \int d^2 y \int d^2 y' \, \delta^d[\mathbf{r}(y) - \mathbf{r}(y')]. \tag{5.25}$$

The second term assigns a positive energetic penalty b whenever two elements of the surface occupy the same position in the three-dimensional embedding space.

To make analytical progress with the statistical mechanics associated with (5.25), it is useful to generalize it by considering *manifolds* $\mathbf{r}(x)$ with a D-dimensional flat internal space embedded in a d-dimensional external space [46–49]. The associated free energy is

$$F = \frac{1}{2} K \int d^D x \left(\frac{\partial \mathbf{r}}{\partial x} \right)^2 + \frac{1}{2} b \int d^D y \int d^D y' \, \delta^d[\mathbf{r}(y) - \mathbf{r}(y')], \tag{5.26}$$

or

$$F = \frac{1}{2} \int d^D x \left(\frac{\partial \mathbf{R}}{\partial x} \right)^2 + \frac{1}{2} b K^{d/2} \int d^D y \int d^D y' \, \delta^d[\mathbf{R}(y) - \mathbf{R}(y')], \tag{5.27}$$

where we have introduced the d-dimensional rescaled variable,

$$\mathbf{R}(x^1, x^2) = \sqrt{K} \, \mathbf{r}(x^1, x^2). \tag{5.28}$$

When $b = 0$, we have a free-field theory and it is easy to show that the mean squared distance between points with internal coordinates x_A and x_B is

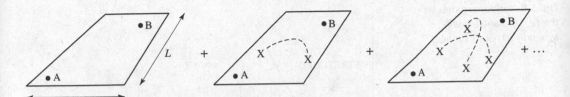

$$\langle |r(x_A) - r(x_B)|^2 \rangle_{x_{AB} \to \infty} \simeq \frac{2dS_D}{(2-D)K}(|x_{AB}|^{2-D} - a^{2-D}), \tag{5.29}$$

where $S_D = 2\pi^{D/2}/\Gamma(D/2)$ is the surface area of a D-dimensional sphere, $x_{AB} = x_A - x_B$ and a is a microscopic cutoff. If we take x_A and x_B to be close to opposite sides of the manifold (in the internal space), Eq. (5.29) becomes a measure of the squared radius of gyration. When $D = 1$, we are dealing with a linear polymer chain and we see that the size R_G increases as the square root of the linear dimension $L \sim |x_{AB}|$, i.e., $R_G \sim L^{1/2}$. The same argument, however, shows that the characteristic membrane size R_G increases only as the square root of the logarithm of the linear dimension L for $D = 2$,

$$R_G \sim \frac{1}{K} \ln^{1/2}\left(\frac{L}{a}\right). \tag{5.30}$$

To see how self-avoiding corrections affect Eq. (5.29), we can carry out perturbation theory in the excluded volume parameter. Each term can be represented as in Fig. 5.7, where the dotted lines represent self-avoiding interactions between different pieces of the manifold. Dimensional analysis using the rescaled free energy Eq. (5.27) shows that this perturbation theory becomes singular in the limit of large internal linear dimension L: The correction to Eq. (5.29) must take the form

$$\langle |r(x_A) - r(x_B)|^2 \rangle \simeq \frac{2dS_D}{(2-D)K}|x_{AB}|^2$$

$$(1 + \text{constant} \times bK^{d/2}L^{2D-(2-D)d/2} + \cdots). \tag{5.31}$$

Whenever

$$2D > (2-D)d/2 \tag{5.32}$$

the corrections to the free-field result (5.31) diverge as $L \to \infty$, signaling a breakdown of perturbation theory. If this inequality is reversed, however, we expect self-avoidance to be asymptotically irrelevant in large systems. This is the case for polymers ($D = 1$) when $d > 4$. Note, however, that the perturbative correction in (5.31) is always large for membranes, i.e., for $D = 2$ [50].

Fig. 5.8. Various regimes in the (d, D)-plane for self-avoiding tethered surfaces.

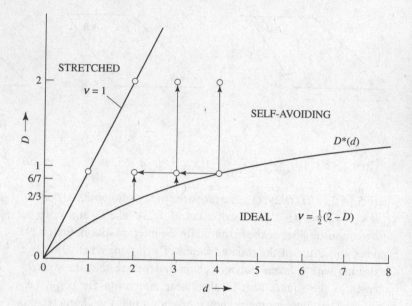

Figure 5.8 shows the critical curve

$$D^*(d) = \frac{2d}{4+d} \tag{5.33}$$

which separates ideal from self-avoiding behavior in the (d, D)-plane. Also shown is the line $D = d$, along which the manifold becomes fully stretched due to self-avoidance. The critical line $D^*(d)$, of course, passes through the point $(d^* = 4, D^* = 1)$, which is the basis for epsilon expansions of polymers [30], but in fact *any point* on this line is an equally good expansion candidate. We could, for example, stay in three dimensions $(d = 3)$ and change the dimensionality of the manifold D. Self-avoidance dominates for solid elastic cubes $(D = 3)$, but is less important for elastic surfaces $(D = 2)$. It produces relatively small corrections to the Gaussian result for linear manifolds $(D = 1)$ and becomes formally negligible when $D < D^* = \frac{6}{7}$. This idea forms the basis for a $\frac{6}{7} + \varepsilon$ expansion for tethered surfaces [46–49]. The result is that the radius of gyration scales with the linear dimension according to

$$R_G \sim L^\nu, \tag{5.34}$$

where

$$\nu = \frac{2-D}{2} + 0.469\left(D - \frac{6}{7}\right) + \mathcal{O}\left(D - \frac{6}{7}\right)^2. \tag{5.35}$$

This novel epsilon expansion gives excellent results for linear polymers in three dimensions ($\varepsilon = \frac{1}{7}$, $\nu = 0.567$), but is not very accurate for polymerized membranes ($\varepsilon = \frac{8}{7}$, $\nu = 0.536$).

Early computer simulations [46] suggested that $\nu \approx 0.8 \pm 0.03$ for

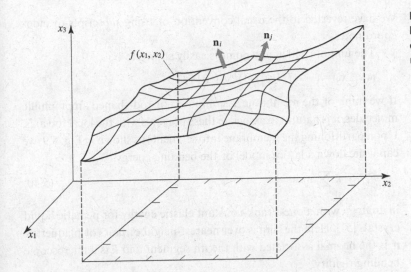

Fig. 5.9. A liquid membrane broken up into plaquettes, each characterized by a unit normal.

self-avoiding membranes. A good approximation for ν for a D-dimensional manifold embedded in d dimensions follows from applying the Flory approximation to Eq. (5.26). Proceeding as in the case of linear polymers, we find

$$F \approx \frac{1}{2} K R_G^2 L^{D-2} + \frac{1}{2} b \left(\frac{L^D}{R_G^d}\right)^2 R_G^d. \tag{5.36}$$

Minimization with respect to R_G leads to $R_G \sim L^\nu$, with [45, 46]

$$\nu = \frac{2 + D}{2 + d}, \tag{5.37}$$

i.e., $\nu = \frac{4}{5}$ for membranes ($D = 2$, $d = 3$). As discussed in the preface to this chapter, the existence of a crumpled phase satisfying (5.37) in computer simulations is still an open question.

5.3 Normal–normal correlations in liquid membranes

The analysis in the previous subsection is restricted to floppy membranes, without an appreciable bending rigidity. The conformations of untethered liquid membranes, however, are dominated by such bending energies. Before discussing how bending energies affect tethered membranes, we first review results for liquid membranes.

Figure 5.9 shows a fragment of a liquid membrane, which we assume is approximately parallel to the (x_1, x_2)-plane so that we can use a Monge representation for its position,

$$\mathbf{r}(x_1, x_2) = (x_1, x_2, f(x_1, x_2)). \tag{5.38}$$

We have reverted to the usual convention of using *sub*scripts to index coordinates.

The unit normal at any point is easily shown to be

$$\mathbf{n}(x_1, x_2) = (-\partial_1 f, -\partial_2 f, 1)/\sqrt{1 + |\nabla f|^2}. \tag{5.39}$$

If we think of the membrane as composed of rod-shaped amphiphillic molecules, it is natural to associate these normals with the local rod axis. Upon partitioning the membrane into segments as shown in Fig. 5.9, we can write down a lattice model of the bending energy,

$$F_b = -\bar{\kappa} \sum_{\langle i,j \rangle} \mathbf{n}_i \cdot \mathbf{n}_j, \tag{5.40}$$

in analogy with a one-Frank-constant elastic energy for nematic liquid crystals [51]. Here the sum is over nearest-neighbor pairs of plaquettes, $\mathbf{n}_i$ is the normal associated with the ith segment and $\bar{\kappa}$ is a microscopic bending rigidity.

Equation (5.40) resembles the energy of a classical Heisenberg ferromagnet on a two-dimensional lattice [52, 53]. The normals are like spin vectors and the rigidity $\bar{\kappa}$ is like a Heisenberg exchange constant. Rotational symmetry is broken at $T = 0$ by a "ferromagnetic" flat surface, with a uniformly aligned normal field. As discussed in Chapter 2, however, long-range order is destroyed at any *finite* temperature in the two-dimensional Heisenberg model by spin-wave fluctuations. In surfaces, we might also expect long range-order to be destroyed, in this case by surface undulations. The analogy with spin waves is not perfect, however, because the normals are constrained to be part of a surface, which forces them to be expressible as in Eq. (5.39). For small undulations, this restriction means that "spin waves" in the normals must be purely longitudinal; transverse "spin waves" would tear the surface.

To determine how undulations affect correlations in the normals, it is useful to take the continuum limit of Eq. (5.40). To leading order in an expansion in gradients of $f(x_1, x_2)$, we can neglect the factor $\sqrt{g} = \sqrt{1 + |\nabla f|^2}$ in the measure as well as the $|\nabla f|^2$ term in the denominator of (5.39) and find

$$F_b \approx \frac{1}{2}\kappa \int d^2x \, |\nabla \mathbf{n}|^2 \approx \frac{1}{2}\kappa \int d^2x \, [(\partial_1^2 f)^2 + 2(\partial_1 \partial_2 f)^2 + (\partial_2^2 f)^2]$$

$$\approx \frac{1}{2}\kappa \int d^2x \, [(\nabla^2 f)^2 - 2\det(\partial_i \partial_j f)] \tag{5.41}$$

where κ is proportional to $\bar{\kappa}$. The last two terms of (5.41) are just the mean-curvature and Gaussian-curvature pieces of the Helfrich [54] bending energy of a liquid membrane. The Gaussian curvature is a perfect derivative, which we can see by writing the second term as

$$2\det(\partial_i \partial_j f) = -\varepsilon_{im}\varepsilon_{jn}\partial_m \partial_n [(\partial_i f)(\partial_j f)]. \tag{5.42}$$

Upon neglecting the contribution from this surface term, we can write

$$F_b \approx \frac{1}{2}\kappa \int d^2x (\nabla^2 f)^2. \tag{5.43}$$

Following de Gennes and Taupin [55], we can now estimate fluctuations in the normals. The angle $\theta(x_1, x_2)$ which the normal $\mathbf{n}(x_1, x_2)$ makes with respect to the $\hat{\mathbf{x}}_3$ axis is given by

$$\mathbf{n}\cdot\hat{\mathbf{x}}_3 = \cos\theta = 1/\sqrt{1 + |\nabla f|^2}. \tag{5.44}$$

If there is a broken symmetry such that the normals point on average along the $\hat{\mathbf{x}}_3$ axis, fluctuations in $\theta^2 \approx |\nabla f|^2$ should be small at low temperatures. Because Eq. (5.43) is a quadratic form, we can calculate $\langle\theta^2\rangle$ by passing to Fourier space and using the equipartition theorem,

$$\langle \theta^2(x_1, x_2)\rangle \approx k_B T \int \frac{d^2q}{(2\pi)^2}\frac{1}{\kappa q^2} \approx \frac{k_B T}{\kappa}\ln\left(\frac{L}{a}\right). \tag{5.45}$$

Just as in many other systems with continuous symmetries in two dimensions [56], there is a logarithmic divergence with the system's size L, signaling the breakdown of long-range order in the normals. More sophisticated calculations by Peliti and Leibler [57] show that the renormalized wave-vector-dependent rigidity $\kappa_R(q)$ is softened by these fluctuations

$$\kappa_R(q) = \kappa - \frac{3k_B T}{4\pi}\ln\left(\frac{1}{qa}\right). \tag{5.46}$$

Note that, if we replace κ by its renormalized value $\kappa_R(q)$ in Eq. (5.45), this only makes the divergence worse. The renormalization-group calculations of [57] are consistent with exponential decay of the normal–normal correlation function,

$$\langle \hat{\mathbf{n}}(\mathbf{x})\cdot\hat{\mathbf{n}}(0)\rangle \propto e^{-x/\xi}, \tag{5.47}$$

with a correlation length that diverges at low temperatures

$$\xi \approx a e^{4\pi\kappa/(3k_B T)}. \tag{5.48}$$

The low-temperature behavior of the two-dimensional Heisenberg model [53] is very similar.

5.4 Tethered surfaces with bending energy

Figure 5.10 shows a surface in which bending energy and tethering are present simultaneously [58, 59]. If $\mathbf{r}_i$ denotes the position of the ith vertex and $\mathbf{n}_\alpha$ is the normal to the αth triangular plaquette, a microscopic model Hamiltonian would be

$$H = -\bar{\kappa}\sum_{\langle\alpha,\beta\rangle}\mathbf{n}_\alpha\cdot\mathbf{n}_\beta + \sum_{\langle i,j\rangle} V(|\mathbf{r}_i - \mathbf{r}_j|), \tag{5.49}$$

Fig. 5.10. Unit normals defined on the triangular plaquettes of a tethered surface.

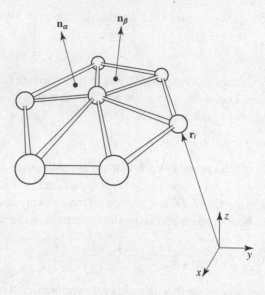

where $V(r)$ is a tethering potential between nearest-neighbor vertices. We have just seen that liquid membranes, which are subject only to bending energy, crumple at finite temperatures, in the sense that long-range order in the normals is destroyed. As we saw in Section 5.2, tethered surfaces which are subject only to the constraint of fixed bonding connectivity, also crumple, like polymers in a good solvent [45, 46]. We shall now argue that there is a fundamental incompatibility when these two energies are simultaneously present, which stabilizes a flat phase at sufficiently low temperatures.

As in our discussion of liquid membranes, we start with a locally flat surface and ask whether a state with long-range order in the normals is stable with respect to thermal fluctuations. We assume that the tethering potential induces nonzero elastic constants in approximately planar membranes, so that there is some elastic stretching energy, as well as the bending energy characteristic of liquid membranes.

As shown in Fig. 5.11, a three-component displacement field is necessary to describe the deformation of an initially flat membrane. If $\mathbf{r}_0(x_1, x_2) = (x_1, x_2, 0)$ describes the undistorted membrane at zero temperature, an arbitrary membrane configuration for $T > 0$ is given by

$$\mathbf{r}(x_1, x_2) = \mathbf{r}_0 + \begin{pmatrix} u_1(x_1, x_2) \\ u_2(x_1, x_2) \\ f(x_1, x_2) \end{pmatrix}. \tag{5.50}$$

A small line element $d\mathbf{r}_0 = (dx_1, dx_2, 0)$ in the undistorted membrane is mapped by this transformation into a line element

$$dr = \begin{pmatrix} (1 + \partial_1 u_1)\,dx_1 + (\partial_2 u_1)\,dx_2 \\ (\partial_1 u_2)\,dx_1 + (1 + \partial_2 u_2)\,dx_2 \\ (\partial_1 f)\,dx_1 + (\partial_2 f)\,dx_2 \end{pmatrix}. \tag{5.51}$$

As usual in discussions of continuum elastic theory [60], we describe the stretching of this line element by using a strain matrix $u_{ij}(x_1, x_2)$,

$$d^2r = d^2r_0 + 2u_{ij}\,dx_i\,dx_j, \tag{5.52}$$

in which, using Eq. (5.51), we find

$$u_{ij} = \tfrac{1}{2}[\partial_i u_j + \partial_j u_i] + \tfrac{1}{2}(\partial_i f)(\partial_j f) + \tfrac{1}{2}(\partial_i u_k)(\partial_j u_k). \tag{5.53}$$

To lowest order in gradients of u and f we can neglect the term $\tfrac{1}{2}(\partial_i u_k)(\partial_j u_k)$ and simply write

$$u_{ij} \approx \tfrac{1}{2}[\partial_i u_j + \partial_j u_i + (\partial_i f)(\partial_j f)]. \tag{5.54}$$

The free energy of a nearly flat tethered membrane is the sum of the bending and stretching energies,

$$F[f, u] = \frac{1}{2}\kappa \int d^2x\,(\nabla^2 f)^2 + \frac{1}{2}\int d^2x\,(2\mu u_{ij}^2 + \lambda u_{kk}^2), \tag{5.55}$$

where the elastic stretching energy has been expanded in powers of the strain matrix and μ and λ are elastic constants [60]. The vertical membrane displacement f in Eq. (5.54) introduces an important element of "frustration" into (5.55). To see this, imagine that we are given a vertical displacement field $f(x_1, x_2)$ that has a particularly low bending energy and hence would lead to a low overall energy if the membrane were a liquid. If this is to be a low-energy configuration for tethered membranes as well, it must be possible to choose phonon displacements such that u_{ij} vanishes for the given f. Note that the term $(\partial_i f)(\partial_j f)$ acts like a matrix vector potential in Eq. (5.54). It will in general be impossible to choose the two independent phonon displacement fields $u_1(x_1, x_2)$ and $u_2(x_1, x_2)$ in such a way as to cancel out all *three* distinct components of this symmetric matrix. We conclude that there must be many low-energy configurations of a liquid membrane that will be energetically unfavorable when we introduce the stretching energy.

To treat the stretching energy more quantitatively, it is useful to eliminate the quadratic phonon field in (5.55) and define

$$\bar{F}(f) = -k_B T \ln\left(\int \mathcal{D}\mathbf{u}(x_1, x_2) e^{-F[f,u]/(k_B T)} \right). \tag{5.56}$$

To carry out the functional integral in (5.56), it is essential to separate u_{ij} into its $\mathbf{q}=0$ and $\mathbf{q}\neq 0$ Fourier components

$$u_{ij}(\mathbf{x}) = u_{ij}^0 + A_{ij}^0 + \sum_{\mathbf{q}\neq 0}\left(\tfrac{1}{2}i[q_i u_j(\mathbf{q}) + q_j u_i(\mathbf{q})] + A_{ij}(\mathbf{q}) \right)e^{i\mathbf{q}\cdot\mathbf{x}}. \tag{5.57}$$

Here, $A_{ij}(q)$ is the qth Fourier component of the "vector potential" $A_{ij}(x) = (\partial_i f)(\partial_j f)$,

$$A_{ij}(\mathbf{q}) = \int d^2 x e^{-i\mathbf{q}\cdot\mathbf{x}}(\partial_i f)(\partial_j f), \tag{5.58}$$

while A_{ij}^0 is the corresponding component for $\mathbf{q}=0$. Although there are only two independent phonon degrees of freedom $\mathbf{u}_i(\mathbf{q})$ in the in-plane strain matrix at nonzero wavevectors, the uniform part of the in-plane strain matrix u_{ij}^0 has in fact *three* independent components, reflecting the three independent ways of macroscopically distorting a flat two-dimensional crystal [61]. The $\mathbf{q}\neq 0$ part of A_{ij} can be decomposed, as can any two-dimensional symmetric matrix, into transverse and longitudinal parts [62],

$$A_{ij}(\mathbf{x}) = \tfrac{1}{2}[\partial_i \phi_j(\mathbf{x}) + \partial_j \phi_i(\mathbf{x})] + P_{ij}^{\text{T}}\Phi(\mathbf{x}), \tag{5.59}$$

where P_{ij}^{T} is the transverse projection operator, $P_{ij}^{\text{T}} = \delta_{ij} - \partial_i\partial_j/\nabla^2$. By applying the transverse projector to both sides of (5.59), we find that

$$\Phi(\mathbf{x}) = \tfrac{1}{2}P_{ij}^{\text{T}}(\partial_i f)(\partial_j f). \tag{5.60}$$

The functional integral can now be efficiently performed by integrating over the shifted variables

$$\bar{u}_{ij}^0 = u_{ij}^0 + A_{ij}^0, \tag{5.61a}$$

$$\bar{u}_i = u_i + \phi_i, \tag{5.61b}$$

which leads to an effective free energy

$$F_{\text{eff}} = \frac{1}{2}\kappa \int d^2 x (\nabla^2 f)^2 + \frac{1}{2}K \int{}' d^2 x \left(\frac{1}{2}P_{ij}^{\text{T}}(\partial_i f)(\partial_j f) \right)^2, \tag{5.62}$$

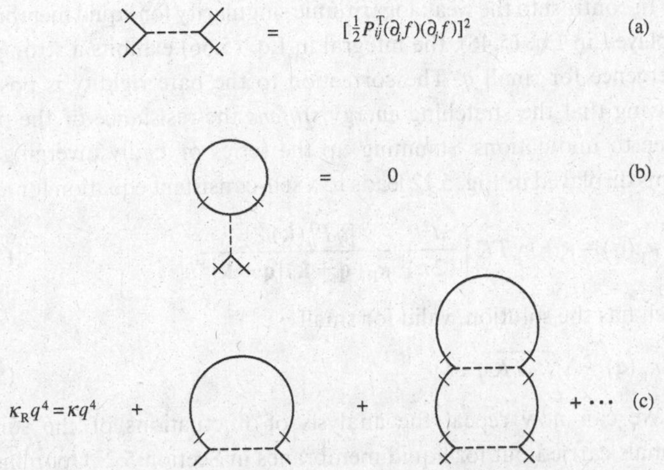

Fig. 5.12. Graphical rules for calculating the renormalized rigidity: (a) the interaction vertex, (b) a graph that vanishes because the $\mathbf{q}=0$ part of the interaction has been eliminated by integrating out the in-plane strain field and (c) the most divergent terms in the perturbation series for κ_R.

where the prime on the integral means that the $q=0$ part of P_{ij}^T has been integrated out, and K is the two-dimensional Young modulus,

$$K = \frac{4\mu(\mu+\lambda)}{2\mu+\lambda}. \tag{5.63}$$

To obtain a physical interpretation of the peculiar form assumed by the stretching energy in (5.62), we note first that the Laplacian of the square root of the integrand is just the Gaussian curvature,

$$-\nabla^2\left(\frac{1}{2}P_{ij}^T(\partial_i f)(\partial_j f)\right) = \det\left(\frac{\partial^2 f}{\partial x_i \partial x_j}\right) = S(\mathbf{x}). \tag{5.64}$$

Thus, the parts of a membrane with nonzero Gaussian curvature act as source terms in the Laplace equation (5.64), leading inevitably to a large positive contribution to the stretching energy. The elastic coupling K penalizes all membrane distortions that are not "isometric," i.e., those with nonzero Gaussian curvature. There are still many low-energy configurations available to the membrane, however, as one can verify by crumpling a piece of paper, which has essentially infinite in-plane elastic constants.

The effect of the nonlinear stretching energy on the renormalized wave-vector-dependent rigidity

$$\kappa_R^{-1}(\mathbf{q}) \equiv q^4\langle|f(\mathbf{q})|^2\rangle \tag{5.65}$$

can be calculated perturbatively in K_0 [58]. Figure 5.12 summarizes the relevant Feynman graphs. The slashes on the interaction vertex denote derivatives of $f(x_1, x_2)$. Note that the "tadpole" graphs vanish identically because we have integrated out the $\mathbf{q}=0$ part of the interaction. The first two terms in the perturbation series for κ_R are

$$\kappa_R(\mathbf{q}) = \kappa + k_B T K \int \frac{d^2 k}{(2\pi)^2} \frac{[\hat{q}_i P_{ij}^T(\mathbf{k})\hat{q}_j]^2}{\kappa|\mathbf{q}+\mathbf{k}|^4}. \tag{5.66}$$

In contrast to the weak logarithmic singularity for liquid membranes displayed in Eq. (5.46), the integral in Eq. (5.66) exhibits a strong $1/q^2$ divergence for small q. The correction to the bare rigidity is positive, showing that the stretching energy *stiffens* the resistance of the membrane to undulations. Summing up the series of badly diverging diagrams displayed in Fig. 5.12 leads to a self-consistent equation for $\kappa_R(\mathbf{q})$,

$$\kappa_R(\mathbf{q}) = \kappa + k_B T K \int \frac{d^2 k}{(2\pi)^2} \frac{[\hat{q}_i P_{ij}^T(\mathbf{k}) \hat{q}_j]^2}{\kappa_R(\mathbf{q}+\mathbf{k})|\mathbf{q}+\mathbf{k}|^4}, \tag{5.67}$$

which has the solution, valid for small q,

$$\kappa_R(\mathbf{q}) \sim \sqrt{k_B T K}\, q^{-1}. \tag{5.68}$$

We can now repeat the analysis of fluctuations of the surface normals carried out for liquid membranes in Section 5.2. Upon inserting the renormalized rigidity (5.68) into Eq. (5.45), we find that the fluctuations in θ^2 are now finite,

$$\langle \theta^2(\mathbf{x}) \rangle = k_B T \int \frac{d^2 q}{(2\pi)^2} \frac{1}{\kappa_R(q)q^2} \simeq \sqrt{\frac{k_B T}{K}} \int \frac{d^2 q}{(2\pi)^2} \frac{1}{q} < \infty. \tag{5.69}$$

The mode-coupling analysis which led to this important result *assumed* that there was no significant renormalization of the elastic coupling K [58]. A recent renormalization-group calculation of Aronovitz and Lubensky [63] suggests that the elastic constants will in fact exhibit weak singularities as $q \to 0$, but that κ_R still diverges strongly enough to make (5.45) finite. Equation (5.69) shows that tethering stabilizes long-range order in the normals at low temperatures. Because entropy favors a crumpled surface at high temperatures, this result suggests the existence of the finite-temperature crumpling transition. There is now strong evidence from computer simulations [59] for this crumpling transition, at least for tethered surfaces without self-avoidance. These arguments suggest that there is a crumpling transition in real self-avoiding membranes as well. The possibility of a crumpling transition due to *long-range forces* was suggested in the paper on liquid membranes by Peliti and Leibler [57]. Although it is possible to rewrite the stretching energy in Eq. (5.26) in terms of a long-range interaction between *Gaussian* curvatures [58], this interaction is rather different from the long-range forces considered in [57]. There are, moreover, no true microscopic long-range forces in a tethered surface: they arise here only to compensate for integrating out the underlying physical phonon field.

5.5 Defects and hexatic order in membranes

In our discussion of tethered surfaces, it has been convenient to consider polymerized membranes, tied together with permanent covalent bonds.

It is interesting to consider instead nominally crystalline membranes bound together by weaker, van der Waals forces. This situation arises, for example, in unpolymerized lipid bilayers at sufficiently low temperatures. If these materials were constrained to be flat, their low-temperature crystalline phase would eventually become unstable with respect to a proliferation of unbound dislocations upon heating. As first elucidated in a famous argument by Kosterlitz and Thouless [64] (and discussed in Chapters 1 and 2), a dislocation with Burger's vector $\mathbf{b}$ (see Fig. 5.13) in a flat two-dimensional crystal of radius R has an energy of order $Kb^2\ln(R/a)$, where K is given by Eq. (5.63) and a is the lattice spacing. This energy cost suppresses the formation of dislocations at low temperatures. However, there is also an entropy of roughly $2k_B\ln(R/a)$ associated with a dislocation, since it can be located at $(R/a)^2$ possible positions. Above the critical melting temperature $k_BT_M \sim Kb^2$, the entropy term dominates, dislocations proliferate and the crystal melts into a hexatic phase [43]. Figure 5.13 also shows another type of defect, the disclination, which has an energy of order KR^2s^2. Here s is the disclination charge, defined as the angle in radians of the wedge which must be removed or added to a perfect crystal to make the defect. Disclination energies diverge so rapidly with a system's size that these defects are extremely unlikely in equilibrated two-dimensional crystals.

In this section we sketch what happens when crystals containing defects are allowed to buckle into the third dimension [58, 65]. For a more detailed discussion, see Chapter 6. As shown in Figure 5.14(a), an initially flat disclination in a triangular lattice (obtained by removing a 60° wedge of material from a perfect crystal) will prefer to buckle into an approximately conical shape in a large enough crystal. A similar, hyperbolic buckling (Fig. 5.14(b)) occurs in a sufficiently large crystal containing a negative disclination, which is obtained by adding a 60° wedge of material. This buckling occurs because the system finds it energetically preferable to trade in-plane elastic energy for bending energy. Suppose for simplicity that the in-plane elastic constants are

Fig. 5.13. Defects of interest in polymerized crystalline membranes: (a) a dislocation with Burger's vector $\mathbf{b}$, (b) $a + 2\pi/6$ disclination and (c) $a - 2\pi/6$ disclination.

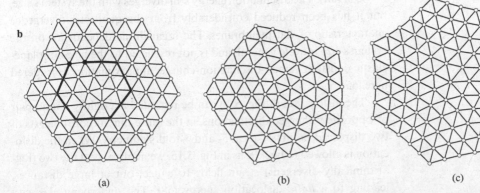

(a)   (b)   (c)

Fig. 5.14. Buckling of (a) positive and (b) negative disclinations.

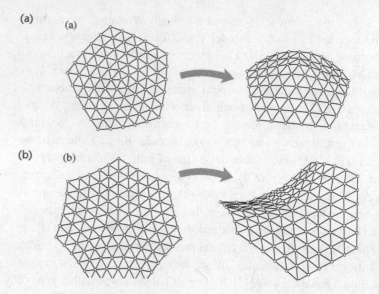

(a)

(b)

very large, such as in an ordinary piece of writing paper. It is easy to check by inserting disclinations into pieces of paper that there are then essentially no elastic distortions in the buckled state. The only remaining energy is now the bending energy, i.e., the first term of Eq. (5.55). It is not hard to show that, for small s, the vertical displacement in polar coordinates (r, θ) is [58]

$$f(r, \theta) \approx \sqrt{\frac{s}{\pi}r} \tag{5.70}$$

and

$$f(r, \theta) \approx \sqrt{\frac{2|s|}{3\pi}}r\cos(2\theta), \tag{5.71}$$

for positive and negative disclinations, respectively. In both cases $\nabla^2 f \propto 1/r$ and the bending energy in Eq. (5.55) diverges logarithmically, $F \sim \kappa \ln(R/a)$. The disclination energy still diverges with the system's size, but it has been reduced considerably from the quadratic divergence characteristic of flat membranes. The logarithmic dependence on the system's size is quite general, and is not restricted to the large-in-plane-elastic-constant, small-disclination-charge approximation considered here [65].

The dislocation in Fig. 5.13 can be regarded a a tightly bound pair of oppositely charged disclinations, in the sense that its core consists of two displaced points of local 5- and 7-fold symmetry. When the dislocation is allowed to buckle, as in Fig. 5.15, we might expect the two (logarithmically divergent) strain fields to cancel out at large distances, leading to a *finite* dislocation energy [58]. The underlying elasticity

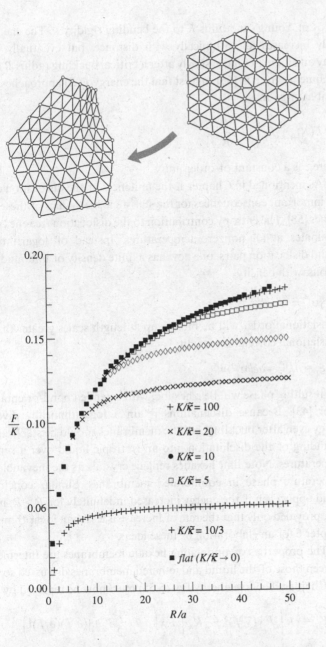

Fig. 5.15. Buckling of a dislocation. Note the pair of $\pm 2\pi/6$ disclinations represented by 5- and 7-coordinated particles at the core of the dislocation.

Fig. 5.16. The energy of a dislocation as a function of the membrane radius R. The energy ceases to increase logarithmically beyond a critical buckling radius.

$\dfrac{F}{K}$

+ $K/\tilde{\kappa} = 100$
× $K/\tilde{\kappa} = 20$
● $K/\tilde{\kappa} = 10$
□ $K/\tilde{\kappa} = 5$

+ $K/\tilde{\kappa} = 10/3$
■ flat ($K/\tilde{\kappa} \to 0$)

R/a

equations for buckled membranes are nonlinear [60], however, so we cannot really apply the superposition principle in this way. Numerical calculations have recently been carried out [65] using the model Hamiltonian (5.49), with

$$V(\mathbf{r}) = \tfrac{1}{2}(|\mathbf{r}|^2 - a^2),\tag{5.72}$$

where a is the lattice constant. The energy of an isolated dislocation at the origin in a system of size R is shown in Fig. 5.16 for a variety of ratios of Young's modulus K to the bending rigidity $\bar{\kappa}$. The energy initially increases logarithmically with distance, but eventually breaks away and increases more slowly after a critical buckling radius $R_c(K_0/\bar{\kappa})$. Arguments given in [65] suggest that the energy $E(R)$ approaches a constant at large R,

$$E(R) \underset{R \to \infty}{\approx} E_D \left(1 - \frac{cR_c}{R} \right),$$

(5.73)

where c is a constant of order unity.

As mentioned in Chapter 1, the finiteness of the dislocation energy has important consequences for the statistical mechanics of these membranes [58]. The entropy contribution to the dislocation free energy now dominates at all nonzero temperatures. Instead of logarithmically bound dislocation pairs, one now has a finite density of unbound dislocations, with density

$$n_D \approx a^{-2} e^{-E_D/(k_B T)}.$$

(5.74)

Translational order will be broken up at length scales greater than the translational correlation length

$$\xi_T \approx n_D^{-1/2} \approx a e^{E_D/(2k_B T)}.$$

(5.75)

The resulting phase will be a hexatic, with extended bond-orientational order [43]. Because disclinations retain a logarithmically divergent energy even after buckling, hexatic membranes should be stable against unbinding of the disclination into an isotropic liquid over a range of temperatures. Note that hexatics replace crystals as the inevitable low-temperature phase in equilibrated membranes. Similar conclusions would apply even if the energy increased indefinitely for $R > R_c$ in Fig. 5.16, provided only that the rate of increase is less than logarithmic. See Chapter 6 for an elaboration of these ideas.

The properties of undulating hexatic membranes are intermediate between those of the liquid and tethered membranes discussed so far in this Chapter. The free energy Eq. (5.55) must now be replaced by [58]

$$F_H = \frac{1}{2}\kappa \int d^2 x (\nabla^2 f)^2 + \frac{1}{2}K_A \int d^2 x \left(\partial_i \theta + \frac{1}{2}\varepsilon_{jk}\partial_k[(\partial_i f)(\partial_j f)] \right)^2$$

(5.76)

where $\theta(x_1, x_2)$ is the local bond-angle field defined within the membrane and K_A is the hexatic stiffness constant [43]. Gradients of θ are accompanied by a "vector potential"

$$A_i = -\frac{1}{2}\varepsilon_{jk}\partial_k[(\partial_i f)(\partial_j f)]$$

(5.77)

in Eq. (5.76). Just as in Eq. (5.54), this vector potential reflects frustration: The bond-angle field cannot return to itself when it is parallel transported around a region of nonzero Gaussian curvature. (The Gaussian curvature is given by the curl of (5.77).) The statistical mechanics associated with Eq. (5.77) was worked out by David *et al.* [66]. Remarkably, there is a low-temperature "crinkled" phase with a radius of gyration R_G controlled by a *continuously variable* Flory exponent

$$R_G \sim L^{\nu(K_A)} \tag{5.78}$$

where L is a characteristic linear dimension of the membrane and

$$\nu(K_A) = 1 - \frac{k_B T}{2\pi K_A} + \mathcal{O}(k_B T)^2. \tag{5.79}$$

Both disclination-unbinding transitions and finite-temperature crumpling transitions are possible in hexatic membranes.

Acknowledgements

The work described here was carried out in collaboration with L. Peliti, Y. Kantor, M. Kardar, S. Seung and M. Paczuski. It is a pleasure to acknowledge these fruitful collaborations, as well as stimulating interactions with F. David and S. Leibler. This chapter is adapted with permission from *Random Fluctuations and Pattern Growth: Experiments and Models*, edited by H. E. Stanley and N. Ostrowsky (Kluwer, Dordrecht, 1988), pp. 193–217.

References

[1] H. P. Duwe, J. Kas and E. Sackmann, *J. Physique* **51**, 945 (1990).

[2] C. F. Schmidt, K. Svoboda, N. Lei, I. B. Petsche, L. E. Berman, C. R. Safinya and G. S. Grest, *Science* **259**, 952 (1993).

[3] T. Hwa, E. Kokufuta and T. Tanaka, *Phys. Rev.* A **44**, R2235 (1991); X. Wen, C. W. Garland, T. Hwa, M. Kardar, E. Kokufuta, Y. Li, M. Orkisz and T. Tanaka, *Nature* **335**, 426 (1992).

[4] M. S. Spector, E. Maranjo, S. Chiruvolu and J. A. Zasadzinski, *Phys. Rev. Lett.* **73**, 2867 (1994).

[5] R. R. Chianelli, E. B. Prestridge, T. A. Pecorado and J. P. De Nefville, *Science* **203**, 1105 (1979).

[6] S. I. Stupp, S. Son, J. C. Lin and L. S. Li, *Science* **259**, 59 (1993).

[7] H. Rehage, B. Achenbach and A. Kaplan, *Ber Bunsenges. Phys. Chem.* **101**, 1683 (1997).

[8] W. T. S. Huck, A. D. Stroock and G. M. Whitesides, *Angew. Chem. Int. Ed.* **39**, 1056 (2000).

[9] L. Radzihovsky and D. R. Nelson, *Phys. Rev.* A **44**, 3525 (1991).

[10] D. C. Morse and T. C. Lubensky, *Phys. Rev.* A **46**, 1751 (1992).

[11] D. C. Morse, T. C. Lubensky and G. S. Grest, *Phys. Rev.* A **45**, R2151 (1992).

[12] D. C. Morse, I. B. Petsche, G. S. Grest and T. C. Lubensky, *Phys. Rev.* A **46**, 6745 (1992).

[13] M. Mutz, D. Bensimon and M. J. Brienne, *Phys. Rev. Lett.* **67**, 923 (1991).

[14] D. R. Nelson and L. Radzihovsky, *Phys. Rev.* A **46**, 7474 (1992).

[15] C. Carraro and D. R. Nelson, *Phys. Rev.* E **48**, 3082 (1993).

[16] F. F. Abraham and D. R. Nelson, *Science* **249**, 393 (1990).

[17] F. F. Abraham and D. R. Nelson, *J. Physique* **51**, 2653 (1990).

[18] M. Bowick, M. Falcioni and G. Thorleifsson, *Phys. Rev. Lett.* **79**, 885 (1997).

[19] L. Radzihovsky and J. Toner, *Phys. Rev. Lett.* **75**, 4752 (1995).

[20] See, e.g., L. Radzihovsky and J. Toner, *Phys. Rev.* E **57**, 1832 (1988); M. Bowick and A. Travesset, *Phys. Rev.* E **59**, 5659 (1999).

[21] G. Gompper and D. M. Kroll, *Phys. Rev. Lett.* **78**, 2859 (1997); *J. Physique* I **11**, 1 (1997).

[22] K. Weise, in *Phase Transitions and Critical Phenomena*, Vol. 19 (Academic Press, New York, 2001), p. 253.

[23] M. Bowick and A. Travesset, in *Renormalization Group Theory at the Turn of the Millennium*, edited by D. O'Connor and C. R. Stephens; *Phys. Rep.* **344**, 255 (2001).

[24] J. Weeks, in *Ordering in Strongly Fluctuating Condensed Matter Systems*, edited by T. Riste (Plenum, New York, 1980).

[25] For an introduction useful to physicists, see B. A. Dubrovin, A. T. Fomenko and S. P. Novikov, *Modern Geometry – Methods and Applications* (Springer, New York, 1984).

[26] J. V. Jose, L. P. Kadanoff, S. Kirkpatrick and D. R. Nelson, *Phys. Rev.* B **16**, 1217 (1977).

[27] J. Fröhlich, in *Applications of Field Theory to Statistical Mechanics*, edited by L. Garido, Lecture Notes in Physics, Vol. 216 (Springer, Berlin, 1985).

[28] P. G. de Gennes, *Rev. Mod. Phys.* **57**, 827 (1985).

[29] E. Brésin, B. I. Halperin and S. Leibler, *Phys. Rev. Lett.* 50, 1387 (1983); R. Lipowsky, D. M. Kroll and R. K. P. Zia, *Phys. Rev.* B **27**, 4499 (1983).

[30] Y. Oono, *Adv. Chem. Phys.* **61**, 301 (1985).

[31] P. G. de Gennes, *Scaling Concepts in Polymer Physics* (Cornell University Press, Ithaca, 1979).

[32] E. Evans and R. Skalak, *Mechanics and Thermodynamics of Biomembranes* (CRC Press, Boca Raton, 1980).

[33] P. G. de Gennes and C. Taupin, *J. Phys. Chem.* **86**, 2294 (1982).

[34] *Physics of Complex and Supermolecular Fluids*, edited by S. A. Safran and N. A. Clark (Wiley, New York, 1987).

[35] *Physics of Amphiphillic Layers*, eds. J. Meunier, D. Langevin and N. Boccara (Springer, Berlin, 1987).

[36] H. Fendler and P. Tundo, *Acc. Chem. Res.* **17**, 3 (1984).

[37] A. Blumstein, R. Blumstein and T. H. Vanderspurt, *J. Colloid Interface Sci.* **31**, 236 (1969).

[38] M. J. Asis, E. Nygren, J. F. Hays and D. Turnbull, *J. Appl. Phys.* **57**, 2233 (1985).

[39] R. Zallen, *The Physics of Amorphous Solids* (Wiley, New York, 1983).

[40] J. Larche, J. Appell, G. Porte, P. Bassereau and J. Marignan, *Phys. Rev. Lett.* **56**, 1700 (1986).

[41] C. R. Safinya, D. Roux, G. S. Smith, S. K. Sinhs, P. Dimon and N. A. Clark, *Phys. Rev. Lett.* **57**, 2718 (1986).

[42] D. R. Nelson and L. Peliti, *J. Physique* **48**, 1085 (1987); S. Seung and D. R. Nelson, *Phys. Rev.* A **38**, 1005 (1988).

[43] D. R. Nelson and B. I. Halperin, *Phys. Rev.* B **19**, 2457 (1979); D. R. Nelson, *Phys. Rev* B **27**, 2902 (1983).

[44] A. M. Polyakov, *Phys. Lett.* B **103**, 207 (1981).

[45] Y. Kantor, M. Kardar and D. R. Nelson, *Phys. Lett.* **57**, 791 (1986).

[46] Y. Kantor, M. Kardar and D. R. Nelson, *Phys. Rev.* A **35**, 3056 (1987).

[47] M Kardar and D. R. Nelson, *Phys. Rev. Lett.* **58**, 1289 (1987); *Phys. Rev.* E **38**, 966 (1988).

[48] J. A. Aronovits and T. C. Lubensky, *Europhys. Lett.* **4**, 395 (1987).

[49] D. Duplantier, *Phys. Rev. Lett.* **58**, 2733 (1987).

[50] There are logarithmic corrections to the result of naïve dimensional analysis in this case. See the appendix of [46].

[51] P. G. de Gennes, *The Physics of Liquid Crystals* (Clarendon, Oxford, 1974).

[52] A. M. Polyakov, *Nucl. Phys.* B **268**, 406 (1986).

[53] A. M. Polyakov, *Phys. Rev. Lett.* B **59**, 79 (1975).

[54] W. Helfrich, *Z. Naturforsch.* C **28**, 693 (1973).

[55] P. G. de Gennes and C. Taupin, *J. Phys. Chem.* B86, 2294 (1982).

[56] D. R. Nelson, in *Phase Transitions and Critical Phenomens*, Vol. 7, edited by C. Domb and J. Lebowits (Academic, New York, 1983), or Chapter 2.

[57] L. Peliti and S. Leibler, *Phys. Rev. Lett.* **54**, 690 (1985); see also W. Helfinch, *J. Physique* **46**. 1263 (1985).

[58] D. R. Nelson and L. Peliti, *J. Physique* **48**, 1085 (1987).

[59] Y. Kantor and D. R. Nelson, *Phys. Rev. Lett.* **58**, 2774 (1987); *Phys. Rev.* A **36**, 4020 (1987).

[60] L. D. Landau and E. M. Lifshitz, *Theory of Elasticity* (Pergammon, New York, 1970).

[61] For an analogous treatment of compressible spin models in d-dimensions, see J. Sak, *Phys. Rev.* B **10**, 3957 (1974).

[62] See, e.g., S. Sachdev and D. R. Nelson, *J. Phys.* C. **17**, 5473 (1984).

[63] J. A. Aronovitz and T. C. Lubensky, *Phys. Rev. Lett.* **60**, 2634 (1988).

[64] J. M. Kosterlitz and D. J. Thouless, *J. Phys.* C **5**, 124 (1972); *J. Phys.* C **6**, 1181 (1973).

[65] S. Seung and D. R. Nelson, *Phys. Rev.* A **38**, 1005 (1988).

[66] F. David, E. Guitter and L. Peliti, *J. Physique* **48**, 2059 (1987).

Chapter 6
Defects in superfluids, superconductors and membranes

Preface

This chapter attempts a unified treatment of the ideas about defects presented in Chapters 2–5. We present as well a more detailed derivation of the well-known universal jump in the superfluid density of helium films, which is probably the most striking prediction to emerge from theories of defect-mediated phase transitions in two dimensions. Of particular importance for this book, we then show how screening changes the interactions between vortices in *charged* superfluids. This discussion provides a useful starting point for understanding vortex lines in high-temperature superconductors, a subject discussed in considerable detail in Chapters 7 and 8. The last part of this chapter discusses related screening phenomena for buckled defects in membranes and gives a more extensive and detailed treatment of solutions of the nonlinear von Karman equations at zero temperature than was possible in Chapter 5. For a discussion of how buckling stimulates and modifies two-dimensional phase transitions for colloidal particles confined between two flat plates, see the articles cited in [1]

Because isolated dislocations in membranes can reduce their energy to a finite value by buckling, crystals are unstable with respect to proliferation of dislocations at any finite temperature (see Chapters 1 and 5). The finite density of unbound dislocations which results leads to a hexatic phase at low temperatures. The interesting question of how the *hexatic* phase of membranes "melts" into an isotropic liquid is addressed briefly at the end of this chapter. Although disclinations can buckle in the hexatic phase, this buckling is not enough to remove their logarithmically diverging energy. Although the energies of isolated 5- and 7-fold disclinations can, in principle, diverge logarithmically with distinct coefficients [2], it appears that, in practice, both types of defect unbind at the same critical temperature in membranes [3]. For a discussion of how this defect-unbinding transition might occur in the fluctuating background geometry of a thermally excited membrane, see the articles cited in [4].

The configurations of buckled defects displayed in this chapter were obtained via numerical minimizations using a useful discretization of tethered-surface elasticity theory discussed in [5]. Witten and his collaborators have recently used similar numerical methods coupled with scaling ideas to discuss stress concentrations in mechanically crumpled sheets with nonzero shear moduli [6–8].

Instead of studying how defects are generated in the fluctuating geometry of a membrane, one can *freeze* the geometry and examine how well defects are able to screen out the elastic strains associated with a quenched-in distribution of Gaussian curvature [9]. There has recently been a renewal of interest in this problem, with efforts to solve the so-called "Thomson problem," that this, to find the ground state of identical particles packed on the surface of a sphere [10–12].

6.1 Introduction

As discussed in Chapters 1 and 2, phase transitions and spontaneous breakings of symmetry abound in modern condensed matter physics. Of particular interest are the low-energy excitations associated with *continuous* broken symmetries, like the broken translational and orientational invariance of a crystal, the broken rotational symmetry of ferromagnets and the broken gauge symmetry of superfluids and superconductors. By imposing a slow spatial variation of the continuous symmetry operation on the ground state, one generates an important class of modes that can easily be excited near $T = 0$. In ferromagnets for example, these Goldstone modes are called spin waves, or, in quantized form, "magnons." In crystalline solids, the broken translational symmetry leads to new shear-wave phonon modes. Unlike in a nematic liquid crystal, the additional broken rotational symmetry of a solid does *not* lead to any extra low-energy modes: Rotational modes are coupled to phonons in a way reminiscent of the Higgs mechanism and are thus rendered massive (see, e.g., Eq. (2.147) of Chapter 2). In superfluid helium, the broken symmetry selects a particular phase for the order parameter of the condensate and leads to second sound. The physics of superconductors is complicated by the *charge* of the Cooper pairs and by the coupling of the motion of electrons to the underlying ionic lattice. Low-energy phase degrees of freedom are still important, although now it is the vector potential which acquires a mass via the usual Higgs mechanism.

The excitations discussed above are all continuously connected to the ground state. However, point, line and wall defects in the order-parameter texture, which are *not* continuously connected to the ground state, may also be important. Dislocation lines mediate plastic flow and point-like vacancies and interstitials allow rapid diffusion of particles in crystalline solids. Vortex rings control the decay of supercurrents in superfluids and superconductors. Most laboratory ferromagnets are subject to weak crystal fields, which break rotational invariance. In this case, the width of hysteresis loops below the Curie temperature is controlled by the motion of Bloch walls. Under some circumstances, such topological defects *also* play an important role in the phase transition out of the state of broken symmetry.

In this review, we discuss the defects which arise in superfluids, superconductors and membranes. Although the emphasis is on how to incorporate defects into simple Landau–Ginzburg field theories, we also touch upon their role in the statistical mechanics. We concentrate on two-dimensional systems, so that the defects are typically point-like.

The classic example of vortices and the Kosterlitz–Thouless transition in superfluid helium films is discussed first, together with universal jump in the superfluid density. Experimental confirmation of this prediction was an important test of the idea of defect-mediated phase transitions. *Screening* of defect energies in superconducting films is then treated, with particular attention to the spreading of magnetic field lines in the three-dimensional volume outside the sample. When this spreading is ignored, screening efficiently cuts off the logarithmic divergence in the vortex energy. The finite-energy vortices in this "naïve" theory then proliferate for entropic reasons at any finite temperature, thus destroying superconductivity. In realistic samples, however, spreading of magnetic field lines (which arises because supercurrents are confined to a two-dimensional plane) renders screening less efficient, so a Kosterlitz–Thouless transition from a finite-temperature superconducting state becomes possible. The screening which occurs in the "naïve" theory is an important feature of the vortex lines in real three-dimensional superconductors discussed in Chapters 7 and 8.

The remainder of the paper contrasts the behaviors of defects in membranes and monolayers. By "monolayers" we mean crystalline films strongly confined to a plane. Dislocations and disclinations control the statistical mechanics of monolayers in a way similar to that for vortices in neutral superfluids. As discussed in Chapter 2, successive dislocation- and disclination-unbinding transitions lead from crystalline to hexatic to liquid monolayer phases. In contrast to monolayers, "membranes" have a relatively low energy cost for buckling out of the two-dimensional plane, thus escaping into the third dimension. As a result, defects can be screened by out-of-plane displacements much like vortices in the "naïve" model of superconducting films. Dislocations, for example, now have a finite energy. Their entropic proliferation causes crystals to be replaced by *hexatics* as the inevitable low-temperature phase of membranes. We also discuss the buckling of 5- and 7-fold disclinations, vacancies, interstitial and impurity atoms and grain boundaries. The important distinction between membranes and monolayers was already discussed briefly in Chapter 1.

We conclude by emphasizing the fundamental asymmetry between 5- and 7-fold disclinations in membranes. There is an exact symmetry between plus and minus vortices in superfluids and between dislocations with equal and opposite Burgers vectors in two-dimensional crystals, which forces the energies of the defects and anti-defects to be equal. No such symmetry, however, insures that the core energies of plus and minus disclinations are identical. These disclinations represent points of local 5- and 7-fold coordination, where the number of neighboring particles is determined by the Dirichlet construction. This situation is similar to vacancies and interstitials in crystalline solids: Vacancies typically have a lower energy than interstitials, even though these excitations are "anti-defects" of each other.

Asymmetry of disclinations is of little consequence in flat monolayers, in which the concentrations of 5- and 7-fold defects must be equal for topological reasons. (More precisely, Euler's theorem ensures that the average coordination number is six in a sufficiently large system.) In hexatic monolayers, moreover, the logarithmic divergences in the energies of isolated plus and minus disclinations have equal coefficients. The situation is different, however, in *membranes*, which are free to change their topology or with free boundary conditions, which allow them to curl up near the edges. When the bending rigidity which controls out-of-plane fluctuations is small, buckling in hexatic membranes leads to different coefficients in the logarithmically diverging disclination energies. In liquid membranes, moreover, different core energies of plus and minus disclinations will inevitably bias the system toward a net positive or negative average Gaussian curvature. The asymmetry between energies of 5- and 7-fold disclinations in the liquid determines the sign of the Gaussian rigidity, which acts as a chemical potential controlling their population difference. Returning to the comparison with a superconductor, the Gaussian rigidity acts like an external magnetic field, which would bias superconducting films toward a net excess of plus or minus vortices. In membranes, the sign of the Gaussian bending rigidity determines whether vesicle phases (with positive net Gaussian curvature) or "plumber's nightmare" phases (with overall negative Gaussian curvature) are preferred.

Theories and simulations of hexatic and liquid membranes usually avoid this issue by imposing a definite spherical or torroidal topology, which automatically fixes the difference between the numbers of 5- and 7-fold defects. Such constraints are clearly artificial, however, for lipid bilayer membranes, which can change their topology at will and may even have free edges at some intermediate stage of a topological change. The analogous constraint for vacancies and interstitials in solids would be imposed by periodic boundary conditions: These defects can then only be created or destroyed in pairs. In real solids, with free boundary conditions, vacancies and interstitials can enter at the surface to produce

Fig. 6.1. (a) A schematic diagram of an oscillating-substrate experiment, which measures the superfluid density. (b) Propagation of a third-sound excitation. Measuring the dispersion relation also allows one to extract $\rho_s(T)$.

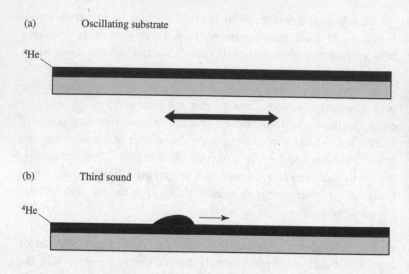

(a)   Oscillating substrate

^{4}He

(b)   Third sound

^{4}He

the concentrations dictated by the microscopic energetics. The simplest probe of disclination bias would be to simulate liquid membranes with free edges and see which defect begins to dominate as the bending rigidity is reduced from large values. See Section 6.3.4 for details.

6.2 Two-dimensional superfluids and superconductors

A detailed and successful theory of two-dimensional superfluidity was constructed in the 1970s. The theory explains the observed properties of thin films of superfluid ^{4}He on various substrates and the physics of thin superconducting films. The basic physical idea of unbinding of defects was proposed by Kosterlitz and Thouless [13], by Berezinski [14] and, in a different physical context, by de Gennes [15]. The first detailed calculations were carried out by Kosterlitz [16], who exploited a pioneering renormalization-group method developed by Anderson and Yuval for the Kondo problem [17]. Because an extensive review is given in Chapter 2 (see also [18–20]), we limit ourselves to a brief sketch of the models and the basic results. Our main goal is to compare and contrast the behavior of superconductors and neutral superfluids. We shall see how vortices control the basic physics and discuss their statistical mechanics. We discuss only equilibrium properties, ignoring the important area of superfluid and superconducting dynamics [21, 22].

6.2.1 Superfluid helium films: Experimental facts

As discussed in Chapter 2, superfluidity arises with decreasing temperature in helium films in a fundamentally different way from how it arises in

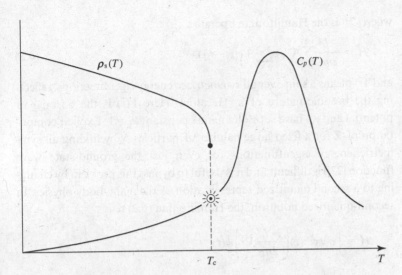

Fig. 6.2. The jump discontinuity in the superfluid density $\rho_s(T)$ as a function of temperature for a fixed thickness of helium film. Also shown is the superfluid specific heat at constant pressure as a function of temperature. There is only an essential singularity, $C_p(T) \approx c_1 + c_2 \exp(-c_3/|T - T_c|^{1/2})$, near the transition, where the c_i are constants.

bulk materials. Two basic experiments that reveal this difference are illustrated in Fig. 6.1. On an oscillating substrate, a finite fraction of the ^{4}He film decouples from the underlying motion below a critical temperature T_c. The film appears lighter than it should be and measuring this deficit allows one to extract the temperature-dependent superfluid density $\rho_s(T)$ [23]. In a related experiment, third-sound excitations similar to capillary waves propagate below T_c with negligible damping through films as thin as a fraction of 1 Å [24]. Such propagating waves would be impossible in a classical fluid, due to viscous damping. The primary restoring force is the van der Waals attraction to the substrate (rather than gravity or surface tension) and again one can extract $\rho_s(T)$ provided that the temperature is not too close to T_c, where the waves become strongly damped.

As illustrated in Fig. 6.2, the superfluid density in films actually jumps discontinuously to zero at T_c, in contrast to the continuous vanishing observed in bulk superfluid ^{4}He. The phase transition is nevertheless *not* of first order, as becomes evident from an examination of the specific heat (see Fig. 6.2). There is only a rounded maximum *above* the transition temperature, in contrast to the delta-function peak expected for a true first-order transition. The detailed theory (see Chapter 2) shows that the specific heat contains only an unobservable essential singularity at the transition itself [25, 26].

6.2.2 Theoretical background

Understanding the equilibrium behavior of superfluid helium films at temperature T requires calculation of quantities such as the quantum partition function,

$$Z = \text{Tr}'\{e^{-\mathcal{H}/T}\}, \tag{6.1}$$

where $\hat{\mathcal{H}}$ is the Hamiltonian operator,

$$\hat{\mathcal{H}} = \frac{-\hbar^2}{2m} \sum_{j=1}^{N} \nabla_j^2 + \sum_{i>j} V(|\mathbf{r}_i - \mathbf{r}_j|), \tag{6.2}$$

and Tr' means a sum over all *symmetrized* energy eigenfunctions, reflecting the bosonic nature of a ^{4}He atom. Here $V(r)$ is the boson pair potential and we have set Boltzmann's constant $k_B = 1$. Explicit computation of Z for a fixed large number of particles N by finding all symmetric energy eigenfunctions (or even just the ground-state wave function [27]) is difficult and it is useful to bypass this problem by changing to a second quantized representation of the many-body physics. In second-quantized notation, the Hamiltonian (6.2) reads

$$\hat{\mathcal{H}} = \int d^3 r \hat{\psi}^+(\mathbf{r}) \left(\frac{-\hbar^2}{2m} \nabla^2 \right) \hat{\psi}(\mathbf{r})$$

$$+ \frac{1}{2} \int d^3 r \int d^3 r' \; \hat{\psi}^+(\mathbf{r}) \hat{\psi}^+(\mathbf{r}') V(|\mathbf{r} - \mathbf{r}'|) \hat{\psi}(\mathbf{r}') \hat{\psi}(\mathbf{r}), \tag{6.3}$$

where $\hat{\psi}^+(\mathbf{r}')$ and $\psi(\mathbf{r})$ are boson creation and destruction operators obeying the commutation relations

$$[\hat{\psi}(\mathbf{r}), \hat{\psi}^+(\mathbf{r}')] = \delta(\mathbf{r} - \mathbf{r}'), \tag{6.4}$$

with all other commutators zero. Within the second-quantization formalism, it is convenient to calculate thermodynamic quantities via the *grand* canonical partition function,

$$Z_{gr} = \text{Tr}' \{ e^{-(\hat{\mathcal{H}} - \mu \hat{N})/T} \}, \tag{6.5}$$

where μ is the chemical potential and the number operator is

$$\hat{N} = \int d^3 r \, \hat{\psi}^+(\mathbf{r}) \hat{\psi}(\mathbf{r}). \tag{6.6}$$

As pointed out by Penrose and Onsager [28], Bose condensation in bulk ^{4}He is associated with off-diagonal long-range order in the correlation function,

$$G(r) = \langle \psi^+(\mathbf{r}) \psi(0) \rangle, \tag{6.7}$$

where $\langle Q \rangle$ means $\text{Tr}' \{ Q \exp[-(\hat{\mathcal{H}} - \mu \hat{N})/T] \}/Z_{gr}$. Specifically, one has

$$\lim_{r \to \infty} G(r) = \langle \hat{\psi}^+(\mathbf{r}) \rangle \langle \hat{\psi}(0) \rangle$$
$$\neq 0. \tag{6.8}$$

A simplified description of superfluidity in terms of a *Landau* theory results from a coarse-graining procedure. In the case of helium films, we start by defining a c-number field

$$\psi(\mathbf{r}) = \langle \hat{\psi} \rangle_{\Omega(r)}, \tag{6.9}$$

where the angle brackets include a spatial average over a volume $\Omega(\mathbf{r})$ centered on $\mathbf{r}$ that is large relative to the spacing between helium atoms but small in comparison with the overall size of the system. The coarse-grained fields $\psi^*(\mathbf{r})$ and $\psi(\mathbf{r})$ (which are nonzero when the ^{4}He liquid is superfluid in view of Eq. (6.8)) behave like complex numbers (instead of operators) for large enough averaging volumes. In helium films, we also take the averaging size to be large relative to the film's thickness. The long-wavelength spatial configurations of the system are now specified by complexions of the classical variable $\psi(\mathbf{r})$, where $\mathbf{r}$ is a *two*-dimensional spatial variable, and the partition function is a functional integral (in units such that $k_B = 1$),

$$Z_{gr} = \int \mathcal{D}\psi(\mathbf{r}) \exp(-F/T). \tag{6.10}$$

The free energy F includes entropic contributions due to the coarse-graining procedure. The Landau expansion of F/T in the order parameter $\psi(r)$ has the same form as $\hat{\mathcal{H}} - \mu\hat{N}$, with $\hat{\psi}(\mathbf{r})$ and $\hat{\psi}^+(\mathbf{r})$ replaced by the c-number fields $\psi(\mathbf{r})$ and $\psi^*(\mathbf{r})$,

$$\frac{F}{T} \approx \int d^2r \left(\frac{1}{2} A|\nabla\psi|^2 + \frac{1}{2} a|\psi|^2 + b|\psi|^4 + \cdots \right), \tag{6.11}$$

where, as usual, a changes sign at the mean-field transition temperature T_c^0, $a \approx a'(T - T_c^0)$.

Because the long-wavelength fluctuations in a helium film are two-dimensional, the actual temperature T_c at which true long-range order develops is suppressed well below T_c^0. For $T \approx T_c$, $a \ll 0$ and the polynomial part of F has a deep minimum, even though thermal fluctuations may nevertheless suppress genuine off-diagonal long-range order at the largest length scales. To a first approximation, we can then neglect fluctuations in the *amplitude* of $\psi(r)$ and set

$$\psi(\mathbf{r}) \approx \psi_0 e^{i\theta(\mathbf{r})}, \tag{6.12}$$

where $\psi_0 \approx \sqrt{-a/(4b)}$ and $\theta(\mathbf{r})$ is a slowly varying phase variable. The free energy becomes

$$\frac{F}{T} = \text{constant} + \frac{1}{2} K_0 \int |\nabla\theta(\mathbf{r})|^2 \, d^2r, \tag{6.13}$$

where $K_0 = A\psi_0^2$. The physical interpretation of K_0 becomes clear if we recall that the superfluid velocity $\mathbf{v}_s$ is given by [29].

$$\mathbf{v}_s(\mathbf{r}) = \frac{\hbar}{m} \nabla\theta(\mathbf{r}), \tag{6.14}$$

and write the contribution to F of the superfluid kinetic energy in terms of the superfluid density ρ_s^0 as

$$F = \text{constant} + \frac{1}{2}\rho_s^0 \int d^2r |\mathbf{v}_s(\mathbf{r})|^2. \tag{6.15}$$

Evidently, K_0 is related to the superfluid density in this "phase-only" approximation by

$$K_0 = \frac{\hbar^2}{m^2 T} \rho_s^0. \tag{6.16}$$

The peculiar nature of superfluid order in two dimensions becomes evident if we now evaluate $G(r)$, by integrating freely over the phase field when evaluating averages weighted by $e^{-F/T}$,

$$G(r) = \langle \psi(r)\psi^*(0) \rangle$$

$$= \psi_0^2 \exp[-\tfrac{1}{2}\langle[\theta(r) - \theta(0)]^2\rangle], \tag{6.17}$$

where the last line is a general property of Gaussian fluctuations. Upon introducing Fourier variables, the remaining average is easily evaluated using the equipartition theorem,

$$\langle[\theta(\mathbf{r}) - \theta(0)]^2\rangle = \frac{2}{K_0} \int \frac{d^2q}{(2\pi)^2} \frac{1}{q^2}(1 - e^{i\mathbf{q}\cdot\mathbf{r}})$$

$$\underset{r\to\infty}{\approx} \frac{1}{\pi K_0} \ln\left(\frac{r}{a_0}\right), \tag{6.18}$$

where a_0 is a microscopic cutoff. We thus find that $G(r)$ decays *algebraically* to zero [30],

$$G(r) \sim 1/r^{\eta(T)}, \tag{6.19}$$

with

$$\eta(T) = 1/[2\pi K_0(T)]. \tag{6.20}$$

Hence, as discussed for a variety of systems in Chapter 2, there is only *quasi*-long-range off-diagonal long-range order at any finite temperature in two dimensions. Nevertheless, this algebraic order is enough to distinguish the low-temperature superfluid from a distinct high-temperature normal liquid film in which $G(r)$ decays exponentially,

$$G(r) \sim \exp[-r/\xi(T)], \tag{6.21}$$

which defines a superfluid coherence length $\xi(T)$. Note that the low-temperature phase still has a nonzero superfluidity density, similar to bulk superfluids. In contrast to bulk superfluids, however, the low-temperature phase is characterized by a *continuously varying* critical exponent $\eta(T)$.

The "phase-only" description used above neglects amplitude fluctuations and ignores the periodic nature of the phase variable. We can account for both effects by introducing a discrete set of vortex singularities in $\psi(\mathbf{r})$. As illustrated in Fig. 6.3(a), vortices represent zeroes of $\psi(\mathbf{r})$, where the phase is undetermined. Near a vortex core, the assumption

(a)

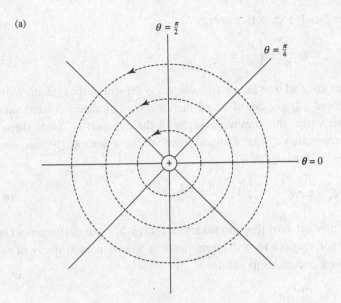

Fig. 6.3. (a) Lines of constant phase associated with a vortex in a two-dimensional superfluid. Dashed circles are the streamlines for the associated supercurrents. (b) The variation in magnitude of the order parameter near a vortex core located at the origin.

(b)

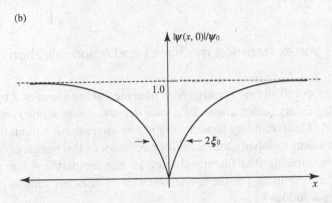

that the amplitude of the order parameter $|\psi(\mathbf{r})|$ is a constant breaks down. Figure 6.3(b) shows how $\psi(\mathbf{r})$ varies in the vicinity of a vortex [31]. $\psi(\mathbf{r})$ rises from zero to its bulk value over a characteristic length ξ_0 related to the coefficients of the first two terms of Eq. (6.11),

$$\xi_0 = \sqrt{A/|a|}. \tag{6.22}$$

To describe a radially symmetric net phase change of $\pm 2\pi$ near a vortex centered at the origin, we set

$$\theta(x, y) = \pm \tan^{-1}(y/x). \tag{6.23}$$

According to Eq. (6.14) we then have a superfluid velocity field

$$\mathbf{v}_s = \frac{\hbar}{m} \frac{\hat{z} \times \mathbf{r}}{r^2}, \tag{6.24}$$

which leads to a vortex energy

$$E_v = \pi \rho_s^0 \frac{\hbar^2}{m^2} \ln\left(\frac{R}{a}\right) + C \tag{6.25}$$

when inserted into Eq. (6.15), where R is the size of the system. We have imposed a lower cutoff a and included the contribution from smaller scales where the amplitude varies in the constant C. Since there are approximately $(R/a)^2$ distinct positions for a vortex, the *free* energy $E_v - TS_v$ for large R is

$$F_v \approx \left(\pi \rho_s^0 \frac{\hbar^2}{m^2} - 2T\right) \ln\left(\frac{R}{a}\right). \tag{6.26}$$

The prediction of this famous argument [13–15] is that it becomes favorable for vortices to proliferate (leading to exponential decay of $G(r)$) above a critical temperature

$$T_c \approx \frac{\pi}{2} \frac{\hbar^2}{m^2} \rho_s^0. \tag{6.27}$$

6.2.3 Vortex statistical mechanics and renormalization of ρ_s

Going beyond the simple argument presented above requires a more detailed theory, which allows for a finite density of interacting vortices near T_c. Understanding superfluidity in two-dimensional helium films is equivalent to solving the statistical mechanics of this vortex gas. We start by noticing that functional integrals like the partition function (6.10) are dominated in the low-temperature "phase-only" approximation by solutions of

$$\nabla^2 \theta = 0. \tag{6.28}$$

To allow for vortices (and thus for amplitude fluctuations), we require that Eq. (6.28) be satisfied "almost everywhere," i.e., everywhere except in the cores of a collection N of vortices located at positions $\{r_\alpha\}$ with integer charges $\{s_\alpha\}$. A vortex singularity has "charge" s_α if the line integral of the phase gradient on any path enclosing the core satisfies

$$\oint \nabla \theta \cdot d\ell = 2\pi s_\alpha. \tag{6.29}$$

The function (6.23) obeys this condition, for example, with $s_\alpha = \pm 1$. More generally we expect, for a contour C enclosing many vortices, that

$$\oint_C \nabla \theta \cdot d\ell = \int_\Omega d^2 r \, n_v(r), \tag{6.30}$$

where Ω is the area spanned by C and the "charge density" of vortices is

$$n_v(\mathbf{r}) = 2\pi \sum_{\alpha=1}^{N} s_\alpha \delta(\mathbf{r} - \mathbf{r}_\alpha). \tag{6.31}$$

Equation (6.30) is a statement about the noncommutivity of derivatives of $\theta(x, y)$. Indeed, by taking the contour C to be a small square loop, we readily find

$$\varepsilon_{ij} \partial_i \partial_j \theta(\mathbf{r}) = \partial_x \partial_y \theta - \partial_y \partial_x \theta$$
$$= n_v(\mathbf{r}), \tag{6.32}$$

where ε_{ij} is the antisymmetric unit tensor in two dimensions, $\varepsilon_{xy} = -\varepsilon_{yx} = 1$. To cast this equation in a more familiar form, we introduce the Cauchy conjugate to the phase field, by writing

$$\partial_i \theta(\mathbf{r}) = \varepsilon_{ij} \partial_j \tilde{\theta}(\mathbf{r}), \tag{6.33}$$

and find that $\tilde{\theta}(x, y)$ satisfies

$$\nabla^2 \tilde{\theta}(\mathbf{r}) = n_v(\mathbf{r}). \tag{6.34}$$

Finding the phase associated with a set of vortex singularities thus requires that we first determine the "electrostatic potential" $\tilde{\theta}(\mathbf{r})$ of a collection of point charges $\{s_j\}$ at positions $\{\mathbf{r}_j\}$ in two dimensions [13]. The solution of Eq. (6.34) is

$$\tilde{\theta}(\mathbf{r}) = 2\pi \sum_\alpha s_\alpha G(\mathbf{r}, \mathbf{r}_\alpha), \tag{6.35}$$

where the Green function satisfies

$$\nabla^2 G(\mathbf{r}, \mathbf{r}_\alpha) = \delta(\mathbf{r} - \mathbf{r}_\alpha), \tag{6.36}$$

For $|\mathbf{r} - \mathbf{r}_j|$ large and both points far from any boundaries, we have

$$G(\mathbf{r}, \mathbf{r}_j) \approx \frac{1}{2\pi} \ln\left(\frac{|\mathbf{r} - \mathbf{r}_j|}{a}\right) + C, \tag{6.37}$$

where C is a constant that contributes to the core energy of the vortex.

We now decompose the phase into a contribution $\theta_v(x, y)$ from vortices, which is obtained by taking the Cauchy conjugate of Eq. (6.35), and a smoothly varying part $\phi(x, y)$,

$$\theta(x, y) = \theta_v(x, y) + \phi(x, y). \tag{6.38}$$

The function $\phi(x, y)$ represents single-valued phase fluctuations superimposed on the extrema of the vortex. Thermodynamic averages are obtained by first integrating over this nonsingular phase field and then summing over all possible complexions of vortex charges and positions. To insert this decomposition into Eq. (6.13), we need

$$\nabla\theta(\mathbf{r}) = 2\pi(\hat{\mathbf{z}}\times\nabla)\int d^2r'\ n(\mathbf{r}')G(\mathbf{r},\mathbf{r}') + \nabla\phi(\mathbf{r}). \qquad (6.39)$$

The resulting free energy takes the form [13]

$$\frac{F}{T} = \text{constant} + \frac{1}{2}K_0\int d^2r|\nabla\phi|^2 + \frac{F_v}{T}, \qquad (6.40)$$

where the vortex part is

$$\frac{F_v}{T} = -\pi K_0\sum_{\alpha\neq\beta}s_\alpha s_\beta\ln\left(\frac{|\mathbf{r}_\alpha-\mathbf{r}'_\beta|}{a}\right) + \frac{E_c}{T}\sum_\alpha s_\alpha^2, \qquad (6.41)$$

and the core energy E_c is usually assumed to be proportional to TK_0. Implicit in the statistical mechanics associated with Eq. (6.41) is a constraint of overall "charge neutrality," $\sum_j s_j = 0$, which is required for a finite energy in the thermodynamic limit.

6.2.4 The renormalization group and the universal jump in the superfluid density

To illustrate the statistical mechanics of the vortex gas described above, consider the renormalized superfluid density $\rho_s^R(T)$ calculated to lowest order in the fugacity of the vortex

$$y = e^{-E_c/T}. \qquad (6.42)$$

The renormalized superfluid density is related to the correlations of the momentum density g(r). On a microscopic level, the momentum-density *operator* $\hat{g}_i(\mathbf{r})$ is given in terms of particle-creation and -destruction operators by

$$\hat{g}_i(\mathbf{r}) = \frac{\hbar}{2i}[\hat{\psi}^+(\mathbf{r})\,\partial_i\hat{\psi}(\mathbf{r}) - \hat{\psi}(\mathbf{r})\,\partial_i\hat{\psi}^+(\mathbf{r})]. \qquad (6.43)$$

When local off-diagonal long-range order is present in helium films, we replace $\hat{\psi}^+(\mathbf{r})$ and $\hat{\psi}(\mathbf{r})$ by coarse-grained two-dimensional classical c-number fields, as usual. In the "phase-only" approximation which led to Eq. (6.13), the coarse-grained momentum density is then

$$g(\mathbf{r}) = \rho_s^0\mathbf{v}_s(\mathbf{r}), \qquad (6.44)$$

where $\rho_s^0 = m|\psi_0|^2$ and $\mathbf{v}_s(\mathbf{r})$ includes possible contributions from vortices. In helium films, the contribution of the *normal* fluid to the momentum vanishes, due to the viscous coupling to the substrate.

The correlation matrix which determines the renormalized super-fluid density is

$$C_{ij}(\mathbf{q}; K, y) \equiv \langle g_i(\mathbf{q}) g^*_j(\mathbf{q}) \rangle, \tag{6.45}$$

where $g_i(\mathbf{q})$ is the Fourier transform of the ith component of $g(\mathbf{r})$. To extract the renormalized superfluid density, we first decompose C_{ij} into transverse and longitudinal parts:

$$C_{ij} = A(q)\frac{q_i q_j}{q^2} + B(q)\left(\delta_{ij} - \frac{q_i q_j}{q^2}\right). \tag{6.46}$$

In an isotropic classical liquid, one would have $A(q) = B(q)$ in the limit $q \to 0$. The momentum fluctuations, moreover, would decouple from the configurational degrees of freedom responsible for most phase transitions in the classical limit. The behavior of quantum fluids is different: The renormalized superfluid density $\rho^R_s(T)$, in particular, is given by the difference between A and B as q tends to zero (see the appendix) [32–34]

$$\rho^R_s(T) = \frac{1}{T}\lim_{q \to 0}[A(q) - B(q)]. \tag{6.47}$$

The momentum density $g(\mathbf{r}) = \rho_s(\hbar/m)\nabla\theta(\mathbf{r})$ is already decomposed into transverse and longitudinal parts in Eq. (6.39). As in Chapter 2, we find that C_{ij} takes the form (6.46), with

$$\frac{\hbar^2}{m^2 T}A(q) = K_0, \tag{6.48}$$

$$\frac{\hbar^2}{m^2 T}B(q) = \frac{4\pi^2 K_0^2}{q^2}\langle \hat{n}_v(\mathbf{q})\hat{n}_v(-\mathbf{q})\rangle, \tag{6.49}$$

where $\hat{n}_v(\mathbf{q})$ is the Fourier transform of the vortex density $n_v(\mathbf{r})$ and the average in (6.49) is to be carried out over the vortex part of the free energy.

When the vortices are dilute, it is straightforward to use these results to obtain a perturbation expansion in the fugacity $y_0 \equiv e^{-E_c/T}$ for the renormalized superfluid density. In terms of $K_R \equiv \hbar^2 \rho^R_s/(m^2 T)$, we have (see Chapter 2)

$$K_R^{-1} = K_0^{-1} + 4\pi^3 y_0^2 \int_a^\infty \frac{dr}{a}\left(\frac{r}{a}\right)^{3 - 2\pi K_0} + O(y_0^4), \tag{6.50}$$

where a is the diameter of the vortex core. At low temperatures, Eq. (6.50) provides a small correction (proportional to $y^2 = e^{-2E_c/T}$) to K_R^{-1}. When $K_0 \lesssim 2/\pi$, however, the integral becomes infrared divergent and perturbation theory breaks down. This is precisely the condition (6.27) required for a vortex-unbinding transition. This potentially divergent perturbation theory can be converted into renormalization-group recursion relations for effective couplings $K(\ell)$ and $y(\ell)$ describing renormalized vortices with effective core diameters $ae^{-\ell}$. See Chapter 2. These differential equations read

$$\frac{dK^{-1}(\ell)}{d\ell} = 4\pi^2 y^2(\ell) + O[y^4(\ell)], \tag{6.51}$$

$$\frac{dy(\ell)}{d\ell} = [2 - \pi K(\ell)]y(\ell) + O[y^3(\ell)]. \tag{6.52}$$

Another important result is the invariance of the superfluid density along a renormalization-group trajectory,

$$K_R(K, y) = K_R(K(\ell), y(\ell)), \tag{6.53}$$

where $K_R(K, y) = [\hbar^2/m^2 T]\rho_s^R(K, y)$. This result also follows from the Josephson scaling relation for ρ_s^R [35].

Equations (6.48) and (6.49) are the famous Kosterlitz recursion relations, which he originally derived using another method [16]. The Hamiltonian trajectories they generate in the (K^{-1}, y)-plane are shown in Fig. 6.4, together with a temperature-dependent locus of initial conditions,

$$y_0 = e^{-cK_0}, \tag{6.54}$$

where c is a constant.

Initial conditions to the left of the incoming separatrix renormalize into the line of fixed points at $y = 0$, which describes the low-temperature phase. At higher temperatures, $y(\ell)$ eventually becomes large, indicating that vortices unbind at long wavelengths. One then expects that order-parameter correlations decay exponentially, as in Eq. (6.21), where the correlation length ξ_+ is related to the density of free vortices n_f:

$$n_f(T) \approx \xi_+^{-2}(T). \tag{6.55}$$

A variety of detailed predictions for the Kosterlitz–Thouless transition follow from these recursion relations and the transformation properties of various correlation functions under the renormalization group [16]. We focus here on the prediction of a *universal* jump discontinuity in the superfluid density [35]. As discussed in Chapter 2, this is a direct consequence of the relation (6.53) and the renormalization-group flows shown in Fig. 6.4. Below T_c, $y(\ell)$ tends to zero for large ℓ and we have

$$K_R(K, y) = \lim_{\ell \to \infty} K_R(K(\ell), y(\ell))$$
$$= \lim_{\ell \to \infty} K(\ell) \quad (T \le T_c). \tag{6.56}$$

Since it is clear from Fig. 6.4 that this limit is just $2/\pi$, at T_c we have

$$\lim_{T \to T_c^-} K_R(K, y) = \lim_{T \to T_c^-} \frac{\hbar^2 \rho_s^R(T)}{m^2 T} = \frac{2}{\pi} \tag{6.57}$$

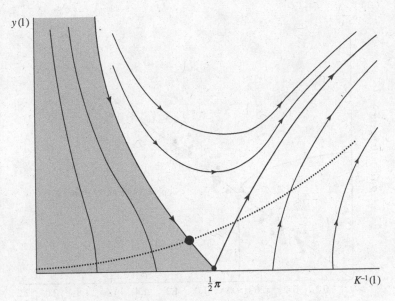

Fig. 6.4. Renormalization flows arising from the Kosterlitz recursion relations. The shaded domain of attraction of the fixed line at $y(\ell) = 0$ is a superfluid. A locus of initial conditions is shown as a dashed line. The superfluid phase is bounded by the incoming separatrix which terminates at $K^{-1} = 2/\pi$.

independent of the way in which the initial locus crosses the incoming separatrix. For helium films of varying thicknesses on varying substrates, one predicts a sequence of curves with jump discontinuities and T_c values all falling on a line with slope

$$\frac{\rho_s(T_c^-)}{T_c^-} = \frac{2m^2}{\pi\hbar^2 k_B T} \approx 3.491 \times 10^{-9} \text{ g cm}^{-2} \text{ K}^{-1}. \tag{6.58}$$

Figure 6.5 shows the locus of jump discontinuities in the superfluid density versus temperature for over 70 experiments on helium films [23]. All points lie on a line with slope close to the universal value (6.58). This remarkable universal jump is related to the universality of the critical exponent $\eta(T)$ at the Kosterlitz–Thouless transition. When vortices are taken into account, $G(r) \sim 1/r^{\eta(T)}$ for $T < T_c$, provided that Eq. (6.20) is replaced by [16]

$$\eta(T) = 1/[2\pi K_R(T)], \tag{6.59}$$

so that (neglecting logarithmic corrections) $\eta(T_c^-) = \frac{1}{4}$.

6.2.5 Two-dimensional superconductors

Superconducting films are similar to substrates coated with superfluid ^{4}He, with the understanding that the complex order parameter $\psi(\mathbf{r})$ now describes Cooper pairs of electrons. Because the bosonic degrees of freedom are now charged, the currents surrounding vortices are screened by a coupling to the vector potential. The interactions between vortices

Fig. 6.5. Jump discontinuities in the superfluid density versus temperature for over 70 different experiments for helium films. Both substrates and film thicknesses are varied. One representative $\rho_s(T)$ curve is shown. Adapted from [23].

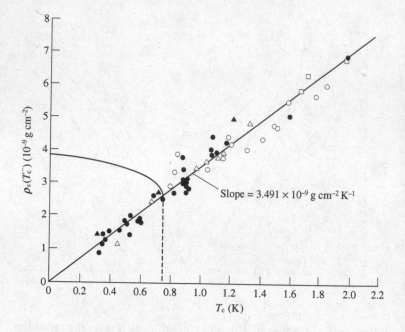

Slope $= 3.491 \times 10^{-9}$ g cm^{-2} K^{-1}

are no longer logarithmic at all distances, but instead die off rapidly at scales larger than the London penetration depth. We first review a "naïve" theory of screening in two-dimensional superconductors. This theory is "two-dimensional" in the sense that all quantities are independent of the z coordinate. It would be directly applicable to situations in which vortices are, in effect, infinitely long rods in a *bulk* material (see Chapters 7 and 8). The conclusion of the naïve theory in two dimensions is that, due to screening, vortices will always unbind at any nonzero temperature [13]. We then discuss real superconducting films, which behave differently, due to the spreading of the vortex magnetic field lines (see Fig. 6.6) as they emerge from the top and bottom of the film [36]. As emphasized by Beasley *et al.* [37], vortices in sufficiently thin superconducting films now interact logarithmically out to an *effective* London penetration depth of order millimeters and the Kosterlitz–Thouless theory for neutral superfluids becomes directly applicable over a wide range of length scales in most samples. The detailed predictions of the Kosterlitz–Thouless theory for superconducting films are discussed in [22].

We begin with the "naïve" theory of two-dimensional superconductors because it provides a simple illustration of the screening of interactions among defects by a vector potential. A closely related, but nonlinear, screening phenomenon arises for dislocations in membranes (see Section 6.3) and leads to the striking conclusion that crystalline membranes must *always* melt into hexatic liquids at any finite temperature.

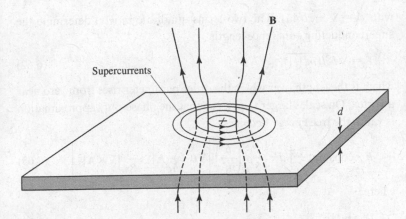

Fig. 6.6. A schematic diagram of currents and magnetic field lines for a vortex in a superconducting film. Currents flow only in a two-dimensional plane, so the screening is weak.

6.2.5.1 The naïve theory

We start with the Ginzburg–Landau theory of superconductivity [38], which generalizes Eq. (6.11) for neutral superfluids to charged Cooper pairs,

$$F = \int d^2r \left[\frac{1}{2m^*} \left| \left(\frac{\hbar}{i} \nabla - \frac{e^*}{c} A \right) \psi \right|^2 + \frac{1}{2} r |\psi|^2 + u |\psi|^4 + \frac{B^2}{8\pi} - \frac{1}{4\pi} \mathbf{H} \cdot \mathbf{B} \right].$$

$$(6.60)$$

Here, $\psi(\mathbf{r})$ is a coarse-grained complex field that represents the microscopic Cooper-pair-destruction operator ($\psi^*(\mathbf{r})$ is the coarse-grained pair-creation operator), $\mathbf{A}(\mathbf{r})$ is the vector potential, $e^* = 2e$ is the charge of the pair and $m^* = 2m$ is the effective mass of the pair. $\mathbf{H}$ is a constant external field and $\mathbf{B}(\mathbf{r}) = \nabla \times \mathbf{A}(\mathbf{r})$ is the fluctuating local magnetic field. Equation (6.60) is incorrect for real superconducting films because it implicitly assumes that *both* $\psi(\mathbf{r})$ and $\mathbf{A}(\mathbf{r})$ vanish outside the thin slab occupied by the superconductor. This is a reasonable assumption for $\psi(\mathbf{r})$, but incorrect for $\mathbf{B}(\mathbf{r})$ and $\mathbf{A}(\mathbf{r})$. This theory does, however, describe a bulk superconductor that is infinitely stiff in the $\hat{z}$ direction, so that all quantities are independent of z. Equation (6.60) should then be interpreted as a free energy per unit length. A related theory provides the correct description of local order in two-dimensional smectic liquid crystals [39].

Let us see what Eq. (6.60) predicts for superconductors, with the external magnetic field $\mathbf{H}$ set to zero. The modifications introduced by the fluctuating magnetic field $\mathbf{B}(\mathbf{r})$ outside the sample will be discussed in the next subsection. Following our treatment of helium films, we assume that thermal excitations have suppressed any possible ordering well below the mean-field transition temperature, so that $r \ll 0$. We again make the "phase-only" approximation and write

$$\psi(\mathbf{r}) = \psi_0 e^{i\theta(\mathbf{r})}, \qquad (6.61)$$

with $\psi_0 = \sqrt{-r/(4u)}$. The two terms quadratic in $\psi(\mathbf{r})$ determine the superconducting coherence length,

$$\xi_0 = \sqrt{\hbar^2/(m^*|r|)}, \tag{6.62}$$

which is the length over which the order parameter rises from zero near a vortex. On scales larger than ξ_0, so that the phase-only approximation is valid, the free energy becomes

$$F = \text{constant} + \int d^2r \left[\frac{1}{2} \rho_s^0 \left(\frac{\hbar}{m^*} \right)^2 \left| \nabla\theta - \frac{e^*}{\hbar c} \mathbf{A} \right|^2 + \frac{1}{8\pi} |\nabla \times \mathbf{A}|^2 \right], \tag{6.63}$$

where

$$\rho_s^0 = m^* |\psi_0|^2 \tag{6.64}$$

is the mass density of the Cooper pairs.

To see qualitatively how the vector potential affects the behavior of vortices, imagine that the vortex solution $\theta(x,y) = \pm\tan^{-1}(y/x)$ is inserted into Eq. (6.63). Since $|\nabla\theta| \sim 1/r$, this would lead to a logarithmically diverging energy, if the vector potential were neglected. The vector potential, however, will cancel out this divergence and reduce the energy, provided that it too falls off like $1/r$ with the correct coefficient far from the vortex. This screening costs gradient energy, however, due to the $|\nabla \times \mathbf{A}|^2$ term. The two terms of Eq. (6.63) quadratic in $\mathbf{A}$ become comparable when $\mathbf{A}(\mathbf{r})$ varies over a length scale λ_0, with

$$\lambda_0 = \sqrt{\frac{(m^*c)^2}{\rho_s^0 4\pi(e^*)^2}}. \tag{6.65}$$

This scale is the London penetration depth, which is of order 1000–10000 Å in most superconductors. Because this is the scale over which screening sets in, we expect the vortex energy to be finite and approximately equal to Eq. (6.25) with R replaced by λ_0, and a replaced by ξ_0,

$$E_v \approx \pi\rho_s^0 \left(\frac{\hbar}{m^*} \right)^2 \ln\left(\frac{c\lambda_0}{\xi_0} \right), \tag{6.66}$$

where c is an undetermined constant. Since the entropy of an isolated vortex is still proportional to $\ln R$ and the energy is now finite, we conclude that vortices will always be unbound in this model [13]. A more detailed analysis, similar to that for finite-energy defects such as vacancies and interstitials in a solid [40], predicts a nonzero density of free vortices

$$n_f \approx \xi_0^{-2} e^{-E_v/T} \tag{6.67}$$

at any temperature $T > 0$.

Because Eq. (6.63) is a simple quadratic form, we can go beyond the arguments sketched above to determine the interactions between screened vortices in more detail. As in helium films, functional integrals weighted by $\exp(-F/T)$ will be dominated by vortex extrema of the free energy. In the Coulomb gauge, $\nabla \cdot \mathbf{A} = 0$, the variational equation $\delta F/\delta\theta = 0$ again leads to $\nabla^2\theta = 0$ away from the vortex cores, so we can immediately write down the vortex contribution to $\nabla\theta$ from N vortices at positions $\{\mathbf{r}_j\}$ with "charges" $\{s_j\}$,

$$\nabla\theta(\mathbf{r}) = 2\pi \sum_\alpha s_\alpha(\hat{z} \times \nabla) G(\mathbf{r}, \mathbf{r}_\alpha), \tag{6.68}$$

where $G(\mathbf{r}, \mathbf{r}_j)$ is the Green function discussed in Section 6.2.3. Note from Eqs. (6.31) and (6.32) that the curl of $\nabla\theta$ does not vanish,

$$\hat{z} \cdot (\nabla \times \nabla\theta) = 2\pi \sum_\alpha s_\alpha \delta^{(2)}(\mathbf{r} - \mathbf{r}_\alpha). \tag{6.69}$$

The vector potential is constrained by vortex singularities, even though $\theta(\mathbf{r})$ and $\mathbf{A}(\mathbf{r})$ appear to decouple in the Coulomb gauge. To see this, consider a closed path C bounding an area Ω in the superconductor far in comparison with λ_0 from any vortex cores. On such a path $\nabla \times \mathbf{A} \approx 0$ and we must minimize the first term in (6.63), insuring that $\nabla\theta = [e^*/(\hbar c)]\mathbf{A}$ and that

$$\oint_c \left(\nabla\theta - \frac{e^*}{\hbar c}\mathbf{A} \right) \cdot d\ell = 0. \tag{6.70}$$

It then follows from Eq. (6.30) that the magnetic flux through Ω is given by the sum of the enclosed vortex "charges,"

$$\int\int_\Omega d^2r\, B_z = \phi_0 \sum_{\{r_\alpha \varepsilon \Omega\}} s_\alpha, \tag{6.71}$$

where $\phi_0 = 2\pi\hbar c/e^*$ is the flux quantum.

A second variational equation follows from $\delta F/\delta\mathbf{A} = 0$, i.e.,

$$\nabla \times [\nabla \times \mathbf{A}(\mathbf{r})] = \nabla \times \mathbf{B}(\mathbf{r}) = 4\pi\rho_s^0 \left(\frac{\hbar}{m^*}\right)^2 \left(\frac{e^*}{\hbar c}\right) \left(\nabla\theta - \frac{e^*}{\hbar c}\mathbf{A}(\mathbf{r})\right). \tag{6.72}$$

Upon casting this equation in the form of Ampère's law $\nabla \times \mathbf{B}(\mathbf{r}) = 4\pi\mathbf{J}(\mathbf{r})/c$, we see that the (gauge-invariant) supercurrent associated with θ and $\mathbf{A}$ is

$$\mathbf{J}(\mathbf{r}) = e^*|\psi_0|^2\frac{\hbar}{m^*}\left(\nabla\theta(\mathbf{r}) - \frac{e^*}{\hbar c}\mathbf{A}(\mathbf{r})\right). \tag{6.73}$$

Note the close analogy with the formula for the superfluid *momentum* density in helium films [41],

$$\mathbf{g}(\mathbf{r}) = m|\psi_0|^2\frac{\hbar}{m}\nabla\theta(\mathbf{r}). \tag{6.74}$$

A closed-form equation for the magnetic field B_z generated by a distribution of vortices follows from taking the curl of Eq. (6.72) and using Eq. (6.69):

$$B_z(\mathbf{r}) - \lambda_0^2 \nabla_\perp^2 B_z(\mathbf{r}) = \phi_0 \sum_{\alpha=1}^{N} s_\alpha \delta^{(2)}(\mathbf{r} - \mathbf{r}_\alpha), \qquad (6.75)$$

where "$\perp$" denotes coordinates in the plane of the film.

Consider one isolated vortex with charge s at the origin. The Fourier-transformed field $B_z(\mathbf{q}_\perp)$ which solves Eq. (6.75) is then

$$B_z(q_\perp) = \frac{s\phi_0}{1 + q_\perp^2 \lambda_0^2}, \qquad (6.76)$$

which leads in real space to

$$B_z(r) = \frac{s\phi_0}{2\pi\lambda_0^2} K_0\left(\frac{r}{\lambda_0}\right), \qquad (6.77)$$

where $K_0(x)$ is the Bessel function $K_0(x) \approx \ln(1/x)$, $x \ll 1$, and $K_0(x) \approx \sqrt{\pi/(2x)}e^{-x}$, $x \gg 1$. The associated supercurrent $\mathbf{J}(\mathbf{r}) = [c/(2\pi)]\nabla \times \mathbf{B}(\mathbf{r})$ is

$$\mathbf{J}(\mathbf{r}) = \frac{sc\phi_0}{8\pi^2\lambda_0^3} K_1\left(\frac{r}{\lambda_0}\right)\hat{\boldsymbol{\theta}}, \qquad (6.78)$$

where $\hat{\boldsymbol{\theta}}$ is a unit vector in the azimuthal direction. Because $K_1(x) \sim 1/x$, $x \ll 1$, $|\mathbf{J}(\mathbf{r})| \sim 1/r$ for $r \ll \lambda_0$, similar to the momentum density in helium films. However, screening sets in for $r \gg \lambda_0$, where $|\mathbf{J}(\mathbf{r})| \sim \exp(-r/\lambda_0)$. Similar manipulations lead straightforwardly to the generalization of the vortex free energy (6.41) for two-dimensional charged superfluids,

$$F_v = \varepsilon_0 \sum_{i \neq j} s_i s_j K_0\left(\frac{|\mathbf{r}_i - \mathbf{r}_j|}{\lambda_0}\right) + E_c \sum_j s_j^2, \qquad (6.79)$$

where $\varepsilon_0 = [\phi_0/(4\pi\lambda_0)]^2$ and E_c is a vortex core energy. Gaussian "spin-wave" fluctuations about these vortex extrema can also be included [39]. The finite range of the interaction potential confirms our earlier conclusion that vortices will unbind for entropic reasons at any finite temperature in this model.

6.2.5.2 Real superconducting films

As illustrated in Fig. 6.6, the magnetic field lines generated by a vortex in real superconducting films spread out into the empty space above and below the plane of the film. This spreading leads to different behavior for the screening currents in the film from that predicted by Eq. (6.78). The true behavior, which was first found by Pearl [36], provides an interesting illustration of how two-dimensional physics can be effected by the outside, three-dimensional environment. We sketch the calculation here, following the treatment of de Gennes [42].

We start by integrating Eq. (6.73) over a film thickness d along $\hat{z}$. Provided that the film is much thinner than the London penetration depth, we may assume that $\theta(\mathbf{r})$ and $\mathbf{A}_{2d}(\mathbf{r})$ remain constant across the thickness of the film. We denote the vector potential by $\mathbf{A}_{2d}(\mathbf{r})$ to emphasize that this quantity is the vector potential *in the two-dimensional film*, other than the full vector potential $\mathbf{A}_{3d}(\mathbf{r}, z)$ which describes the magnetic field in all of three-dimensional space. Note that $\mathbf{r} = (x, y)$ always refers to coordinates in the two-dimensional plane perpendicular to $\hat{z}$. If we imagine that all this current is concentrated in a delta-function sheet in the plane $z = 0$, the full three-dimensional current may be written

$$\mathbf{J}(\mathbf{r}, z) = \frac{c}{4\pi} \frac{\phi_0}{2\pi\lambda_{\text{eff}}} \left(\nabla \theta(\mathbf{r}) - \frac{2\pi}{\phi_0} \mathbf{A}_{2d}(\mathbf{r}) \right) \delta(z), \qquad (6.80)$$

where

$$\lambda_{\text{eff}} = \lambda^2/d. \qquad (6.81)$$

As we shall see, λ_{eff} will play the role of an effective London penetration depth in this problem.

The current (6.80) provides a source for a *three*-dimensional magnetic field $\mathbf{B}(\mathbf{r}, z)$ via Ampère's law, $\nabla \times \mathbf{B}(\mathbf{r}, z) = 4\pi\mathbf{J}(\mathbf{r}, z)/c$. Upon setting $B = \nabla \times \mathbf{A}_{3d}$ and using the Coulomb gauge, Ampère's law becomes

$$-\nabla^2 \mathbf{A}_{3d}(\mathbf{r}, z) + \frac{1}{\lambda_{\text{eff}}} \mathbf{A}_{2d}(\mathbf{r}) \delta(z) = \frac{s\phi_0}{2\pi\lambda_{\text{eff}}} \frac{\hat{z} \times \mathbf{r}}{r^2} \delta(z), \qquad (6.82)$$

where $\nabla^2 = \nabla_\perp^2 + \partial^2/\partial z^2$ and we have inserted the phase gradient for a single vortex of charge s at the origin. We now pass to Fourier-transformed vector potentials

$$\mathbf{A}_{3d}(\mathbf{q}_\perp, q_z) = \int d^2r \int dz \, e^{i\mathbf{q}_\perp \cdot \mathbf{r}} e^{iq_z z} \mathbf{A}_{3d}(\mathbf{r}, z) \qquad (6.83)$$

and

$$\mathbf{A}_{2d}(\mathbf{q}_\perp) = \int d^2r \, e^{i\mathbf{q}_\perp \cdot \mathbf{r}} \mathbf{A}_{2d}(\mathbf{r}) \qquad (6.84)$$

and find that (6.82) becomes

$$\mathbf{A}_{3d}(\mathbf{q}_\perp, q_z) + \frac{1}{\lambda_{\text{eff}}(q_z^2 + q_\perp^2)} \mathbf{A}_{2d}(\mathbf{q}_\perp) = \frac{is\phi_2}{\lambda_{\text{eff}}(q_z^2 + q_\perp^2)} \frac{\hat{z} \times \mathbf{q}_\perp}{q_\perp^2}. \qquad (6.85)$$

The two- and three-dimensional vector potentials must agree on the plane $z = 0$, i.e., $\mathbf{A}_{3d}(\mathbf{r}, z = 0) = \mathbf{A}_{2d}(\mathbf{r})$, so that

$$\mathbf{A}_{2d}(\mathbf{q}_\perp) = \int_{-\infty}^{\infty} \frac{dq_z}{2\pi} \mathbf{A}_{3d}(\mathbf{q}_\perp, q_z). \tag{6.86}$$

After integrating Eq. (6.85) over q_z and using this result, we can solve for $\mathbf{A}_{2d}(\mathbf{q}_\perp)$,

$$\mathbf{A}_{2d}(\mathbf{q}_\perp) = \frac{is\phi_0(\hat{\mathbf{z}} \times \mathbf{q}_\perp)}{q_\perp^2(1 + 2\lambda_{\text{eff}}q_\perp)}. \tag{6.87}$$

Upon inserting Eq. (6.87) into the Fourier transform of Eq. (6.80) after setting $\mathbf{J}(\mathbf{r}, z) \equiv d\mathbf{J}_{2d}(\mathbf{r})\delta(z)$, we have

$$\mathbf{J}_{2d}(\mathbf{q}_\perp) = \frac{c}{2\pi d q_\perp} \frac{is\phi_0(\hat{\mathbf{z}} \times \mathbf{q}_\perp)}{(1 + 2\lambda_{\text{eff}}q_\perp)} \tag{6.88}$$

and so the superconducting current in real space is

$$\mathbf{J}_{2d}(\mathbf{r}) = \frac{is\phi_0}{2\pi d} \int \frac{d^2q_\perp}{(2\pi)^2} \frac{e^{-i\mathbf{q}_\perp \cdot \mathbf{r}}}{(1 + 2\lambda_{\text{eff}}q_\perp)} \frac{\hat{\mathbf{z}} \times \mathbf{q}_\perp}{q_\perp}. \tag{6.89}$$

The behavior of this screening current depends on the effective London penetration depth. For $r \ll \lambda_{\text{eff}}$, we find

$$\mathbf{J}_{2d}(\mathbf{r}) = \frac{sc\phi_0}{8\pi^2 d\lambda_{\text{eff}} r}\hat{\boldsymbol{\theta}}, \tag{6.90}$$

which agrees with the $r \ll \lambda_0$ limit of Eq. (6.78) for our "naïve" model. This unscreened $1/r$ falloff now continues, however, out to a much larger distance $\lambda_{\text{eff}} = (\lambda_0/d)\lambda_0 \gg \lambda_0$. For $r \gg \lambda_{\text{eff}}$, the limiting behavior of (6.89) is

$$\mathbf{J}_{2d}(\mathbf{r}) = \frac{sc\phi_0}{4\pi^2 dr^2}\hat{\boldsymbol{\theta}}. \tag{6.91}$$

The exponential screening of vortex currents in the "naïve" model is thus replaced by a $1/r^2$ power-law falloff.

Similar results hold for the interactions between pairs of vortices [36, 42]. Vortices interact logarithmically, as in neutral superfluids, for $r \ll \lambda_{\text{eff}}$, but exhibit a weaker $1/r$ potential for $r \gg \lambda_{\text{eff}}$. Since λ_{eff} can be of order fractions of a centimeter for film thicknesses $d = 10$–100 Å, the Kosterlitz–Thouless theory becomes directly applicable on essentially all length scales for sufficiently thin films [37]. The relatively weak screening in comparison with that in bulk systems arises because currents are confined to a thin plane instead of forming rings along the entire z axis.

6.3 Defects in membranes and monolayers

Two-dimensional crystals have much in common with two-dimensional superfluids. As discussed in Chapter 2, there are many important experimental examples, including rare gases adsorbed onto periodic substrates like graphite, Langmuir–Blodgett films of amphiphillic molecules at air–water interfaces, freely suspended liquid-crystal films, layers of electrons trapped at the surface of liquid helium and assemblies of colloidal particles confined between two glass plates [43]. Crystals consisting of a few atomic or molecular layers display algebraic decay of a translational order parameter, similar to Eq. (6.19), and the elastic constants of the crystal play a role similar to that of the superfluid density. As emphasized by Kosterlitz and Thouless [13] and by Berezinski [14], dislocations in such crystals are point defects with a logarithmically diverging energy as a function of the system's size, so one might expect them to melt via a dislocation-unbinding mechanism.

The results of the detailed dislocation-unbinding theory were described in Chapter 2 [44, 45]. Melting via dislocations does not lead to an isotropic liquid, as proposed originally [13, 14], but produces instead a new hexatic phase of matter, with residual bond-orientational order [44]. The long-range bond-orientational order in two-dimensional crystals at low temperatures is converted into algebraically decaying correlations in a hexatic fluid by a gas of unbound dislocations. Each dislocation, moreover, contains an embryonic pair of orientational defects called disclinations in its core. These disclinations separate and interact logarithmically in the hexatic phase and themselves unbind via a second phase transition at sufficiently high temperatures. The latent heat associated with the usual first-order melting point can thus be spread out over an entire intermediate phase, separated from the low-temperature crystal and high-temperature liquid by two continuous phase transitions.

We call the experimental systems mentioned above "monolayers," to emphasize that they are constrained to be approximately flat. In all cases, there are nearby substrates, walls or interfaces that force the degrees of freedom to "layer," i.e., to lie in a plane. Excitations out of the plane are strongly disfavored by, for example, a surface tension or substrate potential. As discussed in Chapter 5, there is another important class of two-dimensional materials called "membranes" [46]. Membranes are two-dimensional associations of molecules different from the three-dimensional fluid medium in which they are embedded. Examples include lipid bilayers in water [46] and spectrin protein skeletons extracted from red blood cells [47]. Because membranes are not confined to an interface between two different phases, the surface tension vanishes and they exhibit wild fluctuations out of the plane while retaining a local two-dimensional topology.

In this section, we review the physics of defects such as dislocations, disclinations, grain boundaries, vacancies and interstitials in membranes and monolayers. We discuss crystalline monolayers first and then show in some detail how buckling of defects screens defect energies in membranes [48–50]. The buckling transition of disclinations in hexatic membranes [51] is also described. We point out a fundamental asymmetry between the populations of positive and negative disclinations in liquid membranes with free boundary conditions or with fluctuating topologies. For a discussion of the spin-glass-like statistical mechanics of *polymerized* membranes with *quenched* random distributions of defects such as impurity atoms and grain boundaries, see [52–54].

6.3.1 Landau theory and the elasticity of tethered membranes

Consider first the statistical mechanics of two-dimensional assemblies of atoms and molecules *without* defects. To exclude defects, we assume that we have a perfect triangular lattice of monomers embedded in three dimensions and *tethered* together via an unbreakable network of covalent bonds. Such "tethered surfaces" [55, 56] exhibit interesting fluctuations and phase transitions even in the absence of defects. Once in the low-temperature, broken-symmetry phase of the relevant Landau theory, defects can be introduced both for monolayers and for membranes.

Consider first the high-temperature, crumpled phase of a tethered surface, similar to the crumpled state of a linear polymer chain. We describe the membrane by a function $\mathbf{r}(x_1, x_2)$, where $\mathbf{r}$ is a three-dimensional vector specifying the positions of the monomers as a function of two internal coordinates x_1 and x_2 fixed to the monomers. It turns out that the crumpled membrane can undergo a spontaneously symmetry-breaking transition into a flat phase below a crumpling temperature T_c [48]. The order parameters for this transition are the surface tangents $\{\mathbf{t}_\alpha = d\mathbf{r}/dx_\alpha, \alpha = 1, 2\}$ and the Landau free energy describing the transition is [57],

$$F[\mathbf{r}(x_1, x_2)] = \int d^2x \left(\frac{1}{2} \kappa (\partial_\alpha^2 \mathbf{r})^2 \right.$$

$$\left. + \frac{1}{2} a (\partial_\alpha \mathbf{r})^2 + b(\partial_\alpha \mathbf{r} \cdot \partial_\beta \mathbf{r})^2 + c(\partial_\gamma \mathbf{r} \cdot \partial_\gamma \mathbf{r})^2 + \ldots \right) \quad (6.92)$$

Within this expansion in the tangents, the probability of a surface configuration $\mathbf{r}(x_1, x_2)$ is proportional to $\exp\{-F[\mathbf{r}(x_1, x_2)]/T\}$. Note that F

is invariant under translations and rotations both within the embedding space and in the internal coordinates, as it should be. Self-avoidance of monomers at very different values of $x = (x_1, x_2)$, but close together in the embedding space, is neglected, although it is not hard to incorporate this into the model [57]. We assume that a changes sign at the mean-field crumpling transition, $a \approx a'(T - T_c)$, just as in the more conventional Landau theories for superfluids discussed in Section 6.2.

We discuss here only the low-temperature flat phase. When $a < 0$, F is minimized when

$$\mathbf{r}_0 \equiv \langle \mathbf{r}(x_1, x_2) \rangle$$
$$= m[x_1 \mathbf{e}_1 + x_2 \mathbf{e}_2], \qquad (6.93)$$

where $\mathbf{e}_1$ and $\mathbf{e}_2$ are an arbitrary orthogonal pair of unit vectors in the three-dimensional embedding space and

$$m = \frac{1}{2} \sqrt{\frac{-a}{b + 2c}}. \qquad (6.94)$$

Note that $\langle \mathbf{t}_\alpha \rangle = \langle d\mathbf{r}/dx_\alpha \rangle = m\mathbf{e}_\alpha$, so that m is the "amplitude" of the tangent order parameters. The average of the surface tangents *vanishes* in the high-temperature phase, because the surface is highly crumpled.

Additional physical insight is provided if we rewrite the free energy as

$$F = \text{constant} + \int d^2 x \left[\frac{1}{2} \kappa (\partial_\alpha^2 \mathbf{r})^2 + \mu \left(\frac{\partial \mathbf{r}}{\partial x_\alpha} \cdot \frac{\partial \mathbf{r}}{\partial x_\beta} - m^2 \delta_{\alpha\beta} \right)^2 \right.$$
$$\left. + \lambda \left(\frac{\partial \mathbf{r}}{\partial x_\gamma} \cdot \frac{\partial \mathbf{r}}{\partial x_\gamma} - 2m^2 \right)^2 \right], \qquad (6.95)$$

where

$$\mu = 4bm^4 \qquad (6.96)$$

and

$$\lambda = 8cm^4. \qquad (6.97)$$

This free energy is the sum of a bending energy, controlled by κ, and stretching contributions governed by the elastic moduli μ and λ. The stretching terms provide an energetic penalty whenever the induced metric in the embedding space,

$$g_{\alpha\beta}(x_1, x_2) = \frac{\partial \mathbf{r}}{\partial x_\alpha} \cdot \frac{\partial \mathbf{r}}{\partial x_\beta}, \qquad (6.98)$$

deviates from a flat background value

$$g_{\alpha\beta}^0 = m^2 \delta_{\alpha\beta}. \qquad (6.99)$$

The low-energy Goldstone modes associated with the flat phase are phonons. To study these excitations, we proceed as in the "phase-only" approximation for superfluids [58] and set

$$\mathbf{r}(x^1, x^2) \approx m(x_1 + u_1)\mathbf{e}_1 + m(x_2 + u_2)\mathbf{e}_2 + f\mathbf{e}_3, \tag{6.100}$$

where $\mathbf{u}(x_1, x_2)$ is an in-plane phonon field and $f(x_1, x_2)$ is an out-of-plane displacement along $\mathbf{e}_3 = \mathbf{e}_1 \times \mathbf{e}_2$. With the neglect of an additive constant, the free energy (6.95) becomes

$$F = \int d^2x \left\{ \frac{\mu}{4} [\partial_\alpha u_\beta + \partial_\beta u_\alpha + (\partial_\alpha f)(\partial_\beta f)]^2 \right.$$
$$\left. + \frac{\lambda}{2} \left(\partial_\gamma u_\gamma + \frac{1}{2}(\partial_\gamma f)^2 \right)^2 \right\} + \frac{\kappa}{2} \int d^2x \, (\nabla^2 f)^2. \tag{6.101}$$

This expression is identical to the energy of a bent elastic plate, where κ is the bending rigidity, μ is a shear modulus and $\mu + \lambda$ is the bulk modulus [59]. One often finds the stretching-energy contributions written in terms of the nonlinear strain matrix $u_{\alpha\beta}(x_1, x_2)$,

$$u_{\alpha\beta} = \frac{1}{2}(\partial_\alpha u_\beta + \partial_\beta u_\alpha) + \frac{1}{2}(\partial_\alpha f)(\partial_\beta f). \tag{6.102}$$

We have neglected terms nonlinear in $\mathbf{u}$, which are less important than the nonlinear terms in f which we have kept. In Eq. (6.101), we have recovered from Landau theory the starting point of the analysis of tethered membranes in Chapter 5. Recall that the nonlinear terms in f cause the renormalized bending rigidity to *diverge* at long wavelengths, leading to the remarkable conclusion that the low-temperature flat phase represents a genuine broken continuous symmetry with true long-range order [46, 48]. Recall from Chapter 2 that broken continuous symmetries in two-dimensional systems at finite temperatures are usually impossible and are replaced instead by exponential decay of correlations or by algebraically decaying order such as that found for superfluid helium films in Eq. (6.19).

Note the similarity between Eq. (6.101) and the free energy Eq. (6.63) which arose in our "naïve" theory of superconducting films. The phase gradient $\nabla\theta$ is replaced by the linearized strain matrix $\frac{1}{2}(\partial_\alpha u_\beta + \partial_\beta u_\alpha)$. The superfluid density is replaced by the elastic constants μ and λ. The role of the "vector potential" is played by gradients of f and the contribution by the bending rigidity replaces the field-energy term $[1/(8\pi)]|\nabla \times \mathbf{A}|^2$ for superconductors. Because the "vector-potential" contribution to Eq. (6.102) is nonlinear, the two theories are certainly not identical. Nevertheless, we shall find striking similarities when we relax the constraint of perfect 6-fold coordination and allow for defects: If we first set $f = 0$, then Eq. (6.101) can be used to calculate the energies of defects in planar "monolayers" – see below. Defects in

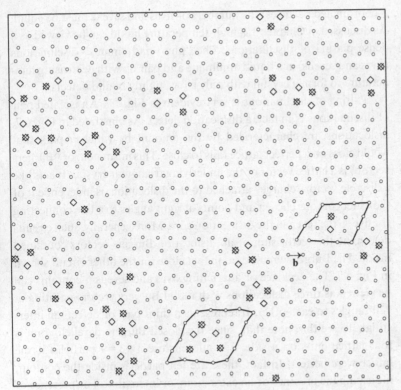

Fig. 6.7. Snapshot of a particle configuration in a computer simulation of a "monolayer." Six-fold coordinated particles are shown as circles. Five-fold and seven-fold disclination defects are shown as diamonds and "crossed diamonds," respectively. Heavy contours surround a dislocation and an interstitial. From a simulation of classical electrons interacting with a repulsive $1/r$ potential by Rudolph Morf.

monolayers are like the vortex singularities discussed earlier for neutral superfluids. When $f \neq 0$, however, the theory applies to the low-temperature phase of "membranes." As we shall see, these defects are then screened by nonzero gradients of f, similar to the screening of vortices by the vector potential in superconductors. The corresponding reduction in energy for membranes is accompanied by buckling of defects out of the plane defined by $f = 0$.

6.3.2 Defects in monolayers

Figure 6.7 shows a typical particle configuration for a thermally excited monolayer in its crystalline phase. Most particles, indicated by circles, have six nearest neighbors, as determined by the Dirichlet or Wigner–Seitz construction: The nearest-neighbor coordination number of a particle is the number of edges of the minimal polygon formed by the bisectors of the lines connecting it to its near neighbors. The 5- and 7-fold coordinated particles (indicated by diamonds and crossed diamonds, respectively) are orientational disclination defects in the otherwise 6-fold-coordinated triangular lattice. An elementary application of Euler's theorem shows that the average coordination number with periodic boundary conditions must be exactly six [60], so (neglecting 3-, 4-, and

8-coordinated particles) the numbers of fives and sevens must be equal. As discussed in Chapter 3, a high-temperature liquid can be viewed as a dense plasma of disclinations. The "plus" and "minus" disclinations (fives and sevens) in a liquid annihilate and pair up with decreasing temperature to form a crystal with only a few, tightly bound defects.

Pairing of disclinations is very evident in Fig. 6.7. An isolated pair of disclinations is, in fact, a dislocation defect. As illustrated in Fig. 6.7, dislocations are characterized by the amount by which a path that would close on a perfect lattice fails to close. This mismatch, or "Burgers vector," is a lattice vector of the underlying triangular crystal. It acts like a discrete vector "charge" attached to the dislocation, is independent of the exact contour chosen and points at right angles to a line connecting the five to the seven in the dislocation core. More generally, such circuits determine the vector sum of all dislocation Burgers vectors contained inside. The Burgers circuit does in fact close for the other circuit shown in Fig. 6.7. The necklace of alternating fives and sevens it contains can be viewed as three dislocations with radial Burgers vectors pointing at 120° angles to each other. It represents a third type of defect, an interstitial. To see this, note that, since the closed Burgers parallelogram is four lattice units on a side, it would enclose $3 \times 3 = 9$ particles if the lattice were perfect. In fact, there are ten particles inside the contour, indicating the presence of an extra atom, or interstitial. The presence of a vacancy, the "anti-defect" of the interstitial, could be detected by similar means.

To calculate the energies of the various monolayer defects discussed above, we use the continuum elastic free energy (6.101) with $f=0$,

$$F = \frac{1}{2} \int d^2r \, (2\mu u_{ij}^2 + \lambda u_{kk}^2), \tag{6.103}$$

where

$$u_{ij} = \tfrac{1}{2}(\partial_i u_j + \partial_j u_i). \tag{6.104}$$

In experimental monolayer systems, the constraint $f=0$ might be imposed by a strong substrate potential. For particles confined at an interface with a surface tension σ, we should add a term

$$\delta F = \frac{1}{2}\sigma \int d^2x \, (\nabla f)^2 \tag{6.105}$$

to Eq. (6.101). Upon integrating out the f field in perturbation theory, we recover a free energy of the form (6.103), with renormalized elastic constants μ and λ [61].

Dislocations, disclinations and other defects can be introduced into the theory in a way similar to the discussion of superfluid vortices in Section 6.2. First, however, we determine the equations satisfied by free-energy extrema away from defect cores. The variation of the monolayer free energy (6.103) with respect to **u** leads to

$$\partial_i \sigma_{ij} = 0, \tag{6.106}$$

where

$$\sigma_{ij} = 2\mu u_{ij} + \lambda u_{kk} \delta_{ij} \tag{6.107}$$

is the stress tensor. Equation (6.106) will be satisfied automatically if we introduce the Airy stress function χ via

$$\sigma_{ij}(\mathbf{r}) = \varepsilon_{im} \varepsilon_{jn} \partial_m \partial_n \chi(\mathbf{r}). \tag{6.108}$$

The individual components of σ_{ij} are thus

$$\sigma_{xx} = \frac{\partial^2 \chi}{\partial y^2}, \quad \sigma_{yy} = \frac{\partial^2 \chi}{\partial x^2}, \quad \sigma_{xy} = -\frac{\partial^2 \chi}{\partial x \, \partial y}. \tag{6.109}$$

The function $\chi(r)$ is similar to the vector potential one uses to insure that $\nabla \cdot \mathbf{B} = 0$ in Maxwell's equations. If we are able to find χ, we know σ_{ij} and hence can obtain the strain matrix by inverting Eq. (6.107),

$$u_{ij} = \frac{1}{2\mu} \sigma_{ij} - \frac{\lambda}{4\mu(\mu + \lambda)} \sigma_{kk} \delta_{ij}$$

$$= \frac{1}{2\mu} \varepsilon_{im} \varepsilon_{jn} \partial_m \partial_n \chi - \frac{\lambda \delta_{ij}}{4\mu(\mu + \lambda)} \nabla^2 \chi. \tag{6.110}$$

So far, we have not used the important fact that u_{ij} is determined by gradients of a displacement field via Eq. (6.104). The corresponding requirement on the superfluid velocity, given by Eq. (6.14), is

$$\nabla \times \mathbf{v}_s = 0, \tag{6.111}$$

which must hold away from vortex cores. The analogous compatibility condition on u_{ij} follows from applying the operator $\varepsilon_{ik} \varepsilon_{j\ell} \partial_k \partial_\ell$ to both sides of Eq. (6.110). This operator vanishes when it is acting on $u_{ij}(\mathbf{r})$, provided that various derivatives of the displacement field commute. Equation (6.110) then simplifies to give a biharmonic equation for $\chi(\mathbf{r})$,

$$\frac{1}{K} \nabla^4 \chi(\mathbf{r}) = \frac{1}{2} \varepsilon_{ij} \varepsilon_{j\ell} \partial_k \partial_\ell (\partial_i u_j + \partial_j u_i)$$

$$= 0, \tag{6.112}$$

with

$$K = \frac{4\mu(\mu + \lambda)}{2\mu + \lambda}. \tag{6.113}$$

Fig. 6.8. Five- (a) and 7-fold (b) disclinations in monolayers (flat) and membranes (buckled).

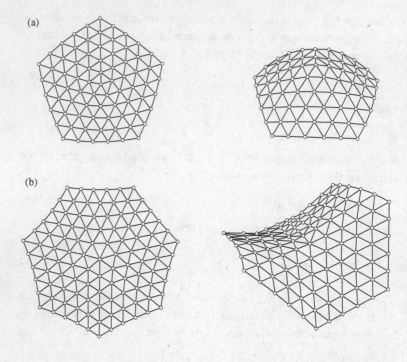

(a)

(b)

Equation (6.112) must be satisfied "almost everywhere," that is away from the cores of defects. Defects introduce source terms on the right-hand side, similar to the vortex density which appears in Eq. (6.34). Just like for superfluid vortices, defects represent points where derivatives fail to commute. Because the line integral of the differential bond-angle field $\theta(\mathbf{r})$ around a disclination in a triangular lattice must be an integral multiple of $2\pi/6 = 60°$,

$$\oint_c d\theta(\mathbf{r}) = s\frac{2\pi}{6}, \tag{6.114}$$

we have

$$\varepsilon_{ij}\,\partial_i\partial_j\theta(\mathbf{r}) = \frac{\pi}{3}\sum_\alpha s_\alpha \delta(\mathbf{r} - \mathbf{r}_\alpha) \tag{6.115}$$

for a collection of disclinations with "charges" $\{s_\alpha = \pm 1, \pm 2, \ldots\}$ at positions $\{\mathbf{r}_\alpha\}$. Isolated 5- and 7-fold disclinations are shown in Fig. 6.8. The local bond angle $\theta(\mathbf{r})$ in a crystalline solid is given by the antisymmetric part of the strain tensor,

$$\theta(\mathbf{r}) = \tfrac{1}{2}[\partial_x u_y(\mathbf{r}) - \partial_y u_x(\mathbf{r})]$$

$$= \tfrac{1}{2}\varepsilon_{k\ell}\,\partial_k u_\ell, \tag{6.116}$$

(a) 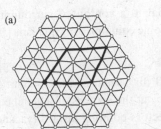   (b)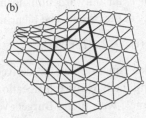

Fig. 6.9. Dislocation defects in (a) a monolayer (flat) and (b) a membrane (buckled). The Burgers construction is shown in both cases.

so the noncommutivity of derivatives for disclinations takes the form

$$= \frac{1}{2}\varepsilon_{ij}\varepsilon_{k\ell}\,\partial_i\partial_j\partial_k u_\ell = \frac{\pi}{3}\sum_\alpha s_\alpha\delta(\mathbf{r}-\mathbf{r}_\alpha). \tag{6.117}$$

We can treat dislocations by regarding them as tightly bound pairs of disclinations, or alternatively by starting with the definition of the Burgers vector in the continuum limit,

$$\oint_c du_i = \oint_c \frac{\partial u_i}{\partial x_j}\,dx_j$$
$$= b_i, \tag{6.118}$$

where (see Fig. 6.9) the circuit C encloses a single dislocation with Burgers vector $\mathbf{b}$. For a collection of Burgers "charges" $\{\mathbf{b}_\alpha\}$ at positions $\{\mathbf{r}_\alpha\}$, the differential statement of the noncommutivity embodied in constraints like (6.118) is

$$\varepsilon_{k\ell}\,\partial_k\partial_\ell u_j = \sum_\alpha b_{\alpha i}\delta(\mathbf{r}-\mathbf{r}_\alpha). \tag{6.119}$$

Upon noting that the right-hand side of Eq. (6.112) may be rewritten

$$\tfrac{1}{2}\varepsilon_{ik}\varepsilon_{j\ell}\,\partial_k\partial_\ell(\partial_i u_j + \partial_j u_i) = \varepsilon_{k\ell}\,\partial_k\partial_\ell\theta + \varepsilon_{ip}\varepsilon_{k\ell}\,\partial_p\partial_\ell\partial_i u_i, \tag{6.120}$$

we have, combining the above results for disclinations and dislocations,

$$\frac{1}{K_0}\nabla^4\chi(\mathbf{r}) = \sum_\alpha \varepsilon_{ij}b_{\alpha i}\,\partial_j\delta(\mathbf{r}-\mathbf{r}_\alpha) + \frac{\pi}{3}\sum_\beta s_\beta\delta(\mathbf{r}-\mathbf{r}_\beta). \tag{6.121}$$

Vacancies and interstitials are special cases of a general class of "impurity defects" that also includes substitutional atoms of the wrong size. If the defect sits in a lattice site of 3-fold or higher symmetry, its stress field is isotropic and it can be characterized by the area deficit Ω_0 (positive or negative) it induces in an otherwise perfect lattice,

$$\Omega_0 = \int (\nabla\cdot\mathbf{u})\,d^2r,$$
$$= \oint_c \hat{\mathbf{n}}\cdot\mathbf{u}(\mathbf{r})\,d\ell \tag{6.122}$$

where $\hat{n}$ is an outward unit normal to a contour C surrounding an impurity at the origin. The biharmonic equation for $\chi(\mathbf{r})$ reads [59]

$$\nabla^4 \chi(\mathbf{r}) = -2\mu\Omega_0 \nabla^2 \delta(\mathbf{r}).$$

(6.123)

A similar equation can be derived for vacancies and interstitials by regarding them as three dislocations at the vertices of a small equilateral triangle with radial Burgers vectors pointing inward (vacancy) or outward (interstitial).

It is straightforward to rewrite the free energy (6.103) in terms of $\chi(\mathbf{r})$,

$$F = \frac{1}{2K} \int d^2r \, [\nabla^2 \chi(\mathbf{r})]^2.$$

(6.124)

Once $\chi(\mathbf{r})$ is known, we can thus calculate both the strain fields (from (6.110)) and the energy of the defect. For an isolated disclination at the origin with angle defect s, we have [62]

$$\chi(\mathbf{r}) = \frac{Ks}{8\pi} r^2 \left[\ln\left(\frac{r}{a}\right) + \text{constant} \right]$$

(6.125)

and

$$F = \frac{Ks^2}{32\pi} R^2,$$

(6.126)

for a circular patch of crystal with radius R. Because disclination strains do not fall off at large distances, the free energy diverges *quadratically* with the size of the system. We have neglected a core-energy contribution that is negligible as $R \to \infty$. For an isolated *dislocation* at the origin, one has

$$\chi(\mathbf{r}) = \frac{K}{4\pi} b_i \varepsilon_{ij} r_j \ln\left(\frac{r}{a}\right)$$

(6.127)

and

$$F = \frac{Kb^2}{8\pi} \ln\left(\frac{r}{a}\right) + \text{constant}.$$

(6.128)

The logarithmic divergence arises because the strains $u_{ij}(r)$ fall off like $1/r$, similar to the superfluid velocity in helium films.

The energy of isolated disclinations is prohibitively large in crystalline monolayers. This energy is reduced to a logarithmic divergence in the hexatic monolayers, however, because of screening by a gas of unbound dislocations [44]. The logarithmic energy of dislocations in two-dimensional crystals leads via the usual energy/entropy argument to an estimate of the dislocation-unbinding temperature [13, 14]. For a more detailed discussion of the statistical mechanics of defect-mediated melting of monolayers, see Chapter 2.

For completeness, we note that the Airy stress function of an isolated impurity at the origin is

$$\chi(\mathbf{r}) = \frac{-\mu\Omega_0}{\pi} \ln\left(\frac{r}{a}\right), \tag{6.129}$$

corresponding to a displacement field

$$\mathbf{u}(\mathbf{r}) = \frac{\Omega_0 \mathbf{r}}{2\pi r^2}. \tag{6.130}$$

Because the strains now fall off as $1/r^2$, the elastic energy is finite and comparable to a core energy.

6.3.3 Defects in crystalline membranes

We now return to the full continuum elastic free energy Eq. (6.101) and determine how the defect energies are reduced in membranes, due to buckling into the third dimension. Variation of F with respect to f and $\mathbf{u}$ leads to the von Karman equations [59]

$$\kappa \nabla^4 f = \frac{\partial^2\chi}{\partial y^2}\frac{\partial^2 f}{\partial x^2} + \frac{\partial^2\chi}{\partial x^2}\frac{\partial^2 f}{\partial y^2} - 2\frac{\partial^2\chi}{\partial x\partial y}\frac{\partial^2 f}{\partial x\partial y} \tag{6.131}$$

$$\frac{1}{K_0}\nabla^4\chi = -\frac{\partial^2 f}{\partial x^2}\frac{\partial^2 f}{\partial y^2} + \left(\frac{\partial^2 f}{\partial x\partial y}\right)^2 + S(\mathbf{r}). \tag{6.132}$$

The Airy stress function is related to the stress as for monolayers, $\sigma_{ij}(\mathbf{r}) = \varepsilon_{im}\varepsilon_{jn}\,\partial_m\partial_n\chi(\mathbf{r})$, where we again have $\sigma_{ij} = 2\mu u_{ij} + \lambda u_{kk}\sigma_{ij}$. Now, however, u_{ij} is the nonlinear stress tensor,

$$u_{ij} = \frac{1}{2}(\partial_i u_j + \partial_j u_i) + \frac{1}{2}\frac{\partial f}{\partial x_i}\frac{\partial f}{\partial x_j} + \frac{1}{2}\frac{\partial\mathbf{u}}{\partial x_i}\cdot\frac{\partial\mathbf{u}}{\partial x_j}$$

$$\approx \frac{1}{2}(\partial_i u_j + \partial_j u_i) + \frac{1}{2}\frac{\partial f}{\partial x_i}\frac{\partial f}{\partial x_j}. \tag{6.133}$$

In the first line of (6.133), we have included nonlinearities in the in-plane displacements $\mathbf{u}$. Although the nonlinearity in f is *always* important for membranes, the term

$$\frac{1}{2}\frac{\partial\mathbf{u}}{\partial x_i}\cdot\frac{\partial\mathbf{u}}{\partial x_j}$$

can often be neglected. This term *does* contribute significantly to disclination energies, however: In monolayers, for example, it leads to a small correction to the coefficient of the R^2 divergence in the disclination energy for monolayers [62]. In contrast to disclinations, the strain fields for dislocations and impurities fall off fast enough to justify neglecting this term in evaluating the far-field elastic energy. We have included a source term on the right-hand side of Eq. (6.132) due to defects. For an isolated defect at the origin, one has

$$
S(\mathbf{r}) = \begin{cases} (\pi/3)s\delta(\mathbf{r}), & \text{disclination,} \\[2mm] \varepsilon_{ij}b_i\partial_j\delta(\mathbf{r}), & \text{dislocation,} \\[2mm] -\dfrac{2\mu+\lambda}{2(\mu+\lambda)}\Omega_0\,\nabla^2\delta(\mathbf{r}), & \text{impurity,} \end{cases} \tag{6.134}
$$

just as in flat space. The additional contribution to the right-hand side of Eq. (6.132) is the Gaussian curvature associated with the out-of-plane membrane displacements. This "curvature charge" can partially cancel out the "topological charges" $S(\mathbf{r})$ of the defects when they buckle. Note that any solution with $f \neq 0$ implies another solution with $f \to -f$. When $f = 0$, we recover the elastic equations for monolayers.

6.3.3.1 Disclinations and dislocations

Equations (6.131) and (6.132) are nonlinear and extremely difficult to solve analytically. Some progress is possible, however, in the inextensional limit $K \to \infty$. All defect energies would be infinite in this case for monolayers. For membranes, however, defects can buckle so as to eliminate the elastic contributions to Eq. (6.101). The only remaining contribution is the bending energy. The required f field is determined by equating the right-hand side of Eq. (6.132) to zero. The inextensional solutions of the von Karman equations representing a 5-fold disclination are

$$
\chi(r) = -\kappa \ln(r/a_0), \quad f(r) = \pm\sqrt{\tfrac{1}{3}}\,r, \tag{6.135}
$$

with energy

$$
E_5 = \tfrac{1}{3}\pi\kappa \ln(R/a_0), \tag{6.136}
$$

for a circular membrane with radius R. A 7-fold inextensional disclination can be represented in polar coordinates approximately by [48, 49]

$$
\chi(r) = 3\kappa \ln(r/a_0), \quad f(\mathbf{r}) = \pm\sqrt{\tfrac{2}{9}}\,r\sin(2\phi), \tag{6.137}
$$

which leads to an energy

$$
E_7 = \pi\kappa \ln(R/a). \tag{6.138}
$$

The R^2 divergence of disclination energies in monolayers (with an *infinite* coefficient when $K \to \infty$) is thus screened considerably by buckling.

Precise numerical solutions of the von Karman equations for arbitrary κ and K can be constructed by minimizing the energy of a triangulated tethered-surface model [49]. The bending rigidity is represented by an interaction between neighboring unit normals to the triangular plaquettes and the in-plane elasticity by nearest-neighbor harmonic springs adjusted to give the correct value of K. This "dynamic-triangulation" method is easier to implement than a direct numerical solution of the von Karman equations and includes automatically all relevant nonlinearities. The numerical results for disclinations in the inextensional limit resemble Eqs. (6.136) and (6.138), with slightly different coefficients [49]. It is still true that $E_5 < E_7$. Buckled disclinations obtained using this approach are shown in Fig. 6.8. Five-fold disclinations buckle into a cone, whereas the 7-fold disclination leads to a saddle surface. For disclinations and dislocations, the crucial physics lies in the existence of a buckling radius [48]: For systems with small radii R, these defects lie flat with no bending. Above a certain critical radius R_c, the defects trade elastic energy for bending energy and buckle out of the plane. Numerical studies [49] show that disclination energies are screened down to a logarithmic divergence with the size of the system beyond the buckling radius, as suggested by results in the inextensional limit. Both disclinations and dislocations behave as if they were inextensional beyond the buckling radius.

Buckling is even more important for dislocations. Beyond the buckling radius R_b, the dislocation energy no longer increases logarithmically with R; in fact, it appears to approach a finite constant [49]. A buckled dislocation is shown in Fig. 6.9. The entropic contribution to the free energy of the dislocation $F_d = E_d - S_d T$, however, still varies logarithmically with the system's size. The finiteness of the dislocation energy (or more precisely, *any* R dependence of the energy increasing more slowly than $\ln(R/a)$) then implies that untethered crystalline membranes must melt at *all* nonzero temperatures [48, 49]. The areal density of dislocations in a hexatic membrane at temperature T is approximately

$$n_d \approx a_0^{-2} e^{-E_d/T}, \qquad (6.139)$$

where E_d is is the (finite) energy of a dislocation and a_0 is the lattice constant. Dislocations in membranes thus behave similarly to vortices in the "naïve" model of superconducting films. Since buckled disclinations still have a logarithmically diverging energy, the hexatic fluid should be separated from a high-temperature isotropic liquid by a finite-temperature disclination-unbinding transition (see below). Melting of membranes confined between two flat plates, which suppresses the buckling of dislocations and leads to a finite melting temperature $T_m(d)$, tending to zero as $d \to \infty$, has been studied by Morse and Lubensky [63]. The implications of a finite dislocation energy for the melting of membranes are discussed in more detail in Chapter 1.

Defects like dislocations and disclinations buckle into the third dimension when the cost in stretching energy to remain flat exceeds the bending energy required to buckle. More generally, buckling occurs whenever [50]

$$K\ell^2/\kappa \gtrsim \gamma \tag{6.140}$$

where ℓ is a characteristic length scale for the defect and γ is a dimensionless constant, typically of order 10^2. In circular membranes of radius R, $\ell = \sqrt{Rb}$ for a dislocation with a Burgers vector of magnitude b and $\gamma \approx 127$. For positive disclinations, we have $\ell = R$ and $\gamma \approx 160$, whereas for negative disclinations $\ell \approx R$ and $\gamma \approx 192$ [49]. Thus isolated dislocations and disclinations *always* buckle in sufficiently large crystalline membranes, irrespective of the values of the elastic constants.

Strictly speaking, these conclusions apply only at $T=0$. In a finite-temperature defect-free flat phase, thermal fluctuations cause the long-wavelength wavevector-dependent bending rigidity to diverge, $\kappa_R(q) \sim q^{-\eta_\kappa}$, while the renormalized elastic parameter $K_R(q)$ tends to zero, $K_R(q) \sim q^{\eta_u}$ [48, 64, 65]. (See Chapters 1 and 5.) It is then appropriate to substitute $K/\kappa \to K_R/\kappa_R \sim 1/R^{\eta_\kappa + \eta_u}$ in Eq. (6.140). Using the values $\eta_\kappa = 0.82$ and $\eta_u = 0.36$ [65], we conclude that disclinations ($\ell \sim R$) will still buckle, whereas dislocations ($\ell \sim R^{1/2}$) become asymptotically flat as $R \to \infty$. Dislocations could still buckle on short scales, before these long-wavelength thermal renormalizations of the elastic parameters set in. The energy of dislocations will in any case tend to a finite constant for large R: Even if the dislocation remains asymptotically flat, we should now replace K by $K_R(R) \sim 1/R^{\eta_u}$ in Eq. (6.128). Thus the conclusion that the hexatic phase is the stable low-temperature phase of membranes is unchanged.

6.3.3.2 Other defects in crystalline membranes

Buckling is not inevitable even at $T=0$ for finite-energy defects such as vacancies, interstitials and impurity atoms. The criterion for buckling is again Eq. (6.140), where ℓ is now related to the excess area induced by the defects, $\ell \approx \sqrt{\Omega_0}$ [50]. Buckling can be triggered in an *infinite* system simply by varying the ratio K/κ. Because of the finite length scale associated with these defects, it is the bare "local" values of K and κ which should appear in (6.140). Since κ usually increases with increasing temperature, whereas K usually decreases, buckling is most likely at low temperatures. The $f \to -f$ symmetry of the von Karman equations insures that buckled defects have at least two degenerate minima, representing displacements on opposite sides of the membrane. The ensemble of two-level systems generated by buckled finite-energy defects will contribute to the specific heat and other equilibrium properties of membranes.

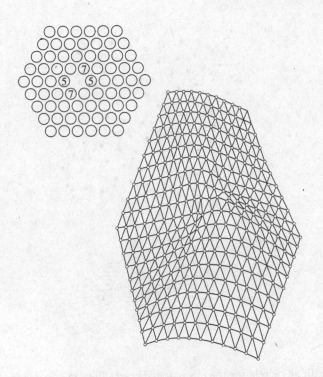

Fig. 6.10. Crushed vacancy defects in a flat monolayer and in a membrane. Note that the initial flat state consists of two fives and two sevens and can be regarded as a tightly bound pair of dislocations.

The buckling of a vacancy is illustrated in Fig. 6.10 [50]. Even in flat space, the vacancy does not have the high symmetry implied by the isotropic impurity source term in Eq. (6.134). For the harmonic-spring nearest-neighbor interaction potential used in [50], an initially 6-fold symmetric vacancy is "crushed" into the lower-symmetry object. There are three energetically equivalent ways in which the symmetry can be broken in this way. The crushed vacancy is equivalent to a tightly bound pair of dislocations indicated by the (5–7) pairs in Fig. 6.10. Such vacancies buckle whenever $Ka_0^2/\kappa \gtrsim 26$ [50]. Figure 6.10 shows a buckled vacancy for $Ka_0^2/\kappa = 92$. Note that the 5-fold disclinations have puckered outwards on opposite sides of the membrane.

It is also interesting to consider buckling of grain boundaries [50]. Three-dimensional crystals often consist of randomly oriented grains separated by defect walls. Similar polycrystalline order may appear in partially polymerized membrane vesicles [50, 54, 66]. In two dimensions, low-angle grain boundaries can be modelled by a row of dislocations, with average Burgers vector perpendicular to the boundary [67]. Consider first a grain boundary in a monolayer. Since all dislocations have the same sign, one might think that the elastic energy would be enormous. However, cancellations in the long-range part of the strain field insure that there is a finite elastic energy per unit length [67]. Let θ

(a)

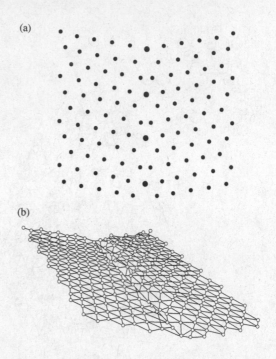

(b)

be the tilt angle relating the mismatched crystallites on either side of the grain. A $\theta = 21.8°$ grain boundary is shown in Fig. 6.11(a). The spacing h between dislocations (associated with the large dots in Fig. 6.11(a)) is given by Frank's law [67], $h = \frac{1}{2}b \sin(\theta/2)$. On length scales large relative to h, the composite stress field of all the dislocations dies off exponentially, provided that the Burgers vectors are strictly perpendicular to the boundary; on shorter scales, however, the behavior of the grain boundary is dominated by the individual dislocations.

Consider now a grain boundary inserted into a *membrane*. We have seen that isolated dislocations buckle whenever $R > R_b \approx 127\kappa/(Kb)$. One might expect a similar buckling in grain boundaries whenever $h > R_b$. Numerical studies [50] show that Eq. (6.140) is again satisfied, with $\ell = \sqrt{bh}$ and $\gamma \approx 120$, consistent with this guess. A buckled 21.8° grain boundary is shown in Fig. 6.11(b) for $Kbh/\kappa = 300$. Note that the membrane remains *flat* far from the boundary.

6.3.4 Defects in hexatic and liquid membranes

As discussed above, a finite concentration of unbound dislocations is present in membranes at any nonzero temperature. Provided that disclinations remain bound together, the resulting phase is a hexatic, similar to the hexatic liquid which arises in the theory of two-dimensional melting of monolayers [44]. Hexatic fluids are characterized by extended correlations in the complex order parameter $\psi_6(\mathbf{r}) = e^{6i\theta(\mathbf{r})}$, where $\theta(\mathbf{r})$ is the angle a bond between neighboring particles centered at $\mathbf{r}$ makes with respect to a local reference axis. Because this reference axis changes when parallel transported on a curved surface, there is an important coupling between the bond-angle field $\theta(\mathbf{r})$ and the geometry of the membrane. This coupling stiffens hexatic membranes relative to their isotropic liquid membrane counterparts, although they still fluctuate more wildly than does a crystalline tethered surface.

The continuum elastic free energy for hexatic membranes was derived in [48]:

$$F_H = \frac{1}{2}K_A \int d^2r \left(\partial_i\theta - \frac{1}{2}\varepsilon_{jk}\partial_k[(\partial_i f)(\partial_j f)] \right)^2 + \frac{1}{2}\kappa \int d^2r \, (\nabla^2 f)^2, \quad (6.141)$$

where K_A is the hexatic stiffness constant controlling fluctuations in $\theta(\mathbf{r})$. Note the similarity between F_H and the crystalline-membrane energy Eq. (6.101). Both free energies contain a bending rigidity, and the out-of-plane displacements act like a "vector potential" coupled to the low-energy Goldstone modes associated with the relevant broken symmetries. Hexatic membranes have many fascinating properties, including a "crinkled" phase, and an interesting literature has developed around them [68–70]. Here we discuss the von Karman equations for hexatics and show that they predict a buckling transition for a positive disclination defect at a critical value of the ratio κ/K_A [51, 69, 71]. We then argue that the energy for negative disclinations must be different and point out that this asymmetry has important consequences for liquid membranes once disclinations proliferate.

The variation of F_H with respect to f and θ leads to

$$\kappa\nabla^4 f = K_A(\partial_i\partial_j f)\varepsilon_{jk}\,\partial_k(\partial_i\theta - A_i) \tag{6.142}$$

and

$$\partial_i(\partial_i\theta - A_i) = 0, \tag{6.143}$$

with

$$A_i = \frac{1}{2}\varepsilon_{jk}\,\partial_k[(\partial_i f)(\partial_j f)]. \tag{6.144}$$

In analogy with Eq. (6.108), we introduce a hexatic stress function χ_H (the Cauchy conjugate function for the hexatic "current") via

$$K_A(\partial_i\theta - A_i) \equiv \varepsilon_{ij}\,\partial_j\chi_H, \tag{6.145}$$

so that Eq. (6.143) is satisfied automatically. Now we apply the operator $\varepsilon_{ik}\partial_k$ to Eq. (6.145) and use Eq. (6.115) to rewrite the derivatives of θ in terms of the density of disclinations $S(\mathbf{r})$. The resulting equation, when it is combined with Eq. (6.139), results in the "von Karman equations for hexatics" [51]

$$\kappa\nabla^4 f = \frac{\partial^2 \chi_H}{\partial y^2}\frac{\partial^2 f}{\partial x^2} + \frac{\partial^2 \chi_H}{\partial x^2}\frac{\partial^2 f}{\partial y^2} - 2\frac{\partial^2 \chi_H}{\partial x \partial y}\frac{\partial^2 f}{\partial x \partial y},$$ (6.146)

$$-\frac{1}{K_A}\nabla^2\chi_H = -\frac{\partial^2 f}{\partial x^2}\frac{\partial^2 f}{\partial y^2} + \left(\frac{\partial^2 f}{\partial x \partial y}\right)^2 + S(\mathbf{r}).$$ (6.147)

Note that these equations are identical to Eqs. (6.131) and (6.132) except that $K \rightarrow K_A$ and $\nabla^4 \rightarrow -\nabla^2$ in Eq. (6.147).

Suppose that $S(\mathbf{r}) = (+\pi/3)\delta(\mathbf{r})$, representing a single positive disclination at the origin. We then look for functions $\chi_H(\mathbf{r})$ and $f(\mathbf{r})$ of the form

$$\chi_H(\mathbf{r}) = -\kappa \ln(r/a_0), \quad f(\mathbf{r}) = \pm\alpha r$$ (6.148)

where α allows for conical dislocation buckling and remains to be determined. It is easy to check that Eq. (6.146) is obeyed for arbitrary α. All three terms in Eq. (6.147) are now proportional to delta functions and equating the coefficients determines α,

$$\alpha^2 = \frac{1}{3} - \frac{2\kappa}{K_A}.$$ (6.149)

For $\kappa/K_A > \frac{1}{6}$, there are no real solutions and the membrane remains flat. For $\kappa/K_A < \frac{1}{6}$, however, the disclination buckles [51]. Using Eq. (6.141), we readily find that the energy of this hexatic disclination in a membrane of radius R is

$$E_5 = \begin{cases} \frac{1}{3}\pi\kappa(1 - 3\kappa/K_A)\ln(R/a_0), & \kappa/K_A < \frac{1}{6}, \\ (\pi K_A/36)\ln(R/a_0), & \kappa/K_A > \frac{1}{6}. \end{cases}$$ (6.150)

When $\kappa/K_A > \frac{1}{6}$, we recover the elastic energy for disclinations in flat hexatic monolayers [44]. When $K_A \rightarrow \infty$, we recover the buckled-disclination energy (6.136) for crystalline membranes in the inextensional limit.

The total free energy of the 5-fold disclination including the positional entropy is

$$\begin{aligned} F_5 &= E_5 - ST \\ &= E_5 - 2T\ln(R/a_0). \end{aligned}$$ (6.151)

This free energy becomes negative above the solid line in Fig. 6.12. A curve of this kind was first presented by Guitter and Kardar [69], who used a slightly more accurate theory to discuss the consequences of buckling of 5-fold disclinations. Their criteria for buckling and for proliferation of disclinations are similar but not identical to ours. Small

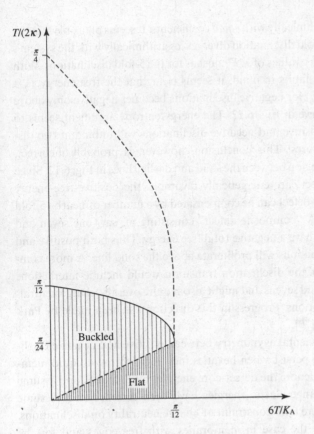

Fig. 6.12. The locus of entropic instabilities for a 5-fold disclination in a hexatic membrane (solid line). The shaded hexatic region is divided into regions where 5-fold disclinations buckle or remain flat, depending on the ratio K_A/κ. The dashed line is an estimate of where an isolated 7-fold disclination becomes unstable. The region above and to the right of the solid line is an isotropic fluid.

errors are to be expected in our approach because the Gaussian curvature $G(\mathbf{r})$ in the Monge representation is actually

$$G(\mathbf{r}) = \frac{(\partial_x^2 f)(\partial_y^2 f) - (\partial_x \partial_y f)^2}{(1 + |\nabla f|^2)^{1/2}}, \tag{6.152}$$

in contrast to the small-$|\nabla f|$ approximation which appears in Eq. (6.147) [72]. A factor of $(1 + |\nabla f|^2)^{1/2}$ should also be included in the measure [72]. Nevertheless, the essential predictions of the two approaches are identical: The hexatic becomes unstable with respect to an entropically driven proliferation of 5-fold disclinations whenever κ/T or K_A/T becomes sufficiently small.

What about *negative* disclinations? When $K_A \to \infty$ the similar approximations lead to a result identical to the inextensional crystalline 7-fold defect described by Eq. (6.138). Although the energies *both* of 5- and of 7-fold defects diverge logarithmically, the coefficient for negative disclinations is *larger*. This asymmetry persists (but is reduced slightly) in more accurate numerical computations [49]. When $\kappa \to \infty$, we approach the monolayer limit and the energies of 5- and 7-fold disclinations will diverge logarithmically with *equal* coefficients. It seems plausible that the energy of a 7-fold disclination diverges logarithmically with the system's size for *arbitrary* values of κ/K_A, just as for the 5-fold disclinations. With the two above limits in mind, it seems clear that the free energy $F_7 = E_7 - 2T \ln(R/a_0)$ for negative disclinations becomes negative only above the dashed curve in Fig. 6.12. The energy/entropy argument seems to predict that positive and negative disclinations will unbind at two distinct temperatures! This conclusion, however, is probably incorrect. Consider the region between the solid and dashed lines in Fig. 6.12. Since the 5-fold defects are energetically favorable, the positive-free-energy cost of a 7-fold defect can be compensated by a number of nearby 5-fold disclinations: A "composite defect" consisting of, say, one seven and three fives will have a negative total free energy. Thus both positive and negative disclinations will proliferate above the solid line. A more complete theory of the disclination transition would include interactions between fives and sevens and might also require overall "charge neutrality" of disclinations. Progress in this direction has been made by Park and Lubensky [70].

The fundamental asymmetry between positive and negative disclinations should persist when hexatics melt into isotropic liquid membranes. Let us denote the defect-core energies by E_5 and E_7 and assume that all long-range elastic energies have been screened out. Assume further that there is *no* constraint of charge neutrality on disclinations. This would be the case in membranes with free edges and for the experimentally relevant case of liquid bilayer surfaces that can change their genus freely. The areal densities of disclinations will then be different,

$$n_5 \approx a_0^{-2} e^{-E_5/T}, \quad n_7 \approx a_0^{-2} e^{-E_7/T}. \tag{6.153}$$

Which defect predominates in liquid membranes depends on microscopic details such as interaction potentials, etc. Close to a transition to a hexatic phase, however, the arguments given above suggest that 5-fold disclinations dominate.

Asymmetry of disclinations in liquid membranes parallels the behavior of vacancies and interstitials in conventional crystalline solids [40]. Although vacancies and interstitials are the anti-defects of each other, they nevertheless have very different energies. Unless vacancies and interstitials are created from a perfect crystal with periodic boundary conditions, they will in general occur with different concentrations. Periodic boundary conditions create an artificial constraint that forces vacancies and interstitials to be created in pairs rather than diffusing in from the surface. The concentrations must also be equal if the defects are charged, as in ionic crystals [40]. The flatness of monolayers similarly constrains the densities of *disclinations* to be equal at all stages in the theory of two-dimensional melting [44]. In membranes artificially constrained to have the topology of a torroidal surface, the numbers of 5- and 7-fold disclinations are forced to be equal to those predicted by Euler's theorem. A calculation similar to that for point defects in ionic crystals then leads to

$$n_5 = n_7 \approx a_0^{-2} e^{-(E_5 + E_7)/(2T)}. \tag{6.154}$$

The free energy of liquid membranes is often expressed in the Helfrich form [73],

$$F_L = \frac{1}{2} \kappa \int d^2 r (\nabla^2 f)^2 + \kappa_G \int d^2 r [(\partial_x^2 f)(\partial_y^2 f) - (\partial_x \partial_y f)^2]. \tag{6.155}$$

The first term is the usual bending rigidity. The remaining one is proportional to the integrated Gaussian curvature and its coefficient κ_G is often called the Gaussian rigidity. This second term is a perfect derivative, which integrates to a constant for surfaces of fixed genus [46]. The microscopic origins of κ_G are obscure. Its sign, however, is clearly determined by the disclination asymmetry discussed above: An excess of 5-fold disclinations corresponds to $\kappa_G < 0$ and favors *positive* net Gaussian curvature. Membrane phases with many spherical vesicles will predominate in this case. We can then estimate a preferred vesicle size R via $4\pi R^2 (n_5 - n_7) = 12$, where 12 is the number of excess disclinations required by Euler's theorem for a spherical topology. An excess of 7-fold disclinations means $\kappa_G > 0$ and a bias toward negative Gaussian curvatures. Complex "plumber's nightmare" lipid-membrane phases [46] are then favored.

On a more formal level, it must be the case from Euler's theorem that the Gaussian curvature integrated over a membrane with free edges of area Ω gives the disclination asymmetry

$$\int d^2 r \sqrt{g} G(\mathbf{r}) = \Omega(n_5 - n_7), \tag{6.156}$$

where $\sqrt{g} = 1 + |\nabla f|^2$ and $G(\mathbf{r})$ is given by Eq. (6.152). Equation (6.155) can thus be rewritten as

$$F_L = \frac{1}{2} \kappa \int d^2 r (\nabla^2 f)^2 + \kappa_G (N_5 - N_7), \tag{6.157}$$

where N_5 and N_7 are the total numbers of 5- and 7-fold disclinations in the membrane. The Gaussian rigidity κ_G thus acts as a *chemical potential* that must be adjusted to give the correct asymmetry between the populations of fives and sevens. In this sense, its effect is similar to that of a nonzero magnetic field in Eq. (6.60), which would lead to a net excess of positive or negative vortices in superconducting films. For a related perspective on the physics of metallic glasses, see Chapter 4 and [74].

It would be interesting to search for the disclination asymmetry discussed here in computer simulations of membranes. The most straightforward approach would be to study initially flat (i.e., large κ) liquid membranes with free edges, so that disclinations can enter and exit freely at the boundary. When the bending rigidity is reduced, a bias in the average Gaussian curvature should emerge as the membranes curl up into the third dimension. One could then vary the interparticle potentials to study what factors influence the sign of $E_7 - E_5$. Such an understanding might lead to controlled synthesis of membranes with a predetermined sign of κ_G.

The issue of asymmetric disclination-unbinding temperatures in hexatic membranes was considered more carefully in [75]. The calculations conform the qualitative picture shown in Fig. 6.12, but additional renormalization-group arguments suggest that positive and negative defects unbind simultaneously when thermal fluctuations are taken into account. See also Ref. [76].

Acknowledgements

I am grateful for the advice of E. Guitter, M. Kardar, T. C. Lubensky and J. D. Reppy while preparing this chapter. This chapter is adapted with permission from *Fluctuating Geometries in Statistical Mechanics and Field Theory*, edited by F. David, P. Ginsparg and J. Zinn-Justin (North-Holland, Amsterdam, 1966), pp. 423–477.

Appendix A. Superfluid density and momentum correlations

Consider a simplified version of the oscillating substrate experiment discussed in Section 6.2.1. Imagine that the substrate is wrapped around to form a cylinder, coated uniformly with ^{4}He. We neglect substrate inhomogeneities, as is appropriate near the transition or if the films are sufficiently thick [77]. Now imagine that the substrate oscillations are around the cylinder axis and slow enough that the motion of the substrate is essentially a uniform translation with velocity **u**. As illustrated in Fig. 6.13(a), we choose a coordinate system such that the substrate moves along the x axis. The cylinder has height L_y along the y axis and there are barriers at $y = \pm L_y/2$ that prevent the film from escaping in this direction. Its circumference is L_x. In two *or* three dimensions, the superfluid density measures the response of liquids to moving walls [32–34]. In Fig. 6.13(a), the "walls" are provided by the substrate itself plus the barriers at the top and bottom of the cylinder. Instead of solving a complicated statistical mechanics problem with moving boundaries, it is easier to make a Galilean transformation

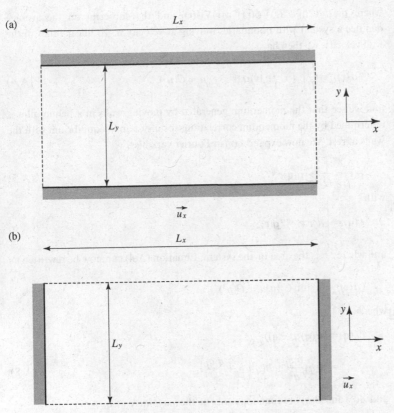

(a)

L_x

L_y

$\vec{u}_x$

(b)

L_x

L_y

$\vec{u}_x$

Fig. 6.13. Two experiments that lead to the superfluid density. In (a), there are periodic boundary conditions in the x direction and impenetrable walls at $y = \pm L_y/2$. These boundary conditions are reversed in (b). In both cases, the walls and substrate move at velocity u_x in the x direction.

to a coordinate system that moves at the velocity of the substrate, so that the boundaries are fixed. In the original laboratory frame of reference, the superfluid fraction of the film will remain at rest. Hence, it appears to be moving with average velocity $-\mathbf{u}$ after the Galilean transformation.

If $\hat{\mathcal{H}}$ is the Hamiltonian in the laboratory frame, averages after the Galilean transformation must be computed with respect to the Hamiltonian [78]

$$\hat{\mathcal{H}}' = \hat{\mathcal{H}} - \mathbf{u} \cdot \mathbf{P} + \tfrac{1}{2} M u^2, \tag{A.1}$$

where $\mathbf{P}$ is the total momentum of the film and M is the total mass. Upon coarse graining the film, the free energy which appears in Eq. (6.10) is correspondingly replaced by

$$F' = F - \mathbf{u} \cdot \int g(\mathbf{r}) \, d^2r + O(u^2), \tag{A.2}$$

where $g(\mathbf{r})$ is given by Eq. (6.43). The relative weights of various configurations of $\psi(\mathbf{r})$ in the moving frame with stationary boundary conditions, given by $\exp(-F'/T)$, must be the same as those in the laboratory frame with moving boundary conditions. The average value of the momentum in the laboratory frame is thus given by

$$\langle g_i(\mathbf{r}) \rangle_u = \frac{\int \mathcal{D}\psi(\mathbf{r}) \, g_i(\mathbf{r}) e^{-F'/T}}{\int \mathcal{D}\psi(\mathbf{r}) e^{-F'/T}}, \tag{A.3}$$

where $g = [\hbar/(2i)][\psi^*(\mathbf{r})\nabla\psi(\mathbf{r}) - \psi(\mathbf{r})\nabla\psi(\mathbf{r})]$ and the subscript on the average denotes a system with boundaries moving at velocity $\mathbf{u}$. To linear order in the wall velocity, we then have

$$\langle g_i(\mathbf{r})\rangle_u = \frac{1}{T}\int d^2r' \, \langle g_i(\mathbf{r})g_j(\mathbf{r}')\rangle_{u=0} u_j + O(u^2) \tag{A.4}$$

and we see that the momentum generated by moving walls in a helium film is determined by the momentum correlations for a system in equilibrium with the walls at rest. We now expand $g(r)$ in Fourier variables,

$$g(\mathbf{r}) = \frac{1}{\Omega_0}\sum_q g(\mathbf{q})e^{i\mathbf{q}\cdot\mathbf{r}}, \tag{A.5}$$

with

$$g(\mathbf{q}) = \int d^2r \, e^{-i\mathbf{q}\cdot\mathbf{r}}g(\mathbf{r}), \tag{A.6}$$

and where Ω_0 is the area of the system. Equation (A.4) can now be rewritten as

$$\langle g_i(\mathbf{r})\rangle_u = \frac{1}{T}\lim_{q\to0}C_{ij}(\mathbf{q})u_j + O(u^2), \tag{A.7}$$

where

$$C_{ij}(\mathbf{q}) = \langle g_i(\mathbf{q})g_j(-\mathbf{q})\rangle_{u=0}$$

$$\equiv A(q)\frac{q_iq_j}{q^2} + B(q)\left(\delta_{ij} - \frac{q_iq_j}{q^2}\right), \tag{A.8}$$

and $A(q)$ and $B(q)$ are functions only of the *magnitude* of $\mathbf{q}$.

The coefficient of u_j Eq. (A.7) involves a delicate limiting procedure whenever $A(q) \neq B(q)$ [32–34]. Consider first the situation outlined above, with the periodic boundary conditions in the x direction appropriate to a smooth cylindrical substrate. For any finite cylinder circumference L_x, there are always Fourier components of the $g(\mathbf{r})$ at $q_x = 0$, projected out by the x' integration in Eq. (A.4). Along the y direction, however, the $q_y = 0$ mode appears only in the limit $L_y \to \infty$, because $g(\mathbf{r})$ must vanish at the top and bottom of the cylinder. Thus, the correct order of limits for large sample sizes in this experiment is the limit $q_x \to 0$, followed by the limit $q_y \to 0$. The x component of the momentum of the film induced by the moving substrate is evidently

$$\langle g_x(\mathbf{r})\rangle_u = \frac{1}{T}\lim_{q_y\to0}\lim_{q_x\to0}C_{xx}(\mathbf{q})u_x$$

$$= \frac{B(0)}{T}u_x. \tag{A.9}$$

Since only the normal component of the film moves with the substrate, we identify $\rho_n(T)$ with the coefficient of u_x,

$$\rho_n(T) = B(0)/T. \tag{A.10}$$

Now consider a related, but fundamentally different, experiment: We repeat the cylinder periodically along the y axis, but erect impenetrable barriers along the x axis at fixed positions $x = \pm L_x/2$. See Fig. 6.13(b). The new barrier (equivalent to gouging a slit in the substrate parallel to the axis of the cylinder) prevents the film from circulating completely around the cylinder. The correct limiting procedure is now the limit $q_y \to 0$, reflecting the periodic boundary conditions along y, followed by the limit $q_x \to 0$, describing the new barriers in the limit $L_x \to \infty$,

$$
\begin{aligned}
\langle g_x(\mathbf{r}) \rangle &= \frac{1}{T} \lim_{q_x \to 0} \lim_{q_y \to 0} C_{xx}(q) u_x \\
&= \frac{A(0)}{T} u_x.
\end{aligned}
\tag{A.11}
$$

In this experiment, *all* of the film must clearly move at the velocity of the substrate since it is pushed around by the barrier. Now we identify ρ_{tot}, the total density of the film, with the coefficient of u_x, so that

$$
\rho_{tot} = A(0)/T.
\tag{A.12}
$$

In a normal liquid, these two different limiting procedures would lead to identical results. In a superfluid, however, there is a nonzero superfluid density, defined by

$$
\rho_s(T) = \rho_{tot} - \rho_n(T).
\tag{A.13}
$$

The formula for the superfluid density is thus

$$
\rho_s(T) = \frac{1}{T} \lim_{q \to 0} [A(q) - B(q)].
\tag{A.14}
$$

References

[1] T. Chou and D. R. Nelson, *Phys. Rev.* E **48**, 4611 (1993); T. Chou and D. R. Nelson, *Phys. Rev.* E **53**, 2560 (1996).

[2] E. Guitter and M. Kardar, *Europhys. Lett.* **13**, 441 (1990).

[3] M. W. Deem and D. R. Nelson, *Phys. Rev.* E **53**, 2551 (1996).

[4] J. M. Park and T. C. Lubensky, *Phys. Rev.* E **53**, 2648 (1996); J. M. Park and T. C. Lubensky, *Phys. Rev.* E **53**, 2665 (1996).

[5] S. Seung and D. R. Nelson, *Phys. Rev.* A **38**, 1005 (1988).

[6] A. E. Lobkovsky, S. Gentges, H. Li, D. Morse and T. A. Witten, *Science* **270**, 1482 (1995).

[7] A. E. Lobkovsky and T. A. Witten, *Phys. Rev.* E **55**, 1577 (1997).

[8] E. M. Kramer and T. A. Witten, *Phys. Rev. Lett.* **78**, 1303 (1997).

[9] S. Sachdev and D. R. Nelson, *J. Phys.* C **17**, 5473 (1984).

[10] M. J. W. Dodgson, *J. Phys.* A **29**, 2499 (1996); M. J. W. Dodgson and M. A. Moore, *Phys. Rev.* B **55**, 3816 (1997).

[11] A. Perez-Garrido, M. J. W. Dodgeson and M. A. Moore, *Phys. Rev.* B **56**, 3640 (1997); A. Perz-Garrido and M. A. Moore, *Phys. Rev.* B **60**, 15628 (1999).

[12] M. Bowick, D. R. Nelson and A. Travesset, *Phys. Rev.* B **62**, 8738 (2000).

[13] J. M. Kosterlitz and D. J. Thouless, *J. Phys.* C **5**, L124 (1972); *J. Phys.* C **6** 1181 (1973).

[14] V. L. Berezinski, *Zh. Éksp. Teor. Fiz.* **59**, 907 (1970) [*Sov. Phys. JETP* **32**, 493 (1971)]; *Zh. Éksp. Teor. Fiz.* **61**, 1144 (1972) [*Sov. Phys. JETP* **34**, 610 (1972)].

[15] P. G. de Gennes in *Proceedings of the Faraday Symposium on Liquid Crystals* (Oxford University Press, London, 1971).

[16] J. M. Kosterlitz, *J. Phys.* C **7**, 1046 (1974).

[17] P. W. Anderson and G. Yuval, *J. Phys.* C **4**, 607 (1971).

[18] J. M. Kosterlitz and D. J. Thouless, *Prog. Low Temp. Phys.* **78**, 371 (1978).

[19] B. I. Halperin, in: *Physics of Low-Dimensional Systems*, edited by Y. Nagaoka and S. Hikami (Publication Office, Progress in Theoretical Physics, Kyoto, 1979).

[20] D. R. Nelson, in *Phase Transitions and Critical Phenomena*, edited by C. Domb and J. L. Lebowitz (Academic, New York, 1983), Vol. 7.

[21] V. A. Ambegaokar, B. I. Halperin, D. R. Nelson and E. D. Siggia, *Phys. Rev.* B **21**, 1806 (1980).

[22] B. I. Halperin and D. R. Nelson, *J. Low. Temp. Phys.* **36**, 599 (1979).

[23] D. J. Bishop and J. Reppy, *Phys. Rev. Lett.* **40**, 1727 (1978); J. D. Reppy, in *Phase Transitions in Surface Films*, edited by J. G. Dash and J. Ruvalds (Plenum, New York, 1980).

[24] I. Rudnick, *Phys. Rev. Lett.* **40**, 1454 (1978).

[25] See, e.g., A. N. Berker and D. R. Nelson, *Phys. Rev.* B **19**, 2488 (1979).

[26] F. M. Gasparini, G. Agnolet and J. D. Reppy, *Phys. Rev.* B **29**, 138 (1984).

[27] G. D. Mahan, *Many Particle Physics* (Plenum, New York, 1981), Chapter 10.

[28] O. Penrose and L. Onsager, *Phys. Rev.* **104**, 576 (1956).

[29] I. M. Khalatnikov, *An Introduction to the Theory of Superfluidity* (Benjamin, New York, 1965).

[30] F. Wegner, *Z. Phys.* **206**, 465 (1967); J. W. Kane and L.P. Kadanoff, *Phys. Rev.* B **155**, 80 (1967).

[31] M. P. Kawatra and R. K. Pathria, *Phys. Rev.* **151**, 132 (1966).

[32] P. C. Hohenberg and P. C. Martin, *Ann. Phys.* **34**, 291 (1965).

[33] G. Baym, in *Mathematical Models in Solid State and Superfluid Theory*, edited by R. C. Clark and G. H. Berrick (Plenum, New York, 1968).

[34] D. Forster, *Hydrodynamic Fluctuations, Broken Symmetry, and Correlation Functions* (Benjamin, Reading, MA, 1975).

[35] D. R. Nelson and J. M. Kosterlitz, *Phys. Rev. Lett.* **39**, 1201 (1977).

[36] J. Pearl, *Appl. Phys. Lett.* **5**, 65 (1964).

[37] M. R. Beasley, J. E. Mooij and T. P. Orlando, *Phys. Rev. Lett.* **42**, 1165 (1979).

[38] M. Tinkham, *Introduction to Superconductivity* (McGraw-Hill, New York, 1975).

[39] J. Toner and D. R. Nelson, *Phys. Rev.* B **23**, 316 (1981).

[40] N. W. Ashcroft and N. D. Mermin, *Solid State Physics* (Holt Rinehart and Winston, Philadelphia, 1976), Chapter 30.

[41] For a detailed comparison of superfluidity in helium and in superconductors, see W. F. Vinen, in *Superconductivity*, edited by R. D. Parks (Marcel Dekker, New York, 1969), Vol. 2.

[42] P. G. de Gennes, *Superconductivity of Metals and Alloys* (Addison-Wesley, New York, 1989), Chapter 3.

[43] See, e.g., the articles in *Bond Orientational Order in Condensed Matter Systems*, edited by K. Strandburg (Springer, New York, 1992).

[44] B. I. Halperin and D. R. Nelson, *Phys. Rev. Lett.* **41**, 121 (1978); *Phys. Rev.* E **41**, 519 (1978); D. R. Nelson and B. I. Halperin, *Phys. Rev.* B **19**, 2457 (1979).

[45] A. P. Young, *Phys. Rev.* B **19**, 1855 (1979).

[46] See also the articles in D. R. Nelson, T. Piran and S. Weinberg, *Statistical Mechanics of Membranes and Surfaces* (World Scientific, Singapore, 1989).

[47] C. Schmidt, H. E. Warriner, P. Davidson, N. L. Slack, M. Schellhorn, P. Eiselt, S. H. J. Idziakc, H.-W. Schmidt and C. R. Safinya, *Science* **259**, 952 (1993).

[48] D. R. Nelson and L. Peliti, *J. Physique* **48**, 1085 (1987).

[49] S. Seung and D. R. Nelson, *Phys. Rev.* A **38**, 1005 (1988).

[50] C. Carraro and D. R. Nelson, *Phys. Rev.* E **48**, 3082 (1993).

[51] S. Seung, *Ph. D. Thesis*, Harvard University, 1990, unpublished.

[52] L. Radzihovsky and D. R. Nelson, *Phys. Rev.* A **44**, 3525 (1991).

[53] D. C. Morse and T. C. Lubensky, *Phys. Rev.* A **46**, 1751 (1992); D. C. Morse, T. C. Lubensky and G. S. Grest, *Phys. Rev.* A **45**, R2151 (1992).

[54] D. R. Nelson and L. Radzihovsky, *Phys. Rev.* A **46**, 7474 (1992).

[55] Y. Kantor, M. Kardar and D. R. Nelson, *Phys. Rev.* A **35**, 3056 (1987).

[56] Y. Kantor and D. R. Nelson, *Phys. Rev.* A **36**, 4020 (1987).

[57] M. Paczuski, M. Kardar and D. R. Nelson, *Phys. Rev. Lett.* **60**, 2638 (1988).

[58] This similarity becomes clearer if one rewrites Eq. (6.12) as $\psi(\mathbf{r}) \approx \psi_0[1 + i\theta(\mathbf{r})]$ for small $\theta(r)$.

[59] L. D. Landau and E. M. Lifshitz, *Theory of Elasticity* (Pergamon, New York, 1970).

[60] See, e.g., L. K. Runnels in *Phase Transitions and Critical Phenomena*, Vol. 2, edited by C. Domb and M. S. Green (Academic, New York, 1972); J. P. McTague, D. Frenkel and M. P. Allen, in *Ordering in Two Dimensions*, edited by S. Sinha (North-Holland, New York, 1980).

[61] The calculations are similar to those for a *quenched f* field in S. Sachdev and D. R. Nelson, *J. Phys.* C **17**, 5473 (1984).

[62] See, e.g., [49].

[63] D. C. Morse and T. C. Lubensky, *J. Physique* II, **3**, 531 (1993).

[64] J. A. Aronovitz and T. C. Lubensky, *Phys. Rev. Lett.* **60**, 2634 (1988).

[65] P. Le Doussal and L. Radzihovsky, *Phys. Rev. Lett.* **69**, 1209 (1992).

[66] M. Mutz, D. Bensimon and M. J. Brienne, *Phys. Rev. Lett.* **67**, 923 (1991).

[67] J. P. Hirth and J. Lothe, *Theory of Dislocations* (Wiley, New York, 1992).

[68] F. David, E. Guitter and L. Peliti, *J. Phys.* **49**, 2059 (1987); see also M. Bowick and A. Travesset, *Phys. Rep.* **344**, 255 (2001).

[69] E. Guitter and M. Kardar, *Europhys. Lett.* **13**, 441 (1990).

[70] J.-M. Park and T. C. Lubensky, *Phys. Rev.* E **53**, 2648 (1996).

[71] E. Guitter, *Ph. D. Thesis*, Saclay (1990); see also Ref. [69].

[72] The *exact* energy for large R of an inextensional 5-fold crystalline disclination is $(11/30)\pi\kappa \ln(R/a_0)$ instead of Eq. (6.136) for similar reasons. See Ref. [51].

[73] W. Helfrich, *Z. Naturforsch.* C **28**, 693 (1973).

[74] D. R. Nelson and F. Spaepen, in *Solid State Physics*, Vol. 42, edited by H. Ehrenreich and D. Turnbull (Harcourt Brance Jovanovich, New York, 1989).

[75] M. Deem and D. R. Nelson, *Phys. Rev.* E **53**, 2551 (1996).

[76] J.-M. Park and T. C. Lubensky, *J. Physique* I **6**, 493 (1996).

[77] Substrate disorder introduces quenched random fluctuations into the coefficient of $|\psi|^2$ in Eq. (6.11). Such randomness is known to be irrelevant if the specific-heat exponent α is negative. The essential singularity in the specific heat at T_c may be interpreted as $\alpha = -\infty$.

[78] E. M. Lifshitz and L. P. Pitaevskii, *Statistical Physics, Part 2* (Pergamon, New York, 1980) Section 23.

Chapter 7
Vortex-line fluctuations in superconductors from elementary quantum mechanics

Preface

The decade of the 1990s was an extraordinary one for superconductivity, much research being stimulated by the discovery of new type-II super-conductors with critical temperatures of order 100 K in the late 1980s. With the discovery of superconductivity below 40 K in MgB_2, a readily available "off-the-shelf" compound, the flood of research continues unabated. Type II superconductors are an especially important class of materials. Although their critical temperatures can be very large, these substances also allow magnetic fields to penetrate in the form of quantized vortex filaments above a (small) lower critical field H_{c1}. Understanding thermally excited configurations of these vortices and how to pin them effectively to prevent dissipation of supercurrents is crucial for many of the proposed applications of superconductors.

Although a consensus on the basic microscopic mechanism for these high critical temperatures remains elusive, considerable progress has been made on understanding the physics of vortices subjected to high temperatures. Fortunately, in the spirit of the "hydrodynamic" renormalization-group ideas of Chapter 1, the results outlined in the next two chapters for vortex physics are independent of whether the action of phonons, magnons, holons, excitons, anyons or some other mechanism is responsible for the underlying superconductivity. All that is required is an underlying Ginzburg–Landau theory with a BCS-like order parameter.

For a popular introduction to the new "vortex physics" which has emerged over the past decade, see [1]. For an authoritative and remarkably detailed account of many of the theoretical developments, see the review by Blatter et al. [2]. All that is needed to understand this chapter, however, is the discussion of point vortices in Chapter 5, possibly augmented by reading the relevant sections on vortex *lines* in the books on superconductivity by de Gennes [3] and Tinkham [4]. The de Gennes book, first published in 1966, is a classic and still very valuable today. Tinkham's book, another classic, was revised in 1996 in light of the new discoveries engendered by the high-T_c materials. Nevertheless, the basic phenomenological description of type-II materials from the 1975 edition has survived intact.

The approach taken here relies on a formal analogy of vortex statistical mechanics with the Feynman path-integral formulation of quantum mechanics in imaginary time. Although much can be learned from application of this approach to isolated vortices, we also provide an elementary illustration of why two or more long flexible vortices, regarded as directed, interacting elastic strings, behave like quantum *bosons* in thick samples, as opposed to fermions or some other statistical entities. For an alternative "back-of-the-envelope" estimate of the flux-lattice melting temperature using the harmonic-oscillator model of Section 7.3, see [5]. The resulting first-order melting curve in the temperature–magnetic-field plane has a *negative* slope, indicative of a crystal phase coexisting with a denser liquid phase. The unusual jump discontinuity in magnetization implied by this "ice-like melting" has been observed experimentally [6]. For an application of related ideas to entanglement and melting of vortex arrays subjected both to point disorder and to thermal fluctuations, see [7].

7.1 Introduction

As discussed above, the 1990s were a time of great ferment in the study of high-temperature superconductors. There is now considerable understanding of the remarkable behavior of these materials in a magnetic field. The theoretical analysis of vortex fluctuations in these extreme type-II materials requires only an underlying Ginzburg–Landau theory with a BCS-like order parameter. We emphasize again that the basic conclusions are *independent* of the precise microscopic mechanism, relying instead on the remarkably different Ginzburg–Landau parameters (coherence length, temperature range and anisotropy) which distinguish the cuprates from their low-T_c counterparts.

Figure 7.1 shows a schematic temperature–magnetic-field phase diagram for cuprate superconductors subjected only to weak point disorder in the form of oxygen vacancies. Throughout this chapter we assume for simplicity that we have a field oriented along the c axis, perpendicular to the copper-oxide planes. The magnetic field B (proportional to the vortex density) is plotted because demagnetizing corrections in the usual slab-like experimental geometry insure that B rather than the magnetic induction H is held fixed in most experimental situations. The famous Abrikosov flux lattice, which exists for all temperatures below the upper critical field $B_{c2}(T)$ in mean-field theory [8], appears here only below a much lower "melting line" $T_m(B)$. Above this line, melted vortex arrays entangle in a novel "flux liquid." The line $B_{c2}^0(T)$ marks the onset of enhanced diamagnetism but is not expected

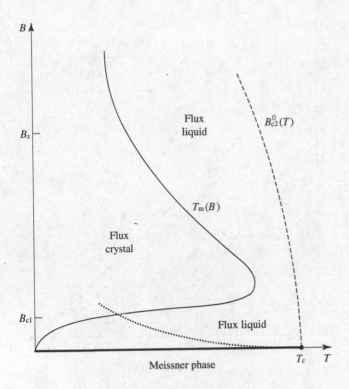

Fig. 7.1. A schematic phase diagram of a high-temperature superconductor in the limit of weak disorder. Although quantized vortex lines appear below $B_{c2}^0 (T)$, the Abrikosov crystal phase appears only below $T_m(B)$. The Meissner phase collapses to the heavy line along the temperature axis in this representation. Point disorder affects the flux liquid only below the dotted line.

to be a sharp phase boundary. The crossover fields B_{c1} and B_x are discussed in Section 7.3. There is considerable evidence [9–12] that the Abrikosov lattice in clean (twin-free) single crystals of yttrium barium copper oxide (YBCO) melts at $T_m(B)$ via a first-order phase transition [13] into a flux liquid in which the quantized vortex filaments presumably wander and entangle in a complicated fashion. Weak point disorder significantly alters the properties of the flux liquid only below the dotted line [14]. Note that a sliver of flux liquid may exist above this line and below the melting curve down to quite low temperatures. It is still not known whether point disorder alone is sufficient to produce a distinct thermodynamic "vortex-glass" phase [15] below the dotted line or within the crystalline region. Another possibility is the "Bragg-glass" phase (with algebraic decay of translational order) mentioned in the preface to Chapter 8.

Figure 7.2 shows another striking development – a remarkable shift in the "irreversibility line" of a highly anisotropic thallium-based compound engineered via the deliberate introduction of columnar pins [16]. Similar pinning centers have been injected by many groups via irradiation with heavy ions of sufficient energy to produce long tracks of damaged material [17–20]. The "irreversibility line" $T_{ir}(B)$ is the boundary below which the dynamics of field-cooled materials slows down

Fig. 7.2. The effect of columnar pins on the irreversibility line of $Tl_2Ba_2Ca_2Cu_3O_{10}$ [16]. In the absence of irradiation, the resistivity becomes unmeasurably small only in the shaded region below $T_{ir}(B)$. The reentrant low-field vortex-liquid regime of Fig. 7.1 appears for $B \lesssim 10^{-2}$ T and is hence not visible on this scale. Figure courtesy of R. Budhani.

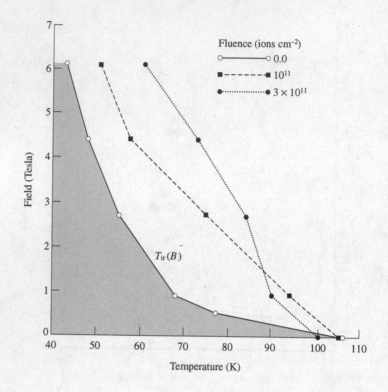

drastically [21]. In samples *without* correlated disorder the melting curve and irreversibility line may in fact be almost identical [15].

Since $T_{ir}(B)$ coincides approximately with the temperature at which the resistivity becomes unmeasurably small, the large upward shift in Fig. 7.2 is of considerable technological as well as intellectual interest. Note, for example, that the effective critical temperature for superconductivity at $B = 2.7$ T shifts from 53 to 87 K upon irradiation. The critical current at liquid-nitrogen temperature (77 K) increases by three orders of magnitude [16]. It is believed that, after irradiation, the irreversibility line becomes a locus of thermodynamically sharp "Bose-glass" transitions, below which the flux lines are localized on the columnar defects [22]. Other forms of correlated disorder may be important even in unirradiated samples. Twin boundaries, for example, appear to be responsible for the apparently continuous transition observed in twinned YBCO samples with the field aligned parallel to the c axis [12].

Here, we highlight these developments by describing how flux lines interact with correlated pinning centers and with each other. To account for thermal fluctuations, one must average over vortex configurations. Because the configuration sums bear a strong resemblance to the imaginary-time path-integral formulation of quantum mechanics [23], wave functions and binding energies that appear in simple quantum problems (such as the square well and the harmonic oscillator) have important implications for flux lines. With this identification we can quickly compute the free energy and localization length of vortices in the Bose-glass phase and the corresponding critical currents and determine as well the decoupling field and line-of-melting temperatures in clean systems [22]. We can also readily estimate the degree of entanglement of the wandering vortex lines in the flux liquid. If entanglement occurs in temperature and field ranges within which flux-cutting barriers are high, it may prevent crystallization and lead to a polymer-like glass transition even in the absence of strong pinning disorder [24]. The experimental consequences of this scenario are reviewed at the end of this brief, elementary chapter.

The idea of using Schrödinger's equation to solve problems in classical statistical mechanics is not new. It was used by S. F. Edwards, P. G. de Gennes and others starting in the 1960s to treat the conformations of flexible polymer chains [25]. One can go further with this analogy for flux lines, however, because these are equivalent to a system of *directed* polymers, with a common average orientation. As a result, distant self-interactions along the filaments can usually be neglected, in contrast to the case for isotropic self-avoiding polymer solutions. The essential physics is captured if one introduces a line tension to control vortex wandering and allows interactions between *different* polymers as well as interactions with the relevant pinning centers. Although we focus here primarily on simple one- and two-line problems, many-flux-line statistical mechanics (with multiple pins) is in fact equivalent to the many-body quantum mechanics of bosons in two dimensions [22, 24]. This system differs from the otherwise closely related problem of helium films on disordered substrates [26] because vortices behave dynamically like *charged* bosons and are easily manipulated in experiments by the injection of supercurrents. A simple explanation of why vortex probability distributions are boson-like in a thick sample can be found in Section 7.2. The richness and complexity of the *many*-flux-line problem rivals the physics of correlated electrons in semiconductors and metals. For details, see [22, 24, 27] and Chapter 8.

7.2 Correlated pinning and quantum bound states

7.2.1 The model free energy

We start with a model free energy F_N for N flux lines in a sample of thickness L, defined by their trajectories $\{\mathbf{r}_j(z)\}$ as they traverse a sample with

Fig. 7.3. A schematic
representation of columnar
pins and pinned vortex lines.

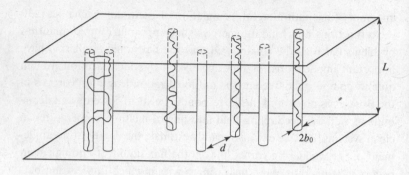

Fig. 7.3. A schematic representation of columnar pins and pinned vortex lines.

both columnar pins and an external magnetic field aligned with the
z-axis, i.e., in the direction perpendicular to the CuO_2 planes [22],

$$F_N = \frac{1}{2}\tilde{\varepsilon}_1 \sum_{j=1}^{N} \int_0^L \left| \frac{d\mathbf{r}_j(z)}{dz} \right|^2 dz + \frac{1}{2} \sum_{i \neq j} \int_0^L V(|\mathbf{r}_i(z) - \mathbf{r}_j(z)|) dz$$

$$+ \sum_{j=1}^{N} \int_0^L V_D[\mathbf{r}_j(z)] dz \tag{7.1a}$$

with

$$V_D(\mathbf{r}) = \sum_{k=1}^{M} V_1(\mathbf{r} - \mathbf{R}_k). \tag{7.1b}$$

Here $V(\mathbf{r}) = 2\varepsilon_0 K_0(r/\lambda_{ab})$ is the interaction potential between lines with
in-plane London penetration depth λ_{ab} and the random potential $V_D(\mathbf{r})$
arises from a z-independent set of M disorder-induced columnar
pinning potentials $V_1(\mathbf{r})$ centered on sites $\{\mathbf{R}_k\}$. The tilt modulus is
$\tilde{\varepsilon}_1 \approx (M_\perp/M_z)\varepsilon_0 \ln(\lambda_{ab}/\xi_{ab})$, where the material anisotropy is embodied in
the effective mass ratio $M_\perp/M_z \ll 1$, and $\varepsilon_0 \approx [\phi_0/(4\pi\lambda_{ab})]^2$ is the energy
scale for the interactions. The potential $V_D(\mathbf{r})$ arises from identical cylin-
drical traps, which are assumed for simplicity to pass completely
through the sample with a well depth per unit length of U_0 and an effec-
tive radius b_0. The parameter $b_0 \approx \max\{c_0, \xi_{ab}\}$, where $c_0 \approx 25$–40 Å, is
the radius of the columnar pins and ξ_{ab} is the superconducting coher-
ence length in the ab-plane.

A complete analysis of the many-line statistical mechanics associated with Eqs. (7.1) requires multiple path integrals over vortex trajectories weighted by $e^{-F_N/T}$ and subject to a complicated random pinning
potential. See Fig. 7.3. Our goal here is to illuminate the essential
physics by studying a few simple problems involving one or two flux
lines and only a few columnar pins. See Chapter 8 for an analysis of the
many-flux-line problem associated with Eqs. (7.1).

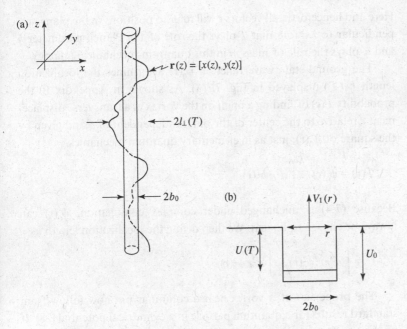

Fig. 7.4. One columnar pin and one vortex line, indicating (a) the localization length $\ell_\perp(T)$ and (b) the thermally renormalized pinning potential $U(T)$.

7.2.2 One flux line and one columnar pin

Consider a vortex line trapped near a single columnar pin with well depth U_0 and radius b_0 parallel to z in an otherwise defect-free sample of thickness L, as shown in Fig. 7.4. A quantity of considerable physical interest is the binding *free* energy per unit length.

$$U(T) = U_0 - TS, \tag{7.2}$$

where S is the reduction in entropy due to confinement. This free energy is given by a path integral,

$$e^{U(T)L/T} = \frac{\int \mathcal{D}\mathbf{r}(z)\exp\left[-\frac{\tilde{\varepsilon}_1}{2T}\int_0^L\left(\frac{d\mathbf{r}}{dz}\right)^2 dz - \frac{1}{T}\int_0^L V_1[\mathbf{r}(z)]\,dz\right]}{\int \mathcal{D}\mathbf{r}(z)\exp\left[-\frac{\tilde{\varepsilon}_1}{2T}\int_0^L\left(\frac{d\mathbf{r}}{dz}\right)^2 dz\right]}, \tag{7.3}$$

where the denominator is required in order to subtract off the entropy of an unconfined line far from the pin. The cylindrically symmetric confining potential $V_1(r)$, indicated in Fig. 7.4(b), tends to zero as $|\mathbf{r}| \to \infty$. The path integrals in (7.3) follow from standard statistical methods, which express them in terms of the eigenvalues of a transfer matrix. In the limit $L \to \infty$, the smallest eigenvalue dominates and $U(T) = -E_0(T)$, where $E_0(T)$ is the ground-state energy of a two-dimensional "Schrödinger equation" (see Appendix A),

$$\left(-\frac{T^2}{2\tilde{\varepsilon}_1}\nabla_\perp^2 + V_1(r)\right)\psi_0(\mathbf{r}) = E_0\psi_0(\mathbf{r}). \tag{7.4}$$

Here and henceforth, all vectors **r** will refer to positions in the plane perpendicular to $\hat{z}$. Note that T plays the role of the Planck parameter $\hbar$ and $\tilde{\varepsilon}_1$ plays the role of mass m in this quantum-mechanical analogy.

The ground-state wave function $\psi_0(\mathbf{r})$ determines the localization length $\ell_\perp(T)$ displayed in Fig. 7.3(a). As shown in Appendix B, the probability $P(\mathbf{r})$ of finding a point on the vortex at a transverse displacement **r** relative to the center of the pin is independent of z and given by the square of $\psi_0(\mathbf{r})$, just as in elementary quantum mechanics,

$$\mathcal{P}(\mathbf{r}) = \psi_0^2(\mathbf{r}) \bigg/ \int d^2r\, \psi_0^2(\mathbf{r}). \tag{7.5}$$

Because (7.4) is unchanged under complex conjugation, $\psi_0(\mathbf{r})$ can always be chosen to be real. We then define the localization length as

$$\ell_\perp^2 = \int d^2r\, r^2 \psi_0^2(r) \bigg/ \int d^2r\, \psi_0^2(r). \tag{7.6}$$

The properties of a vortex near a columnar pin now follow from standard results for a quantum particle in a cylindrical potential [28]. If we assume for simplicity a cylindrical square well ($V_1(r) \equiv -U_0$, $r < b_0$, and $V_1(r) = 0$, $r > b_0$) the binding free energy $U(T) = -E_0$ takes the form (see Chapter 8)

$$U(T) = U_0 f(T/T^*), \tag{7.7}$$

where T^* is an important characteristic temperature defined by

$$T^* = \sqrt{\tilde{\varepsilon}_1 U_0}\, b_0. \tag{7.8}$$

When $T \ll T^*$, the well depth is effectively infinite and we find the usual particle-in-a-box result,

$$U(T) \approx U_0 - c_1 \frac{T^2}{2\tilde{\varepsilon}_1 b_0^2}, \tag{7.9}$$

where c_1 is a constant related to the first zero of the Bessel function $J_0(x)$ which solves Eq. (7.4) in this limit. The localization length is then

$$\ell_\perp(T) \approx b_0 \left[1 + \mathcal{O}\!\left(\frac{1}{\kappa b_0} \right) \right], \tag{7.10}$$

where $\kappa^{-1} \approx T/\sqrt{2\tilde{\varepsilon}_1 U(T)} \ll b_0$ is the distance the "particle" penetrates into the classically forbidden region. The low-temperature correction in Eq. (7.9) represents the entropy lost each time a wandering flux line is reflected off the confining walls of the binding potential [29]. At the crossover temperature T^*, this "zero-point energy" of confinement becomes comparable to the well depth.

When $T \gg T^*$, the flux line is only weakly bound, although a strictly localized ground state *always* exists in this effectively two-dimensional problem [28]. In the limit, the analysis of Chapter 8 leads to

$$U(T) \approx \frac{1}{2} U_0 \left(\frac{T}{T^*} \right)^2 e^{-2(T/T^*)^2},$$
(7.11)

and a localization length $\ell_\perp(T) \approx \kappa^{-1} = T / \sqrt{2\tilde{\varepsilon}_1 U(T)}$, so that

$$\ell_\perp(T) \approx b_0 e^{(T/T^*)^2}.$$
(7.12)

The flux line now "diffuses" within a confining tube with radius of order $\ell_\perp(T)$ as it crosses the sample. The length along $\hat{z}$ required to "diffuse" across this tube is $\ell_z \approx \ell_\perp^2/(T/\tilde{\varepsilon}_1)$, i.e.,

$$\ell_z = (b_0^2 \tilde{\varepsilon}_1 / T) e^{2(T/T^*)^2}.$$
(7.13)

These results imply a strong thermal renormalization of the critical current $J_c(T)$ [22]. $J_c(T)$ is the current necessary to produce a Lorentz force $f_c = J_c f_0/c$ so strong that thermal activation is unnecessary to tear a flux line away from its columnar pin. At low temperatures, one would expect that $f_c \approx U_0/b_0$, i.e.,

$$J_c \approx c U_0 / (\phi_0 b_0).$$
(7.14)

Here we assume that the confining potential $V_1(r)$ does not really jump abruptly at $r = b_0$, but instead rises smoothly to zero over a distance of order b_0. Results such as (7.11–7.13) are in any case independent of the precise form of the microscopic potential [28]. To account for line wandering, we should replace U_0 by $U(T)$ and b_0 by $\ell_\perp(T)$. For $T \gg T^*$, this leads to

$$J_c(T) \approx J_c(0) e^{-3(T/T^*)^2}.$$
(7.15)

7.2.3 One flux line and two columnar pins

The localization length discussed above is like the Bohr radius of an isolated "atom" consisting of one columnar pin and one vortex line. Now consider a vortex line that is able to hop between *two* nearby identical columnar pinning series at $\mathbf{R}_1$ and $\mathbf{R}_2$ (analogous to a H_2^+ molecule) as it traverses the sample – see Fig. 7.5. For now, we ignore the additional dashed flux line. The binding potential $V_1(r)$ in Eq. (7.4) in this case should be replaced by

$$V_2(\mathbf{r}) = V_1(\mathbf{r} - \mathbf{R}_1) + V_1(\mathbf{r} - \mathbf{R}_2),$$
(7.16)

where identical wells of depth U_0 and radius b_0 separated by d are assumed. As in a quantum double well, the dominant configurations are those in which the particle delocalizes further by "tunneling" from one

Fig. 7.5. Two columnar pins
with a single vortex (solid
line) tunneling between
them. The dashed line shows
a second vortex, which
requires the introduction of a
simple Hubbard model to
account for intervortex
interactions.

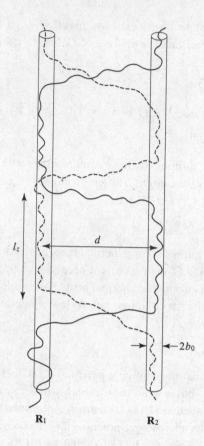

well to another. In the subspace spanned by the isolated pin ground-state wave functions $\psi_0(\mathbf{r} - \mathbf{R}_1)$ and $\psi_0(\mathbf{r} - \mathbf{R}_2)$, the effective Hamiltonian takes the form

$$\mathcal{H}_2 = \begin{pmatrix} E_0 & t \\ t & E_0 \end{pmatrix}, \tag{7.17}$$

where E_0 is the ground-state energy determined in the previous subsection and $t > 0$ is a tunneling-matrix element. The two lowest-energy eigenvalues for the double well are then

$$E_\pm = E_0 \pm t. \tag{7.18}$$

The partition function in this approximation is $Z_{\text{tot}} = \sum_{i,j=1}^2 \langle i | e^{-\mathcal{H}L/T} | j \rangle$, where $|1\rangle$ and $|2\rangle$ are states localized on pins 1 and 2, respectively.

To determine t, we proceed variationally and minimize

$$E(\psi) \equiv \frac{\int d^2\mathbf{r}\left(\dfrac{T^2}{2\tilde{\varepsilon}_1}|\nabla\psi(\mathbf{r})|^2 + V_2(\mathbf{r})|\psi(\mathbf{r})|^2\right)}{\int d^2\mathbf{r}\,|\psi(\mathbf{r})|^2}, \tag{7.19}$$

with the trial function.

$$\psi(\mathbf{r}) = \alpha\psi_0(\mathbf{r} - \mathbf{R}_1) + \beta\psi_0(\mathbf{r} - \mathbf{R}_2), \tag{7.20}$$

where $\psi_0(\mathbf{r})$ is the ground state for an isolated well. We assume that we have widely separated wells, i.e., $d \gg \ell_\perp(T)$. The minimum occurs for the symmetric case $\alpha = \beta$ and has energy $E = E_0 - t$, with

$$t \approx \text{constant} \times \frac{U(T)}{\sqrt{E_k/T}} e^{-E_k/T}, \tag{7.21}$$

where $U(T) = -E_0(T) > 0$ and

$$E_k = \sqrt{2\tilde{\varepsilon}_1 U(T)}\, d \tag{7.22}$$

is similar to a WKB tunneling exponent.

The flux line has now delocalized a distance $\sim d$ in the transverse direction. Delocalization proceeds via wandering in a tube of radius $\ell_\perp(T)$ and occasional tunneling across to a neighboring tube, as indicated in Fig. 7.5. When $b_0 \ll \ell_\perp \ll d$ the spacing between such tunneling events along the z axis is of order

$$\ell_z(T) \approx \left(\frac{b_0^2\tilde{\varepsilon}_1}{T} e^{2(T/T^*)^2} \right) e^{E_k/T}, \tag{7.23}$$

where the prefactor (an inverse "attempt frequency" in imaginary time) comes from the isolated-pin result Eq. (7.13). Note the close analogy between Fig. 7.5 and the configurations of a classical one-dimensional Ising model with exchange constant $J = E_k$ disrupted by kinks along the z axis.

The flux line will not be delocalized much further by adding a third, more distant columnar pin to the problem, because the new available state is not in resonance with the double-well ground-state energy calculated above. One isolated flux line will always be localized by a random array of columnar pins [22].

7.2.4 Hubbard model: Two flux lines and two columnar pins

Interactions are crucial for determining vortex configurations when correlated pinning is present. In the absence of a repulsive pair potential all vortices would pile up at $T = 0$ in a single deep minimum of the random pinning potential produced, for example, by an unusually dense region of columnar pins. By allowing *two* vortices to wander simultaneously between a pair of columnar pins, we can study interactions in a particularly simple context. This elementary model also illustrates why flux lines behave like bosons in thick samples. A related treatment describes the H_2 molecule in real quantum mechanics [30].

Figure 7.5 shows two vortices (solid and dashed lines) hopping back and forth between two columnar pins as we trace their trajectories along the z axis. In the absence of interactions, the probability distribution of each vortex would be described by the symmetric double-well ground-state wave function discussed in the previous section. The vortices would find themselves on the same columnar pin approximately half the time. Introducing a repulsive energy between vortices will lead to *correlated* hopping as the fluxons exchange places.

To model this situation, we introduce a tight-binding Hamiltonian similar to Eq. (7.17), which operates on a set of four normalized orthogonal basis states, $|12\rangle$, $|21\rangle$, $|11\rangle$ and $|22\rangle$. The state $|12\rangle$ means that the first fluxon line occupies pin 1 and the second occupies pin 2; $|21\rangle$ is the state in which the vortices have exchanged places. Both fluxons are localized on pin 1 and on pin 2 in the states $|11\rangle$ and $|22\rangle$, respectively. The classical partition function which gives the probability of making a transition from one of these four states $|a\rangle$ to a final state $|b\rangle$ across a slab of thickness L is then

$$Z(a, b; L) = \langle b | e^{-\mathcal{H}L/T} | a \rangle, \tag{7.24}$$

where the tight-binding model is defined by the 4×4 matrix Hamiltonian

$$\mathcal{H} = \begin{array}{c} \\ |12\rangle \\ |21\rangle \\ |11\rangle \\ |22\rangle \end{array} \overset{\displaystyle |12\rangle \quad |21\rangle \quad |11\rangle \quad\quad |22\rangle}{\begin{pmatrix} 2E_0 & 0 & t & t \\ 0 & 2E_0 & t & t \\ t & t & 2E_0 + V_{\text{int}} & 0 \\ t & t & 0 & 2E_0 + V_{\text{int}} \end{pmatrix}} \tag{7.25}$$

The diagonal terms include the one-vortex binding energy E_0 discussed in Section 7.2.2 and a "Hubbard-repulsion" term V_{int}, which represents the energy which arises when two vortices occupy the same columnar pin. A reasonable estimate of V_{int} when $d \ll \lambda_{ab}$ is [22]

$$V_{\text{int}} \approx 2\varepsilon_0 \ln(d/\xi_{ab}), \tag{7.26}$$

reflecting the energy cost of one doubly quantized vortex as opposed to two singly quantized vortices separated by d. For $d \gg \lambda_{ab}$, $V_{\text{int}} \approx 2\varepsilon_0 \ln(\lambda_{ab}/\xi_{ab})$. The off-diagonal terms in (7.25) reflect transitions between the four basis states caused by hops of a single vortex. The Hamiltonian breaks into symmetric and antisymmetric subspaces,

$$\mathcal{H} = \begin{array}{c} \\ |-\rangle \\ |+\rangle \\ |11\rangle \\ |22\rangle \end{array} \overset{\displaystyle |-\rangle \quad\quad |+\rangle \quad\quad |11\rangle \quad\quad |22\rangle}{\left(\begin{array}{c:ccc} 2E_0 & 0 & 0 & 0 \\ \hdashline 0 & 2E_0 & \sqrt{2}t & \sqrt{2}t \\ 0 & \sqrt{2}t & 2E_0 + V_{\text{int}} & 0 \\ 0 & \sqrt{2}t & 0 & 2E_0 + V_{\text{int}} \end{array}\right)}, \tag{7.27}$$

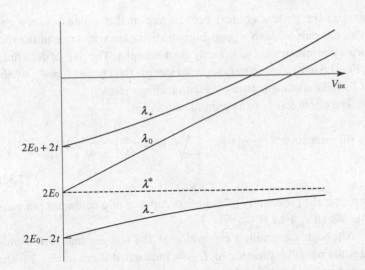

Fig. 7.6. Energy eigenvalues for the four-state Hubbard model. Solid lines represent the symmetric "boson" subspace, while the dashed line corresponds to the antisymmetric "fermion" excitation.

when $|12\rangle$ and $|21\rangle$ are eliminated in favor of the symmetrized states

$$|-\rangle = \frac{1}{\sqrt{2}}(|12\rangle - |21\rangle)) \qquad (7.28a)$$

and

$$|+\rangle = \frac{1}{\sqrt{2}}(|12\rangle + |21\rangle)). \qquad (7.28b)$$

The one-dimensional subspace of states antisymmetric under vortex interchange would describe spinless fermions in conventional quantum mechanics. The three-dimensional symmetric subspace is appropriate to boson quantum mechanics. As $L \to \infty$, partition functions such as (7.24) will be dominated by the smallest eigenvalue of $\mathcal{H}$. The four eigenvalues of (7.27) are easily found to be

$$\lambda^* = 2E_0,$$

$$\lambda_0 = 2E_0 + V_{int},$$

$$\lambda_+ = 2E_0 + \tfrac{1}{2}V_{int} + \tfrac{1}{2}\sqrt{V_{int}^2 + 16t^2},$$

$$\lambda_- = 2E_0 + \tfrac{1}{2}V_{int} - \tfrac{1}{2}\sqrt{V_{int}^2 + 16t^2}. \qquad (7.29)$$

The eigenvalue λ^* belongs to the fermion subspace and is unaffected by interactions. The remaining eigenvalues are bosonic. As shown in Fig. 7.6, the lowest eigenvalue is always λ_-, in the boson subspace. It differs from the fermion eigenvalue $2E_0$ by an amount proportional to t^2/V_{int} as $V_{int} \to \infty$.

The boson character of the ground state is a general feature of Hamiltonians that are symmetric under interchange of particles [23]. It arises in the present context because repeated hopping between the columnar pins leads to a probability distribution symmetric in the two vortex coordinates for sufficiently thick samples. The pair of flux lines in Fig. 7.5 will behave like bosons for sample thicknesses $L \gg \ell_z$, where $\ell_z(T)$ is the spacing between these tunneling events.

The eigenvector of the ground state is

$$|0\rangle = \text{constant} \times \left\{ \frac{1}{\sqrt{2}} |+\rangle + \frac{1}{2} \left[\sqrt{1 + \left(\frac{V_{\text{int}}}{4t}\right)^2} - \frac{V_{\text{int}}}{4t} \right] (|11\rangle + |22\rangle) \right\}.$$

(7.30)

Note that the probability of double occupancy of a columnar pin vanishes like $(t/V_{\text{int}})^2$ as $V_{\text{int}} \to \infty$.

Although the smallest eigenvalue of the transfer matrix V dominates the partition function as $L \to \infty$, the excited states in Fig. 7.6 are important for correlation functions connecting different values of z and when $L \lesssim \ell_z$. Only states in the *bosonic* subspace contribute even in this case, however. To understand this, note that Eq. (7.24) must be summed over states $|a\rangle$ and $|b\rangle$ describing the entry and exit points of the vortices. The total partition function is thus

$$Z_{\text{tot}} = |\alpha\rangle e^{-\mathcal{H}L/T} |\alpha\rangle,$$

(7.31)

where $|\alpha\rangle$ is some linear combination of $|11\rangle$, $|22\rangle$, $|12\rangle$ and $|21\rangle$. Because the flux lines are indistinguishable, the initial state must take the form

$$|\alpha\rangle = a|11\rangle + b|22\rangle + c(|12\rangle + |21\rangle),$$

(7.32)

where a, b and c are constants. We allow for $a \neq b$ because the two columnar pins could in fact have slightly different binding energies (both at the surface and in the interior), as in the Anderson model of localization [22]. Even in this case, the initial state $|\alpha\rangle$ lies completely within the *boson* subspace of Eq. (7.27). Thus the fermion eigenstate never contributes to the statistical mechanics. Similar arguments show that only boson excitations are relevant to arbitrarily large assemblies of interacting flux lines [24, 27]. (See Chapter 8.)

7.3 Flux melting and the quantum harmonic oscillator

Consider one representative fluxon in the confining potential "cage" provided by its surrounding vortices in a triangular lattice. The partition function for a fixed entry point $\mathbf{0}$ and exit point $\mathbf{r}_\perp$ in a sample of thickness L is

$$Z_1(\mathbf{r}_\perp, \mathbf{0}; L) = \int_{\mathbf{r}(0)=\mathbf{0}}^{\mathbf{r}(L)=\mathbf{r}_\perp} \mathcal{D}\mathbf{r}(z)$$

$$\exp\left\{-\frac{1}{T}\int_0^L \left[\frac{1}{2}\tilde{\varepsilon}_1\left(\frac{d\mathbf{r}(z)}{dz}\right)^2 + V_1[\mathbf{r}(z)]\right]dz\right\}, \qquad (7.33)$$

where $V_1[\mathbf{r}(z)]$ is now a one-body potential chosen to mimic the interactions in Eq. (7.1a). We assume clean samples and high temperatures so that both correlated and point disorder can be neglected.

Three important field regimes for fluctuations in vortex crystals are easily extracted from this simplified model. Following the approach in the previous section, we rewrite this imaginary-time path integral as a quantum-mechanical matrix element,

$$Z(\mathbf{r}_\perp, \mathbf{0}; L) = \langle \mathbf{r}_\perp | e^{-L\mathcal{H}/T} | \mathbf{0} \rangle, \qquad (7.34)$$

where $|\mathbf{0}\rangle$ is an initial state localized at $\mathbf{0}$, $\langle \mathbf{r}_\perp |$ is a final state localized at $\mathbf{r}_\perp$ and the "Hamiltonian" $\mathcal{H}$ is the operator which appears in Eq. (7.4)

$$\mathcal{H} = -\frac{T^2}{2\tilde{\varepsilon}_1}\nabla_\perp^2 + V_1(\mathbf{r}). \qquad (7.35)$$

Recall that the probability of finding the flux line at a transverse position $\mathbf{r}$ within the crystal is $\psi_0^2(\mathbf{r})$, where $\psi_0(\mathbf{r})$ is the normalized ground-state eigenfunction of (7.35)

When $B \gg B_{c1} \equiv \phi_0/\lambda_{ab}^2$, the pair potential $V(r_{ij}) = 2\varepsilon_0 K_0(r_{ij}/\lambda_{ab})$ is logarithmic, $K_0(x) \approx \ln x$, and we expand $V_1(\mathbf{r}_\perp)$ about its minimum at $\mathbf{r}_\perp = 0$ to find

$$\left(-\frac{T^2}{2\tilde{\varepsilon}_1}\nabla_\perp^2 + \frac{1}{2}kr_\perp^2\right)\psi_0 = E_0\psi_0, \qquad (7.36)$$

where (neglecting logarithmic corrections to ε_0 and constants of order unity)

$$k \approx \frac{d^2V}{dr^2}\bigg|_{r=a_0}$$

$$\approx \varepsilon_0/a_0^2 \qquad (7.37)$$

and a_0 is the mean spacing between vortices. Equation (7.36) is the Schrödinger equation for a two-dimensional quantum oscillator, with $\hbar \to T$ and mass $m \to \tilde{\varepsilon}_1$. The ground-state wave function is

$$\psi_0(r_\perp) = \frac{1}{\sqrt{2\pi}r_*}e^{-r^2/(4r_*^2)}, \qquad (7.38)$$

with spatial extent

$$r_* = \left(\frac{T^2 a_0^2}{\varepsilon_0 \tilde{\varepsilon}_1}\right)^{1/4}. \tag{7.39}$$

Melting occurs when $r_* = c_L a_0$, where c_L is the Lindemann constant, so the melting temperature is

$$T_m = c_L^2 \sqrt{\varepsilon_0 \tilde{\varepsilon}_1}\, a_0 \qquad (B_{c1} < B \lesssim B_x), \tag{7.40}$$

in agreement with other estimates [31]. Vortices in the crystalline phase will travel across their confining tube of radius $r_\perp^*$ in a "time" along the z axis of order ℓ_z^0, where [24]

$$\ell_z^0 \approx r_*^2/(T/\tilde{\varepsilon}_1)$$

$$\approx \sqrt{\frac{\tilde{\varepsilon}_1}{\varepsilon_0}}\, a_0. \tag{7.41}$$

A new high-field regime arises when $\ell_z^0 \lesssim d_0$, where d_0 is the average spacing of the copper-oxide planes, i.e., for $B \gtrsim B_x$, with decoupling field

$$B_x \approx \frac{\tilde{\varepsilon}_1}{\varepsilon_0}\frac{\phi_0}{d_0^2}, \tag{7.42}$$

again in agreement with earlier work [15, 31]. Above this field, the planes are approximately decoupled and T_m may be estimated from the theory of two-dimensional dislocation-mediated melting described in Chapter 2 [15, 32],

$$T_m \approx \frac{\varepsilon_0 d_0}{8\pi\sqrt{3}} \qquad (B \gtrsim B_x). \tag{7.43}$$

The estimate (7.40) also breaks down at low fields $B \lesssim B_{c1}$, for which the logarithmic interaction potential must be replaced by an exponential repulsion. The two-dimensional harmonic-oscillator model again applies, with the replacement

$$k \rightarrow \frac{\varepsilon_0}{\lambda_{ab}^2} e^{-a_0/\lambda_{ab}}. \tag{7.44}$$

The transverse wandering distance is now

$$r_* \approx \left(\frac{T^2 \lambda_{ab}^2}{\varepsilon_0 \tilde{\varepsilon}_1}\right)^{1/4} e^{a_0/(4\lambda_{ab})} \tag{7.45}$$

and takes place over a longitudinal distance

$$\ell_z^0 = \sqrt{\frac{\tilde{\varepsilon}_1}{\varepsilon_0}}\, \lambda_{ab} e^{a_0/(2\lambda_{ab})}. \tag{7.46}$$

Table 7.1. *Estimates for the flux-lattice melting temperature determined for the three regimes discussed in the text*

Regime	$T_m(B)$	
$B_x \lesssim B$	$\dfrac{\varepsilon_0 d_0}{8\pi\sqrt{3}}$	$B_x \approx \dfrac{\tilde{\varepsilon}_1}{\varepsilon_0}\dfrac{\phi_0}{d_0^2}$
$B_{c1} < B \lesssim B_x$	$c_L^2 \sqrt{\varepsilon_0\tilde{\varepsilon}_1}(\phi_0/B)^{1/2}$	$B_{c1} \approx \phi_0/\lambda_{ab}^2$
$B \lesssim B_{c1}$	$c_L^2 \sqrt{\varepsilon_0\tilde{\varepsilon}_1}\,\lambda_{ab}(B_{c1}/B)e^{-(B_{c1}/B)^{1/2}}$	

The low-field melting temperature becomes

$$T_m \approx c_L^2 \sqrt{\varepsilon_0\tilde{\varepsilon}_1}\,\frac{a_0^2}{\lambda_{ab}}e^{-a_0/(2\lambda_{ab})} \qquad (B \lesssim B_{c1}), \tag{7.47}$$

consistent with earlier predictions [15, 24]. Although we have retained the distinction between $\tilde{\varepsilon}_1$ and ε_0 in these formulas, note that $\tilde{\varepsilon}_1 \approx \varepsilon_0$ in this regime [15].

The predictions (7.40), (7.43) and (7.47) are combined to give the reentrant phase diagram for melting shown in Fig. 7.1. Analytical estimates and boundaries for melting in the various regimes are summarized in Table 7.1.

7.4 Vortex entanglement in the liquid phase

Above the melting line in Fig. 7.1, weak point disorder due to oxygen vacancies can usually be neglected [14] and we must consider a liquid of wandering, essentially unconfined, lines. To estimate the degree of entanglement, we consider a *single* flux line $\mathbf{r}(z)$ and determine how far it wanders perpendicular to the z axis as it traverses the sample. The relevant path integral is

$$\langle |\mathbf{r}(z) - \mathbf{r}(0)|^2 \rangle = \frac{\displaystyle\int \mathcal{D}\mathbf{r}(s)|\mathbf{r}(z) - \mathbf{r}(0)|^2 \exp\left[-\frac{\tilde{\varepsilon}_1}{2T}\int_0^L\left(\frac{d\mathbf{r}}{ds}\right)^2 ds\right]}{\displaystyle\int \mathcal{D}\mathbf{r}(s) \exp\left[-\frac{\tilde{\varepsilon}_1}{2T}\int_0^L\left(\frac{d\mathbf{r}}{ds}\right)^2 ds\right]},$$

which, when discretized as in Appendix A, yields

$$\langle |\mathbf{r}(z) - \mathbf{r}(0)|^2 \rangle = \frac{2T}{\tilde{\varepsilon}_1}|z|, \tag{7.48}$$

which shows that the vortex "diffuses" as a function of the time-like variable z,

$$\langle|\mathbf{r}(z) - \mathbf{r}(0)|^2\rangle^{1/2} = (2Dz)^{1/2}, \tag{7.49}$$

with diffusion constant

$$D = \frac{T}{\tilde{\varepsilon}_1} = \frac{M_z}{M_\perp} \frac{4\pi T}{\phi_0 H_{c1}}. \tag{7.50}$$

At $T = 77$ K, we take $H_{c1} \approx 10^2$ G and $M_z/M_\perp \approx 10^2$ and find $D = 10^{-6}$ cm, so that vortex lines wander a distance of order 1 μm while traversing a sample of thickness 0.01 cm.

Close encounters between neighboring vortex lines will thus occur quite frequently in fields of order 1 T or more, where vortices are separated by distances of order 500 Å or less. The "entanglement length" ℓ_z is defined as the distance along the z axis such that $\langle|\mathbf{r}(\ell_z) - \mathbf{r}(0)|^2\rangle = a_0^2 = B/\phi_0$, i.e. [24]

$$\ell_z = \frac{a_0^2}{2D} = \frac{\tilde{\varepsilon}_1 a_0^2}{2T}. \tag{7.51}$$

Collisions and entanglement of vortex lines will be important for flux liquids whenever

$$L > \ell_z, \tag{7.52}$$

i.e., for $B > B_x \approx (M_\perp/M_z)[\phi_0 \varepsilon_0/(LT)]$.

The discrete flux filaments which comprise the flux liquid form when a superconductor is cooled through the mean-field-transition line $B_{c2}^0(T)$. These fluxons are "phantom vortices" for $B \lesssim B_{c2}^0(T)$ in the sense that the barriers to line crossing are expected to be negligible. The barriers will increase, however, as the flux liquid is cooled toward the fluctuation-induced-melting temperature of the Abrikosov flux crystal. Entanglement can then cause the flux liquid to become very viscous [24, 33].

As described in more detail in Chapter 8, the consequences for transport of vortices in the vicinity of a few strong pinning centers such as twin boundaries can be quite striking. Assume that a supercurrent $\mathbf{j}_s$ flows along $\hat{\mathbf{y}}$, $\mathbf{j}_s = j_s\hat{\mathbf{y}}$, and that the magnetic field, as usual, is parallel to $\hat{\mathbf{z}}$. Vortices will then be subjected to a constant driving Lorentz force,

$$\mathbf{f}_L = \frac{1}{c} n_0 \phi \hat{\mathbf{z}} \times \mathbf{j}_s, \tag{7.53}$$

where $n_0 \approx a_0^{-2}$ is the density of vortices. The equation of motion which describes a z-independent flux-liquid velocity field in the vicinity of, say, a twin boundary in the xz-plane is then [33]

$$-\gamma\mathbf{v} + \eta\nabla_\perp^2\mathbf{v} + \mathbf{f}_L = 0. \tag{7.54}$$

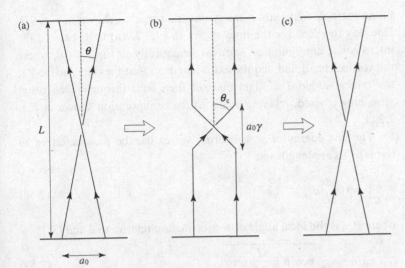

Fig. 7.7. A mechanism for vortex line crossing. Vortices in the initial (a) and final (c) configurations are displaced out of the plane of the figure by a distance $\sim a_0$. An intermediate saddle-point configuration connecting these states is shown in (b). Here $\gamma = (M_\perp/M_z)^{1/2}$.

Equation (7.54) is a hydrodynamic description of flux flow, which is valid on length scales large relative to the intervortex spacing. The Bardeen–Stephen parameter γ is a "friction" coefficient that represents the resistance to motion of vortices encountered by the normal electrons in the core. The combination of the drag and viscous terms in Eq. (7.54) introduces an important new length scale into the problem,

$$\delta = \sqrt{\eta/\gamma}. \tag{7.55}$$

The length δ is the scale over which the velocity rises to its bulk value from the center of the twin boundary, where it is small or vanishes entirely. (See Chapter 8.)

The flux-line viscosity has been estimated, e.g., by Cates [34], who finds a remarkably simple formula for δ,

$$\delta \approx a_0 e^{U_\times/T}, \tag{7.56}$$

where $U_\times$ is the barrier to line crossing. We estimate this barrier as shown in Fig. 7.7, following the treatment of Obukhov and Rubinstein [35]. The projections of two vortices (displaced initially by $\sim a_0$ perpendicular to the plane of the figure) intersect near the center of the sample and approach the intervortex spacing at $z = \pm L/2$. Crossing is very difficult at low angles θ, because of the extra energy associated with a doubly quantized filament formed near the crossing region [24]. Crossing is easier at high angles, however. In an isotropic superconductor, the crossing energy goes to zero when $2\theta = 90°$, for example [36]. A simple rescaling argument shows that this critical angle for the zero-crossing energy becomes

$$\theta_c \approx \tan^{-1}(\sqrt{M_z/M_\perp}) \tag{7.57}$$

for anisotropic materials in the symmetric situation shown in Fig. 7.7. One way to exchange the lines shown in Fig. 7.7(a) is to pass to the intermediate configuration shown schematically in Fig. 7.7(b), where the vortices bend and acquire extra length so that they are inclined at the critical angle. The filaments can then pass through each other remaining crossed, relaxing finally to the configuration shown in Fig. 7.7(c).

The line energy of a wandering vortex can be parameterized in terms of its arc length s as

$$\int_0^L \varepsilon_1(\theta)\frac{ds}{dz}dz, \tag{7.58}$$

where $\theta(z)$ is the local angle of the inclination relative to z and [37]

$$\varepsilon_1(\theta) = \varepsilon_1\sqrt{\cos^2\theta + \frac{M_\perp}{M_z}\sin^2\theta}, \tag{7.59}$$

with $\varepsilon_1 = \varepsilon_0\ln(\lambda_{ab}/\xi_{ab})$. For the configuration in Fig. 7.7(a), this formula leads immediately to an energy

$$E_0 = 2\varepsilon_1 L[1 + \mathcal{O}(a_0^2/L^2)]. \tag{7.60}$$

The saddle-point energy shown in Fig. 7.7(b), on the other hand, has the energy

$$E_0' = 2\varepsilon_1\left(L - \sqrt{\frac{M_\perp}{M_z}}a_0\right)$$

$$+ 2\varepsilon_1\sqrt{\cos^2\theta_c + \frac{M_\perp}{M_z}\sin^2\theta_c}\sqrt{a_0^2 + (\cot^2\theta_c)a_0^2}, \tag{7.61}$$

which simplifies using Eq. (7.57) to

$$E_0' = 2\varepsilon_1 L + 2(\sqrt{2}-1)\sqrt{\frac{M_\perp}{M_z}}a_0\varepsilon_1. \tag{7.62}$$

The change in the energy of interaction between the lines due to the sharp bends and nonzero tilt will add a correction to the coefficient of the second term, which we neglect. Note also that the terms neglected in (7.60) are of higher order in a_0/L than are the terms kept in (7.62). The crossing energy $U_\times \approx E_0' - E_0$ is thus approximately [35]

$$U_\times \approx 2(\sqrt{2}-1)\sqrt{\frac{M_\perp}{M_z}}a_0\varepsilon_1. \tag{7.63}$$

We see that the crossing energy tends to zero as $M_\perp/M_z \to 0$ and has essentially the same functional form as the intermediate-field melting temperature displayed in Eq. (7.40). The key parameter which determines the viscous length scale (7.56) at the melting temperature is then

$$\frac{U_\times}{T_m} \approx \frac{c_\times}{c_L^2},\tag{7.64}$$

where $c_\times$ is a dimensionless constant, $c_\times = (2\sqrt{2}-1)\ln(\lambda_{ab}/\xi_{ab})$, for the simple model discussed above. This dimensionless ratio should be independent of the field strength in the range $B_{c1} \ll B \ll B_\times$. If $U_\times/T_m \lesssim 1$, the barriers to crossing that are associated with entanglement will not interfere with crystallization into an Abrikosov flux lattice. (This should *always* be the case for $B \gg B_\times$.) Note, however, that it is the Lindemann ratio squared which enters the denominator of (7.64). Assume for concreteness that $c_\times = 0.75$. If $c_L = 0.3$, then $U_\times/T_m \approx 8$ and $\delta = a_0 e^{U_\times/T_m} \approx 4 \times 10^3 a_0$. If $c_L = 0.15$, as indicated in the most recent experiments on untwinned YBCO [10], then $U_\times/T_m = 33$ and $\delta \approx 3 \times 10^{14} a_0$. If $U_\times/T_m$ is really this large, the kinetic barriers associated with entanglement will *preclude* crystallization on experimental time scales in samples thick enough or fields high enough to allow multiple entanglements of the vortex lines. The flux liquid will instead form a "polymeric-glass" phase upon cooling [24, 35]. It is interesting to note for comparison purposes that $U_\times/T_m \approx 75$ in a *real* polymer such as polyethyene. More accurate estimates of the numerator of (7.64) would, of course, be highly desirable.

Some evidence for the "polymer-glass" scenario already exists. Recent experiments by Safar *et al.* find that the first-order melting transition in untwinned YBCO goes away in sufficiently high magnetic fields, $B \gtrsim 10$ T. One possibility is that point disorder somehow becomes more important and causes a "vortex-glass" transition [15] at high fields [38]. An alternative explanation, however, is that the flux liquid simply becomes more entangled and viscous as its density increases with increasing field, leading eventually to undercooling and a polymer-glass transition. Simulations of a lattice superconductor have revealed large crossing barriers and are never able to recover the crystalline phase upon cooling once the flux lattice has melted [39], in agreement with this picture. In real experiments, the hysteresis loops in the resistivity associated with first-order freezing would slowly go away as crystallization became more difficult for a fixed rate of cooling. The low-temperature dynamics of this polymer glass in the presence of point disorder should be similar to that predicted by the collective pinning theory [40], with a *polymeric* shear modulus replacing the usual crystalline flux-lattice elastic constant c_{66}.

Fig. 7.8. The discretized path-integral representation of a vortex line passing through $N+1$ CuO_2 planes with average spacing $\delta = \ell/N$.

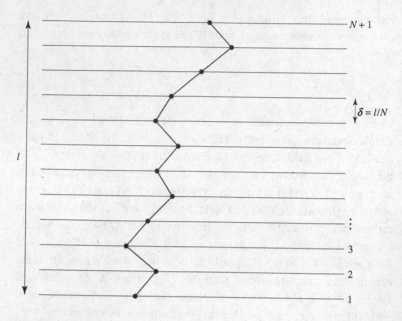

Acknowledgements

Much of this work is the result of a stimulating collaboration with V. M. Vinokur. Discussions with D. Bishop, R. Budhani, L. Civale, G. Crabtree, D. S. Fisher, P. L. Gammel, T. Hwa, P. Le Doussal and M. C. Marchetti are also gratefully acknowledged. This chapter is adapted with permission from *Phase Transitions and Relaxation in Systems with Competing Energy Scales*, edited by T. Riste and D. Sherrington (Kluwer, Dordrecht, 1993), pp. 95–117.

Appendix A. The transfer-matrix representation of the partition function

We review here how path integrals like those represented in Eq. (7.3) can be written in terms of a transfer matrix, which is the exponential of the Schrödinger operator which appears in elementary quantum mechanics [23].

We first consider

$$Z(\mathbf{r}_\perp, \mathbf{0}; \ell) = \int_{\mathbf{r}(0)=\mathbf{0}}^{\mathbf{r}(\ell)=\mathbf{r}_\perp} \mathcal{D}\mathbf{r}(z) \exp\left[-\frac{\tilde{\varepsilon}_1}{2T} \int_0^\ell \left(\frac{d\mathbf{r}}{dz}\right)^2 dz - \frac{1}{T} \int_0^\ell V_1[r(z)]dz \right] \quad (A.1)$$

and discretize this path integral as indicated in Fig. 7.8, where the planes of constant z are separated by a small parameter δ. This is precisely the situation which arises in the high-T_c superconductors with a field parallel to the c axis, provided that δ represents the mean spacing between CuO_2 planes. The discretized path integral reads

$$Z(\mathbf{r}_\perp, \mathbf{0}; \ell) \approx \left(\prod_{j=2}^{N} \frac{\tilde{\varepsilon}_1}{2\pi T\delta} \int d^2 r_j \right) \exp\left(\frac{-\tilde{\varepsilon}_1}{2T\delta} \sum_{j=2}^{N+1} (\mathbf{r}_j - \mathbf{r}_{j-1})^2 - \frac{\delta}{T} \sum_{j=1}^{N+1} V_1(\mathbf{r}_j) \right),$$

(A.2)

where it is understood that $\mathbf{r}_1 = \mathbf{0}$, $\mathbf{r}_{N+1} = \mathbf{r}_\perp$ and the normalization of the integrals is chosen so that $Z(r_\perp, 0; \ell) = 1$ when $V_1(\mathbf{r}) = 0$.

As usual in statistical mechanics, the effect of adding one copper-oxide plane to the system can be represented in terms of a transfer matrix.

$$Z(\mathbf{r}, \mathbf{0}; \ell + \delta) = \int d^2 r'\, T(\mathbf{r}, \mathbf{r}') Z(\mathbf{r}', \mathbf{0}; \ell),$$

(A.3)

where (neglecting small edge effects at the top and bottom of the sample)

$$T(\mathbf{r}, \mathbf{r}') = \frac{\tilde{\varepsilon}_1}{2\pi T\delta} \exp\left(\frac{-\tilde{\varepsilon}_1}{2T\delta} |\mathbf{r} - \mathbf{r}'|^2 - \frac{\delta}{2T} [V_1(\mathbf{r}) + V_1(\mathbf{r}')] \right).$$

(A.4)

Ryu et al. have studied the spectrum of $T(\mathbf{r}, \mathbf{r}')$ for finite δ [41]. Here, we shall instead consider the limit of small δ. We can then expand the potential term and derive a differential equation for Z,

$$T\partial_\ell Z(\mathbf{r}, \mathbf{0}; \ell) = \left(\frac{T^2}{2\tilde{\varepsilon}_1} \nabla_\perp^2 - V_1(\mathbf{r}) \right) Z(\mathbf{r}, \mathbf{0}; \ell).$$

(A.5)

A formal expression for the partition function results from integrating Eq. (A.5) across a thickness L of sample with initial state $|\mathbf{0}\rangle$ and final state $|\mathbf{r}_\perp\rangle$,

$$Z(\mathbf{r}_\perp, \mathbf{0}; \ell) = \langle \mathbf{r}_\perp | e^{-\mathcal{H}L/T} | \mathbf{0} \rangle,$$

(A.6)

where $\mathcal{H} = [-T^2/(2\tilde{\varepsilon}_1)]\nabla_\perp^2 - V_1(\mathbf{r})$. The partition function $Z(0, r_\perp; L)$ has been defined to be the ratio of the path integrals which appear in Eq. (7.3). Upon inserting a complete set of normalized eigenfunctions $\{\psi_n(\mathbf{r})\}$ of $\mathcal{H}$ with energies $\{E_n\}$ into (A.6), we obtain

$$e^{U(T)L/T} = Z(\mathbf{r}_\perp, \mathbf{0}; L)$$

$$- \sum_n \psi_n(\mathbf{r}_\perp) \psi_n(\mathbf{0}) e^{-E_n L/T},$$

(A.7)

which leads, when $L \to \infty$, to the result

$$U(T) = -E_0(T)$$

(A.8)

quoted in the text.

Appendix B. Vortex probability distributions

We show here that the probability of finding an individual vortex line at height z with position $\mathbf{r}_\perp$ in an arbitrary binding potential $V_1(\mathbf{r}_\perp)$ is related to the ground-state wave function of the corresponding Schrödinger equation. The probability distribution at a free surface is proportional to the wave function itself, whereas the probability far from the surface is proportional to the wave function squared.

Consider first a fluxon that starts at the origin $\mathbf{0}$ and wanders across a sample of thickness L to position $\mathbf{r}_\perp$. As discussed in Appendix A, the partition function associated with this constrained path integral may be written as a quantum-mechanical matrix element

$$Z(\mathbf{r}_\perp, \mathbf{0}; L) = \int_{\mathbf{r}(0)=\mathbf{0}}^{\mathbf{r}(L)=\mathbf{r}_\perp} \mathcal{D}\mathbf{r}(z) \exp\left(-\frac{\tilde{\varepsilon}_1}{2T} \int_0^L \left|\frac{d\mathbf{r}}{dz}\right|^2 dz - \frac{1}{T} \int_0^L V_1[r(z)]\, dz \right)$$

$$\equiv \langle \mathbf{r}_\perp | e^{-L\mathcal{H}/T} | 0 \rangle, \tag{B.1}$$

where $|0\rangle$ is an initial state localized at $\mathbf{0}$ while $\langle \mathbf{r}_\perp |$ is a final state localized at $\mathbf{r}_\perp$. The "Hamiltonian" $\mathcal{H}$ appearing in (B.1) is the Schrödinger operator,

$$\mathcal{H} = -\frac{T^2}{2\tilde{\varepsilon}_1} \nabla_\perp^2 + V_1(\mathbf{r}). \tag{B.2}$$

The probability distribution $\mathcal{P}(\mathbf{r}_\perp)$ for the position of the tip of the vortex at the upper surface is then

$$\mathcal{P}(\mathbf{r}_\perp) = Z(\mathbf{r}_\perp, \mathbf{0}; L) \Big/ \int d^2 r_\perp\, Z(\mathbf{r}_\perp, \mathbf{0}; L). \tag{B.3}$$

Upon inserting a complete set of (real) energy eigenstates $|n\rangle$ with eigenvalues E_n into Eq. (B.1), we have

$$\mathcal{P}(\mathbf{r}_\perp) = \frac{\sum_n \psi_n(\mathbf{0})\psi_n(\mathbf{r}_\perp) e^{-E_n L/T}}{\sum_n \psi_n(0) \int d^2 r_\perp\, \psi_n(\mathbf{r}_\perp) e^{-E_n L/T}}. \tag{B.4}$$

In the limit $L \to \infty$ the ground state dominates, so the probability $\mathcal{P}(\mathbf{r}_\perp)$ becomes

$$\mathcal{P}(\mathbf{r}_\perp) \approx \frac{\psi_0(\mathbf{r}_\perp)}{\int d^2 r_\perp\, \psi_0(\mathbf{r}_\perp)} [1 + \mathcal{O}(e^{-(E_1 - E_0)L/T})], \tag{B.5}$$

where E_1 is the energy of the first excited state. Because the ground-state wave function is nodeless [28], $\mathcal{P}(r_\perp)$ is always positive and well defined.

Consider now the more general problem of a vortex that enters the sample at $\mathbf{r}_i$, exits at $\mathbf{r}_f$ and passes through $\mathbf{r}$ at a height z that is far from the boundaries. The normalized probability distribution is now

$$\tilde{P}(\mathbf{r}; L) = \tilde{Z}(\mathbf{r}; L) \Big/ \int d^2 r\, \tilde{Z}(\mathbf{r}; L), \tag{B.6}$$

where

$$\tilde{Z}(\mathbf{r}; L) = \int d^2 r_i \int d^2 r_f\, Z(\mathbf{r}_f, \mathbf{r}; L - z) Z(\mathbf{r}, \mathbf{r}_i; z) \tag{B.7}$$

and $Z(\mathbf{r}_2, \mathbf{r}_1; L)$ is given by Eq. (B.1). Upon inserting complete sets of states as before, we find that

$$\tilde{P}(\mathbf{r}; L) = \frac{\psi_0^2(r)}{\int d^2 r_\perp\, \psi_0^2(r)} [1 + \mathcal{O}(e^{-L(E_1 - E_0)/2T})], \tag{B.8}$$

where the correction assumes that $\mathbf{r}$ is at the midplane of the sample.

References

[1] G. W. Crabtree and D. R. Nelson, *Phys. Today* **50**, 38 (1997).

[2] G. Blatter, M. V. Feigel'man, V. B. Geshkenbein, A. I. Larkin and V. M. Vinokur, *Rev. Mod. Phys.* **66**, 1125 (1994).

[3] P. G. de Gennes, *Superconductivity of Metals and Alloys* (Addison-Wesley, Reading, MA, 1989).

[4] M. Tinkham, *Introduction to Superconductivity* (McGraw-Hill, New York, 1996).

[5] D. R. Nelson, *Mol. Cryst. Liq. Cryst.* **288**, 1 (1996).

[6] E. Zeldov, D. Majer, M. Konczykowski, V. B. Geshkenbein, V. M. Vinokur and H. Shtrikman, *Nature* **375**, 373 (1995); see also D. R. Nelson, *Nature* **375**, 356 (1995).

[7] D. Ertas and D. R. Nelson, *Physica* C **272**, 79 (1996).

[8] A. A. Abrikosov, *Zh Éksp. Theor. Fiz.* **32**, 1442 (1957) [*Sov. Phys. JETP* **5**, 1174 (1957)].

[9] M. Charalambous, J. Chaussy and P. Lejay, *Phys. Rev.* B **45**, 45 (1992).

[10] W. K. Kwok, S. Fleshler, U. Welp, V. M. Vinokur, J. Downey, G. W. Crabtree and M. M. Miller, *Phys. Rev. Lett.* **69**, 3370 (1992).

[11] D. E. Farrell, J. P. Rice and D. M. Ginsberg, *Phys. Rev. Lett.* **67**, 1165 (1991).

[12] H. Safar, P. L. Gammel, D. A. Huse, D. J. Bishop, J. P. Rice and D. M. Ginsberg, *Phys. Rev. Lett.* **69**, 824 (1992).

[13] E. Brezin, D. R. Nelson and A. Thiaville, *Phys. Rev.* B **31**, 7124 (1985).

[14] D. R. Nelson and P. Le Doussal, *Phys. Rev.* B **42**, 10112 (1990).

[15] D. S. Fisher, M. P. A. Fisher and D. A. Huse, *Phys. Rev.* B **43**, 130 (1991).

[16] R. C. Budhani, M. Suenaga and H. S. Liou, *Phys. Rev. Lett.* **69**, 3816 (1992).

[17] M. Konczykowski, F. Rullier-Albenque, E. R. Yacoby, A. Shaulov, Y. Yeshurun and P. Lejay, *Phys. Rev.* B **44**, 7167 (1991).

[18] L. Civale, A. D. Marwich, T. K. Worthington, M. A. Kirk, J. R. Thompson, L. Krusin-Elbaum, Y. Sun, J. R. Clem and F. Holtzberg, *Phys. Rev. Lett.* **67**, 648 (1991).

[19] W. Gerhauser *et al.*, *Phys. Rev. Lett.* **68**, 879 (1992).

[20] V. Hardy, D. Groult, M. Hervieu, J. Provost, B. Raveau and S. Bouffard, *Nucl. Instrum. Methods* B **54**, 472 (1991).

[21] A. P. Malozemoff, T. K. Worthington, Y. Yeshurun and F. Holtzberg, *Phys. Rev.* B **38**, 7203 (1988).

[22] D. R. Nelson and V. M. Vinokur, *Phys. Rev. Lett.* **68**, 2392 (1992); *Phys. Rev.* B **48**, 13060 (1993).

[23] R. P. Feynman and A. R. Hibbs, *Quantum Mechanics and Path Integrals* (McGraw-Hill, New York, 1965); R. P. Feynman, *Statistical Mechanics* (Benjamin, Reading, MA, 1972).

[24] D. R. Nelson, *Phys. Rev. Lett.* **60**, 1973 (1988); D. R. Nelson and S. Seung, *Phys. Rev.* B **39**, 9153 (1989).

[25] S. F. Edwards, *Proc. Phys. Soc.* **85**, 613 (1965); P. G. de Gennes, *Rep. Prog. Phys.* **32**, 187 (1969).

[26] M. P. A. Fisher, P. B. Weichman, G. Grinstein and D. S. Fisher, *Phys. Rev.* B **40**, 546 (1989) and references therein.

[27] D. R. Nelson, in *Phenomenology and Applications of High Temperature Semiconductors*, edited by K. Bedell, M. Inui, D. Meltzer, J. R. Schrieffer and S. Doniach (Addison-Wesley, New York, 1991).

[28] L. D. Landau and E. M. Lifshitz, *Quantum Mechanics*, 2nd Edition (Pergamon, New York, 1965).

[29] See, e.g., D. R. Nelson, *J. Statist. Phys.* **57**, 511 (1989).

[30] N. W. Ashcroft and N. D. Mermin, *Solid State Physics* (Sanders College, Philadelphia, 1976), Chapter 32.

[31] See, e.g., L. I. Glazman and A. E. Koshelev, *Phys. Rev.* B **43**, 2835 (1991).

[32] D. S. Fisher, *Phys. Rev.* B **22**, 1190 (1980).

[33] M. C. Marchetti and D. R. Nelson, *Phys. Rev.* B **42**, 9938 (1990); *Physica C* **174**, 40 (1991).

[34] M. Cates, *Phys. Rev.* B **45**, 12415 (1992).

[35] S. Obukhov and M. Rubinstein, *Phys. Rev. Lett.* **66**, 2279 (1991); see also S. Obukhov and M. Rubinstein, *Phys. Rev. Lett.* **65**, 1279 (1990).

[36] E. H. Brandt, J. R. Clem and D. G. Walmsley, *J. Low Tempt. Phys.* **37**, 43 (1979).

[37] V. G. Kogan, *Phys. Rev.* B **24**, 1572 (1981).

[38] H. Safar, P. L. Gammel, D. A. Huse, D. J. Bishop, W. C. Lee, J. Giapintzakis and D. M. Ginsberg, *Phys. Rev. Lett.* **70**, 3800 (1993).

[39] S. Teitel, private communication.

[40] M. V. Feigel'man. V. B. Geshkenbein, A. I. Larkin and V. M. Vinokur, *Phys. Rev. Lett.* **63**, 2303 (1989).

[41] S. Ryu, A. Kapitulnik and S. Doniach, *Phys. Rev. Lett.* **71**, 4245 (1993).

Chapter 8
Correlations and transport in vortex liquids

Preface

This chapter describes in some detail a "boson analogy" that has proven useful for describing the statistical mechanics of *many* interacting vortex lines at finite temperatures. The meaning of off-diagonal long-range order, in terms of the behavior of a pair of well-separated magnetic monopoles superimposed on a vortex configuration, is described. This chapter also discusses a hydrodynamic theory of the motion of viscous, entangled flux liquids in the presence of a few strong macroscopic pinning centers.

The "boson-localization" ideas for vortices have now been developed into a comprehensive theory of a Bose-glass transition for flux lines, including the phenomenon of the transverse Meissner effect [1, 2]. For interesting new effects that arise when columnar defects are *splayed* instead of parallel, see [3]. For a detailed survey of results for flux liquids obtainable via the boson analogy in a variety of contexts, see [4]. In [5] there is an analysis of the response of the Bose glass to a transverse magnetic field in terms of solutions of a (non-Hermitian) Schrödinger equation with a constant imaginary vector potential and [6] analyzes the *continuous* freezing transition associated with evolution of smectic order out of a vortex liquid. Experiments suggest the possibility of such a continuous freezing in a layered vortex crystal for fields oriented nearly parallel to the CuO_2 planes in the cuprate high-temperature superconductors [7].

Considerable activity has been directed toward understanding various vortex-glass phases in impure type-II superconductors, especially when they are subjected to strong point-like disorder. Of particular importance is the idea of a stable, topologically ordered (arbitrarily large dislocation loops are forbidden) "Bragg glass," which has algebraic peaks at the reciprocal-lattice positions usually associated with an Abrikosov flux lattice [8–11], similar to the algebraic translational order discussed for two-dimensional crystals in Chapter 2. There are also fascinating issues involving flowing vortex "phases" driven far from equilibrium by a large supercurrent. Nattermann and

Scheidl [12] provide an extensive review of recent developments in this area.

Kes and collaborators have carried out beautiful experiments [13] designed to probe vortex flow in confined geometries along the lines discussed in this chapter. One result of these experiments is a direct measurement of the intervortex shear viscosity of arrays of vortices near the two-dimensional freezing transition in superconducting films. Lopez *et al.* have recently investigated the motion of vortices in an annular "Corbino-disk" geometry with a radial current, so that edge pinning of flux lines is minimized [14]. For a study of viscous vortex hydrodynamics applicable to this situation, see [15]. Paltiel *et al.* have carried out striking Corbino-disk flow experiments in $NbSe_2$, which strongly suggest that a first-order transition out of a crystalline or Bragg-glass phase occurs for this material [16].

8.1 Introduction

In 1957, Abrikosov presented his remarkable mean-field theory of the mixed state of type-II superconductors [17]. For applied fields H such that $H_{c1} < H < H_{c2}$, the magnetic field penetrates in the form of quantized flux tubes, which, in the absence of disorder, form a regular lattice of parallel defect lines. In conventional low-temperature superconductors, this line lattice was believed to exist at essentially all temperatures up to $H_{c2}(T)$ [18]. Disorder and pinning (which play a crucial role in transport experiments) were described by the extent to which they broke up the translational order embodied in the underlying Abrikosov flux lattice [19]. The flow of flux in the presence of an externally imposed current was assumed to take place via the motion of blobs of flux crystal, with dimensions given by disorder-induced translational correlation lengths parallel and perpendicular to the direction of the field [19].

The new high-temperature (HTC) superconductors can apparently be described by the same *s*-wave Ginzburg–Landau functional as that used by Abrikosov for conventional materials. Striking differences arise, however, because the phenomenological coupling constants have very unusual values and because the theory must be solved in a qualitatively different, high-temperature regime. It has been argued, in particular, that a melted vortex liquid replaces the conventional Abrikosov flux lattice over large regions of the phase diagram because of the weak interplanar couplings, high critical temperatures and short coherence lengths characteristic of the new HTC materials [20–23].

These theoretical suggestions were inspired by early flux-decoration experiments on $YBa_2Cu_3O_7$ (YBCO) [24]. For simplicity, we shall confine our attention throughout this review to results for magnetic fields along the z axis, perpendicular to the CuO_2 planes. Although flux quanta (decorated via the Bitter technique) were observed emerging from a single-crystalline sample at $T = 4.2$ K, no flux patterns have ever been discerned at $T = 77$ K, possibly due to time-dependent flux wandering in an equilibrated flux liquid. Regions of extensive crystallinity were subsequently observed at low temperatures in YBCO by Dolan et al. [25], suggesting that melting into a vortex liquid could proceed from a highly ordered state, as well as from a disorder-dominated "vortex glass" [26]. Extensive decoration studies of $Bi_2Sr_2CaCu_2O_8$ (BSCCO) have recently revealed a transition from disorderly flux arrays at very low fields ($H \lesssim 15$ Oe) to "hexatic-glass" configurations with very-long-range orientational order at higher fields [27, 28]. A broken orientational symmetry – even if translational order is absent – is sufficient to insure that a thermally driven phase transition (hexatic "melting") to an isotropic vortex liquid will occur at higher temperatures.

Because the magnetic field becomes almost uniform when vortex lines are much closer than the in-plane London penetration depth λ_{ab}, decoration experiments are usually restricted to fields of less than a few hundred gauss. The first experimental claim for melting at high fields ($H \gtrsim 10^3$ Oe) was based on a peak observed in vibrating-reed experiments on YBCO and BSCCO by Gammel et al. [29]. The very sharp peak found in recent low-frequency torsional-oscillator experiments by Farrell et al. [30] supports the hypothesis that this effect is an actual phase transition, rather than merely a thermal enhancement of the mobility of vortices [31]. Resistance measurements in very clean YBCO crystals by Worthington et al. [32] have also been interpreted as evidence for an underlying melting transition. The vanishing shear modulus associated with melting would allow the vortex liquid to flow freely around macroscopic pinning centers, leading to an abrupt increase in the resistivity [33].

In real HTC superconductors, any putative melting must take place in the presence of disorder, such as oxygen vacancies or twin boundaries. One can certainly question the interpretation of experiments like those above when the disorder is strong. Strong disorder, varying on scales comparable to the intervortex spacing, can undoubtedly destroy the translational and orientational order associated with the Abrikosov flux lattice and completely wash out a sharp melting transition. We know, moreover, that arbitarily *weak* disorder will *always* destroy translational correlations at sufficiently large length scales [19].

The operative question, however, is that of how large the translational and orientational lengths actually become below the "melting" transition. If the disorder is predominantly due to oxygen vacancies, which pin very weakly at liquid-nitrogen temperatures [34], translational correlation lengths of hundreds, even thousands, of lattice constants may be possible at high fields. The measured translational correlation lengths at low fields in [28] become as large as 20 lattice constants when $B \approx 100$ G and are in fact steadily *increasing* with applied field. The orientational correlation range is *hundreds* of lattice constants and displays the same trend with field. Under these circumstances, large regions of vortex lines at high fields may appear to be crystalline, disrupted only by widely spaced macroscopic defects such as twin boundaries. "Melting" would then be an appropriate description for the destruction of the translational and orientational order within these crystalline "grains," even if different physics arises on scales large relative to the grain spacing. Note that macroscopic planar defects such as twin boundaries are an example of *correlated* disorder. Theoretical treatments based on microscopic Gaussian disorder with only short-range correlations [19] are *not* appropriate at these large length scales, especially if the twins pass completely through the sample.

In this paper, we shall be interested primarily in vortex liquids, regardless of whether they form a "polycrystalline" array or a disorder-dominated glass at low temperature. A vortex liquid has no translational order and can flow appreciably on experimental time scales. To have some idea of when vortex liquids can occur, we first discuss where they appear in the temperature–field phase diagram, both for relatively clean HTC superconductors and for superconductors dominated by disorder.

8.1.1 Phase diagrams

A phase diagram for very pure materials, with the applied field H perpendicular to the copper-oxide planes, is shown in Fig. 8.1(a). The line $H_{c2}^0(T)$ marks the onset of a significant expulsion of flux from the sample and is not a sharp phase transition. The melting transition near H_{c1} at low fields was first predicted [20] by mapping the statistical mechanics of thermally excited vortex lines onto the physics of two-dimensional boson superfluids [35]. Melting occurs from this point of view because bosons with a purely repulsive pair potential (decreasing faster than $1/r^2$) are always melted by zero-point motion at sufficiently low density. When applied to flux lines with a repulsive exponential interaction [20], this argument simply means that entropy (i.e., "zero-point motion" in the boson language) always favors an entangled flux liquid over an ordered vortex crystal at low densities. When the flux lines are very dense, the interaction between neighboring lines is a logarithmic, rather than exponential, repulsion [36]. The boson analogy then

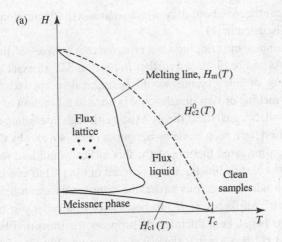

(a) H

Melting line, $H_m(T)$

$H_{c2}^0(T)$

Flux
lattice

Flux
liquid

Clean
samples

Meissner phase

$H_{c1}(T)$ T_c T

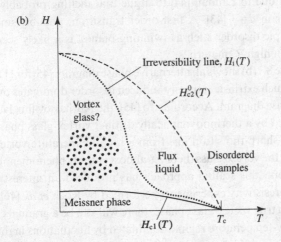

(b) H

Irreversibility line, $H_i(T)$

$H_{c2}^0(T)$

Vortex
glass?

Flux
liquid

Disordered
samples

Meissner phase

$H_{c1}(T)$ T_c T

Fig. 8.1. Phase diagrams for (a) clean and (b) highly disordered high-temperature superconductors. The dashed lines mark the onset of the Meissner effect, which is not a sharp phase transition.

predicts that zero-point motion always melts the crystal at sufficiently *high* densities, which accounts for the reentrant upper branch of the melting curve in Fig. 8.1(a) [37].

A more precise determination of the melting curve for a particular material follows from the Lindemann criterion. As pointed out in [38], the Lindemann integrals are dominated by wavevectors near the zone boundary and wavevector-dependent (i.e., nonlocal) elastic constants should be used to obtain quantitatively accurate estimates. This has been done by Houghton *et al.* [23] who produced reasonable fits to the high-field melting lines proposed by Gammel *et al.* [29] for YBCO and BSCCO materials with a single adjustable Lindemann parameter. The importance of wavevector-dependent elastic constants in superconductors has also been emphasized by Brandt [39]. The tilt and bulk moduli are soft near the zone boundary, which *enhances* the fluctuation-induced melting discussed in [21].

The Lindemann condition is a *criterion*, not a *theory* of flux-lattice melting. As will be discussed further in Section 8.2, an exact theory of the melting of line crystals would be equivalent to understanding quantum melting of two-dimensional bosons as a function of pressure or $\hbar$ at $T=0$. No analytical theory exists, although interesting efforts in this direction have been made by Sengupta *et al.* [40] and by Chui [41]. Dislocation-mediated melting of the flux lattice would lead to a three-dimensional *hexatic* line liquid, as pointed out in [42]. It can be shown that amplitude fluctuations cause the continuous mean-field-theory transition of Abrikosov at H_{c2} to become of first order just below six dimensions [43]. These fluctuations suppress the transition below the mean field H_{c2} line, so that there is a nonzero amplitude of the local order parameter, even in the normal state. One can then invoke a famous argument due to Landau [44] to argue that melting probably remains first order in $d=3$ [43]. A first-order transition, possibly smeared by macroscopic disorder such as twinning planes, is a likely scenario for most clean high-T_c materials.

Figure 8.1(b) shows an alternative phase diagram [45] for HTC materials in which extrinsic impurity-induced disorder dominates over much of the phase diagram. According to [45], the Abrikosov flux lattice may be replaced by a thermodynamically distinct vortex-glass phase, separated by a sharp transition line from a high-temperature vortex liquid. Although the vortex-glass theory is almost entirely phenomenological at the moment, this scaling approach has produced an interesting fit to results of resistivity experiments performed by Koch *et al.* [46]. Even if there is no true vortex-glass phase, there will still be a gradual crossover from a low-temperature regime dominated by fluctuations in the impurity potential to a high-temperature region dominated by thermal fluctuations. In this case, the dotted line in Fig. 8.1 would simply represent a locus of crossover temperatures [31]. If the disorder is weak, there should actually be a deep notch of flux liquid, stable against both crystallization and disorder, cut into this line. See Eqs. (8.73) and (8.75) below.

To provide a point of reference for vortex liquids in three dimensions, it is worth summarizing what is known about *point* vortices in two dimensions, for which there is the well-developed theory of melting discussed in Chapter 2 [47]. The existence of liquid phases of point vortices in conventional low-temperature superconducting films was proposed in [48, 49]. A phase diagram predicted (for clean samples) using the dislocation/disclination theory of two-dimensional melting is shown in Fig. 8.2 [49]. Fluctuations are particularly strong in two dimensions, so a substantial region of the mean-field Abrikosov phase

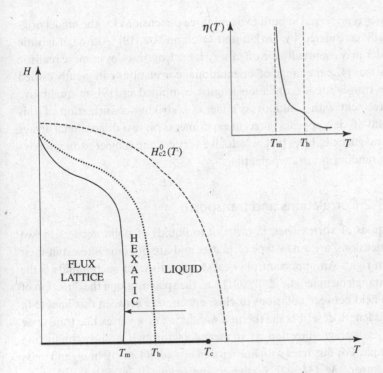

Fig. 8.2. A schematic phase diagram for a thin-film superconductor [49]. H_{c1} is zero in thin films. The inset shows the expected temperature dependence of the viscosity along the path indicated by the arrow [47].

diagram is occupied by a flux liquid, just as for three-dimensional HTC materials. The estimate of the melting line is based on the Kosterlitz–Thouless dislocation criterion [47–49]. This condition is an *upper bound* on the true melting temperature, which insures that there is indeed a large region of melted flux liquid, even if the melting transition occurred via some other mechanism. In addition to an isotropic liquid of point vortices, there is also a region of hexatic order, to be discussed further in Section 8.3.

The inset to Fig. 8.2 shows the shear viscosity [50] of the vortex liquid on the path indicated by an arrow, as predicted by the theory of dislocation-mediated melting. The hexatic-to-crystal transition is continuous and characterized by a strongly diverging intervortex viscosity. As we shall see, the viscosity of a flux liquid in three dimensions can also be large, not only near the melting temperature but also over much of the temperature–field phase diagram [21]. As will be discussed in Section 8.4, these large viscosities allow the effects of a few strong pins to propagate large distances, with important implications for transport experiments [51].

Strong disorder will, of course, eliminate the crystalline phase in Fig. 8.2, just as it does in three dimensions [19]. No vortex-glass transition is expected in two dimensions although sluggish dynamics may accompany a diverging vortex-glass correlation length at zero temperature [45]. As pointed out by Chudnovsky [52], hexatic orientational order is *preserved* in both two and three dimensions for the model originally considered by Larkin and Ovchinnikov [19]. Although hexatic order may eventually die off at *very* large length scales in more realistic systems [42], the range of orientational correlations can greatly exceed the translational correlation lengths computed in [19], in qualitative agreement with Chudnovsky's ideas. A striking confirmation of this point of view appears in recent experiments on two-dimensional arrays of magnetic bubbles [53], which, like vortices, are subject to an underlying random substrate potential.

8.1.2 Correlations and transport

Liquids of vortex *lines*, in contrast to liquids of point vortices in two dimensions, are a new type of matter and are of some interest in their own right. An important physical characteristic of such liquids is the "entanglement length" ℓ_z [20, 21], i.e., the spacing along the direction of the field between collisions or close encounters between flux lines [54]. This length describes the thermal wandering of a vortex line transverse to its average direction as it meanders through a superconducting sample. For flux lines with line tension $\varepsilon_1 = [\phi_0/(4\pi\lambda_{ab})]^2 \ln \kappa_{ab}$ and mass anisotropy $M_\perp/M_z \ll 1$, ℓ_z is given approximately by [55]

$$\ell_z \simeq (M_\perp/M_z)[\phi_0\varepsilon_1/(Bk_B T)]. \tag{8.1}$$

Note that ℓ_z decreases with increasing magnetic field and mass anisotropy.

Consider a vortex liquid confined to a slab of thickness L along the direction of the field with average spacing c_0 between copper-oxide planes. Vortex liquids should then display the three qualitatively different regimes shown in Fig. 8.3, depending on the value of the magnetic field relative to two important crossover fields. The first field is obtained by equating ℓ_z to L [55]:

$$B_{x1} = (M_\perp/M_z)[\phi_0\varepsilon_1/(Lk_B T)]. \tag{8.2}$$

For $B \lesssim B_{x1}$, we have a *disentangled* flux liquid, whose properties should be qualitatively similar to those of the liquid of point vortices indicated in Fig. 8.2. For $L = 1$ mm, this field is about 100 G.

Above $B = B_{x1}$, we have an entangled flux liquid. Entanglement continues to increase until B exceeds B_{x2}, which is the field such that $\ell_z = c_0$,

$$B_{x2} = (M_\perp/M_z)[\phi_0\varepsilon_1/(c_0 k_B T)]. \tag{8.3}$$

When $B > B_{x2}$, flux lines are "superentangled," in the sense that vortices hop more than an intervortex spacing on passing from one CuO_2 plane to the next. It then makes little sense to draw lines connecting them.

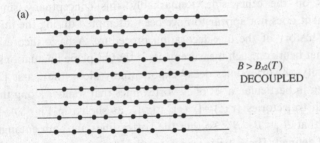

(a)

$B > B_{x2}(T)$
DECOUPLED

Fig. 8.3. A schematic diagram of flux lines relative to the CuO_2 planes in HTC materials, viewed edge on. Three regimes are shown: (a) decoupled, (b) entangled and (c) disentangled.

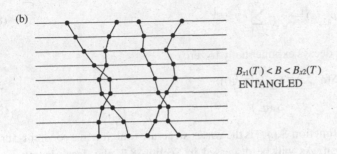

(b)

$B_{x1}(T) < B < B_{x2}(T)$
ENTANGLED

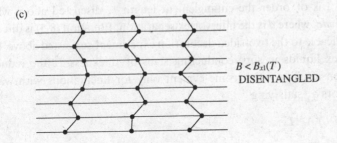

(c)

$B < B_{x1}(T)$
DISENTANGLED

Above B_{x2} (about 50 T for YBCO but only about 1 T for BSCCO), the liquid degenerates into decoupled planes of point vortices. A crossover at $B_{x2}(T)$ appears to be the explanation for the changeover in resistivity observed recently in artificial multilayers by White *et al.* [56]. An identical criterion for decoupling vortices in neighboring copper-oxide planes has been suggested by Glazman and Koshelev [57]. A related condition for decoupling in the *crystalline* phase was also derived by these authors, as well as by Fisher *et al.* [45].

Three-dimensional equal-time correlations in flux-line liquids may be measurable via neutron diffraction [58]. An interesting alternative would be to compare (using magnetic decoration or tunneling microscopy) the configuration of flux lines entering a sample with that which emerges on the other side. Remarkably, this conceptually simple experiment does not appear to have been attempted during the initial 25-year history of the flux-decoration procedure. A basic theoretical result that bears on such measurements is that in-plane fluctuations in the density of vortex lines decay exponentially along the z axis [21]. Consider, in particular, a set of N vortex lines that wander along the z axis with trajectories $\{r_j(z) = [x_j(z), y_j(z)]\}$, as shown in Fig. 8.4. We assume that $B_{x1} < B < B_{x2}$, so that the vortex lines are both entangled and well defined. The Fourier-transformed density in a constant-z cross section,

$$\rho_{q_\perp}(z) = \frac{1}{\sqrt{N}} \sum_{j=1}^{N} e^{iq_\perp \cdot r_j(z)} \tag{8.4}$$

then decays exponentially to zero,

$$\bar{S}(q_\perp, z) \equiv \langle \rho_{q_\perp}(z) \rho_{q_\perp}^*(0) \rangle$$
$$\approx S_2(q_\perp) e^{-|z|/\ell_z(q_\perp)}. \tag{8.5}$$

The function $S_2(q_\perp)$ is the vortex structure function in a constant-z cross section. As will be discussed in Section 8.3, the decay length $\ell_z(q_\perp)$ diverges as $q_\perp \to 0$ and is simply related to the phonon–roton spectrum of the equivalent two-dimensional boson superfluid. We expect that $\ell_z(q_\perp)$ is of order the entanglement length ℓ_z discussed above when $q_\perp \approx \pi/d$, where d is the intervortex separation. Equation (8.5) is directly applicable to the two-sided decoration experiment suggested above. For vortex liquids in superconducting slabs of thickness L, the reduced dimensionality will become evident only for fluctuations with wave-vectors $q_\perp$ satisfying

$$\ell_z(q_\perp) \gtrsim L. \tag{8.6}$$

For a discussion of double-sided flux-decoration experiments that measure $\ell_z(q_\perp)$, see Ref. [1] of Chapter 7.

The above results were first derived in the absence of the pinning disorder shown in Fig. 8.4. Dense flux liquids, however, are *stable* against weak disorder and will persist even if the Abrikosov flux lattice is replaced by a glassy phase at low temperatures [59]. Weak disorder merely produces "Lorentzian squared" corrections to the Fourier transform of Eq. (8.5) – see Section 8.2. This insensitivity of vortex liquids to disorder arises because flux lines collide and "restart" their random walks along the z axis before the effects of randomness can significantly alter their trajectories.

The explicit microscopic derivation of these results [21, 59] assumes the applicability of simplified models that ignore nonlocal interactions between vortex lines. Such interactions produce quantitative changes

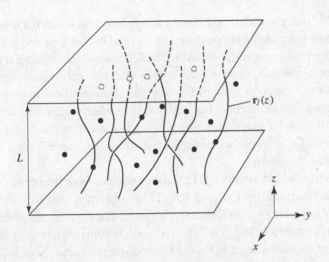

Fig. 8.4. A schematic diagram of flux lines wandering through a sample of thickness L in the presence of random impurities.

in phase boundaries at high fields [23, 39]. Nonlocal interactions, however, are easily incorporated into a long-wavelength "hydrodynamic" version of the theory [59], with essentially no changes in the basic conclusions.

There are also important issues related to the *dynamics* of flux-line liquids. As was suggested in [21, 51], these liquids can become very viscous, allowing the effects of a few strong pinning centers to propagate over long distances. A large viscosity will arise provided that (1) the lines are entangled and (2) there is a large barrier to flux-line crossing [60]. The entanglement criterion simply means that B must exceed $B_{x1} \approx 100$ G. A barrier to line crossing arises because two singly quantized vortices behave like a doubly quantized one in the region where the crossing takes place. The energy stored in the circulating supercurrents which extend a distance λ_{ab} perpendicular to the lines is proportional to the *square* of the effective flux quantum. If ε_1 is the line tension and the crossing takes place over a distance ℓ_0 along the z axis, the crossing energy is then roughly [21]

$$U_x \approx (2^2 - 1^2 - 1^2)\varepsilon_1 \ell_0$$
$$= 2\varepsilon_1 \ell_0 \qquad\qquad (8.7)$$

when the lines are dilute. For flux lines at arbitrary densities, this formula becomes [61]

$$U_x = 2\varepsilon_1 \ell_0 \ln (H_{c2}^0/H)/\ln(H_{c2}^0/H_{c1}). \qquad\qquad (8.8)$$

Although the crossing energy vanishes near the mean-field upper critical field H_{c2}^0, it is more than 20 times $k_B T$ (with $\ell_0 = 10$ Å) at liquid-nitrogen temperatures for $B = 1$ T in YBCO, consistent with a very high viscosity [51].

In isotropic superconductors, some of this interaction energy can be reduced by forcing the lines to cross at 90° [62]. Something like this surely takes place above the decoupling field B_{x2}, at which the requisite large crossing angles are produced by thermal fluctuations and the continuum picture breaks down. At lower fields, however, crossing at large angles requires additional line length and the net crossing energy remains of the order given by Eq. (8.8) [63]. Because the currents prefer to flow in the anisotropic CuO_2 planes even when vortex lines tilt, the critical angle for easy flux cutting is in any case much *larger* than the 90° required for isotropic materials. See Section 4 of Chapter 7 for a more detailed discussion.

Suppose that we cool a HTC superconductor from the normal state at fixed magnetic field, $B_{x1} < B < B_{x2}$. The viscosity is always small near $H_{c2}^0(T)$ at which vortex lines are little more than loci of zeroes of the order parameter. The flux liquid should become significantly more viscous, however, once $k_B T < U_x(T)$, because the vortex lines become more tangible and resist crossing. If this intervortex viscosity becomes large well before the equilibrium freezing line, one might expect regimes in which a vortex liquid drops out of equilibrium at a polymer-like glass transition well before it ever freezes into a flux crystal [21, 61]. Such a "polymeric-glass" state has little to do with oxygen-vacancy disorder, although it will greatly influence the response of the system to a few strong pins. The polymer-glass idea provides an interesting alternative to the "vortex-glass" conjecture [45] as an explanation for experimentally observed "irreversibility lines" [64].

How can we understand transport measurements in vortex-line liquids? Because the underlying Abrikosov flux lattice is melted, it does not make sense to invoke the traditional picture [19] of collective motion of crystalline flux bundles [65]. At the opposite extreme is the idea of "thermally assisted flux flow" (TAFF) [66], which essentially assumes that we have an ideal gas of disconnected flux bits moving in a tilted washboard potential. This approach allows thermally activated depinning to be incorporated into the usual [36] Bardeen–Stephen flux-flow resistivity formulas. The TAFF picture, however, takes little account of intervortex interactions and is not well suited for dealing with phase transitions. To describe transport in strongly interacting vortex liquids, it is essential to include the intervortex viscosity.

To see how this viscosity affects flux-flow experiments, consider the flow of an isotropic flux liquid in a channel contained between two flat twin boundaries in the xz-plane at $y = -W/2$ and $y = W/2$, shown in Fig. 8.5. The magnetic field is in the z direction. The transport current is applied along $\hat{y}$, $\mathbf{j}_T = j_T \hat{y}$, and yields a constant driving Lorentz force [36]

$$\mathbf{f}_T = \frac{1}{c} n_0 \phi_0 \hat{z} \times \mathbf{j}_T \tag{8.9}$$

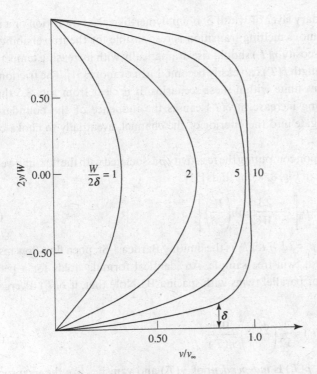

Fig. 8.5. Flux-line-velocity profiles for flow in a channel between two twin boundaries. We have assumed for simplicity that the velocity is forced to vanish on the boundary. The viscous length δ is indicated for the case $W/(2\delta) = 10$.

in the x direction, where n_0 is the areal vortex density. The equation of motion which describes the velocity field of the vortex liquid is then [51]

$$-\gamma \mathbf{v} + \eta \nabla_\perp^2 \mathbf{v} + \mathbf{f}_T = 0. \qquad (8.10)$$

Equation (8.10) is a hydrodynamic description of flux flow, which is valid on scales large relative to the intervortex spacing. The combination of the drag and the viscous term introduces an important new length scale into the problem,

$$\delta = \sqrt{\eta/\gamma}. \qquad (8.11)$$

Here γ is the vortex "friction" coefficient of Bardeen and Stephen [36] augmented by the effects of weak microscopic disorder such as oxygen vacancies. The parameter η is the intervortex viscosity discussed above. We assume for simplicity that $\mathbf{v} = 0$ at the walls, due to strong pinning by the twin boundaries, which are believed to contain exceptionally high concentrations of point disorder. The solution of (8.10) is then

$$v_x(y) = v_\infty \left(1 - \frac{\cosh(y/\delta)}{\cosh[W/(2\delta)]} \right). \qquad (8.12)$$

Here $v_\infty = \mathbf{f}_T/\gamma$ is the usual Bardeen–Stephen limiting flux-line velocity obtained far from any strong pins. The flow velocity drops to zero in a boundary layer of width δ. If a polymer-like glass transition or a nearly continuous melting transition is responsible for the irreversibility line, the viscosity $\eta(T)$ should rise dramatically with increasing temperature. The length $\delta(T)$ can easily become 1 μm or more [51]. The friction $\gamma(T)$ remains finite within these scenarios. It is clear from Fig. 8.5 that the resulting increase in $\delta(T)$ causes the influence of the boundaries to propagate into the interior of the channel, eventually to choke off the flow.

Upon computing the resistivity ρ associated with the flux-line-velocity profile in Fig. 8.4, one finds [51]

$$\rho = \rho_0 \left[1 - \frac{2\delta}{W} \tanh\left(\frac{W}{2\delta}\right) \right],$$

(8.13)

where $\rho_0 = (\phi_0 n_0/c)^2/\gamma$ is the limiting Bardeen–Stephen flux-flow resistivity for a twin-free sample. An identical formula holds for a periodic array of parallel twins with spacing W. Note that, if $\delta(T)$ diverges, we have

$$\rho(T) \approx \frac{1}{12} \left(\frac{\phi_0 n_0}{c} \right)^2 \frac{W^2}{\eta(T)},$$

(8.14)

so that $\rho(T)$ is *independent* of $\gamma(T)$ and vanishes like the reciprocal of the viscosity in this limit.

It is very important to determine the parameters $\gamma(T)$ and $\eta(T)$ separately. Although the bulk resistivity can be interpreted in terms of a diverging intervortex viscosity $\eta(T)$ [51], it is hard to rule out an even stronger divergence in $\gamma(T)$ (as predicted by the vortex-glass hypothesis [45]) to explain the vanishing resistivity in flux-flow experiments. In Section 8.4, we discuss how to perform an independent determination of $\gamma(T)$ and $\eta(T)$ and provide a clean experimental test of the various explanations of the irreversibility line.

8.1.3 Contents

In the remainder of this chapter, we present a more technical review of the ideas sketched above. In Section 8.2, we discuss the statistical mechanics of flexible vortex lines at finite temperature. We describe the thermal wandering of a single line and show that this produces a large, temperature-dependent renormalization of the energy for binding of a flux line to planar and linear defects. The boson analogy, which allows us to gain intuition about flux liquids from two-dimensional many-particle quantum mechanics, is presented in detail. For a more elementary introduction, see Chapter 7. This section concludes with a discussion of when disorder has a significant effect on the results for pure systems.

In Section 8.3, we discuss equilibrium correlations in flux liquids, using a hydrodynamic approach that is more general than the microscopic boson theories. We also discuss the intermediate hexatic phase, which may appear both in glassy form [52] and as an equilibrated flux liquid [42]. We conclude with some comments about vortex liquids when the dominant source of disorder is random planar twin boundaries or linear defects that pass completely through the sample. We argue that conventional models including only microscopic point disorder [19, 45] are then *inappropriate* and that a better description of the vortex-line configurations and dynamics results from an analogy with boson localization [1].

Flux-liquid dynamics is discussed in Section 8.4. We treat motion of flux in the presence of a few strong pins and describe how to distinguish among the four possible explanations of the irreversibility line via an experiment that probes the physics of "viscous electricity."

8.2 Statistical mechanics of flexible lines

8.2.1 The model free energy

Consider the Gibbs free energy for the N flux lines shown in Fig. 8.4, in a slab of thickness L, whose positions with a field $\mathbf{H}$ along the z direction (perpendicular to the CuO_2 planes) in a sample of length L are given by $\mathbf{r}_j(z) = [x_j(z), y_j(z)], j = 1, \ldots, N$. Since the ratio of the penetration depth λ_{ab} and the coherence length ξ_{ab} is typically quite large, $\kappa_{ab} = \lambda_{ab}/\xi_{ab} \approx 10^2$, we shall often work in the London limit. If

$$\varepsilon_1 = \left(\frac{\phi_0}{4x\lambda_{ab}}\right)^2 \ln \kappa_{ab}$$

is the energy per unit length of a single flux line and $\phi_0 = 2\pi\hbar c/(2e) = 2.07 \times 10^{-7} \, G \, cm^2$ is the flux quantum, the energy reads

$$G = \left(\varepsilon_1 - \frac{H\phi_0}{4\pi}\right)NL + \sum_{i>j} \int_0^L V[\mathbf{r}_{ij}(z)] \, dz + \sum_{j=1}^N \int_0^L dz \, V_D[\mathbf{r}_j(z), z]$$

$$+ \frac{1}{2}\tilde{\varepsilon}_1 \sum_{j=1}^N \int_0^L \left|\frac{d\mathbf{r}_j(z)}{dz}\right|^2 dz, \tag{8.15}$$

where $\mathbf{r}_{ij} = \mathbf{r}_i - \mathbf{r}_j$, $V(\mathbf{r}_{ij})$ is a vortex-line pair potential, which in the London approximation takes the form [36]

$$V(\mathbf{r}_{ij}) = \frac{\phi_0^2}{8\pi^2\lambda_{ab}^2} K_0\left(\frac{r_{ij}}{\lambda_{ab}}\right), \tag{8.16}$$

where $K_0(x)$ is the modified Bessel function, $K_0(x) \approx [\pi/(2x)]^{1/2}e^{-x}$ for large x. Although we have neglected interactions along the z axis, this

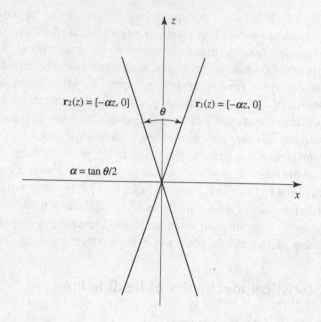

simplified model is still quite nontrivial and informative. Qualitatively similar results, moreover, emerge when these "nonlocal" interactions are taken into account using the hydrodynamic approach of Section 8.3.

The calculations of [21, 59] are in fact easily carried out for *arbitrary* in-plane pair potentials. Some of the calculations in [21], for example, used a cutoff delta-function pseudopotential,

$$V(\mathbf{r}_{ij}) = v_0 \int_{k<\lambda} \frac{d^2k}{(2\pi)^2} e^{i\mathbf{k}\cdot[\mathbf{r}_i(z)-\mathbf{r}_j(z)]}$$

$$\equiv v_0 \delta_\lambda [\mathbf{r}_i(z) - \mathbf{r}_j(z)], \tag{8.17}$$

where $v_0 = \phi_0^2/(4\pi)$. With this potential, it is easy to show that the crossing energy for the lines inclined at angle θ in Fig. 8.6 is finite and given by

$$U_x = \frac{v_0}{2\pi} \int_0^\lambda q\,dq \int_{-\infty}^\infty dz\, J_0(2\alpha qz)$$

$$= \frac{v_0\lambda}{2\pi} \cot\left(\frac{\theta}{2}\right). \tag{8.18}$$

The renormalization-group calculation in [21] is just a resummation of perturbation theory in U_x.

The function $V_D[\mathbf{r}_j(z), z]$ is a Gaussian pinning potential for individual vortex lines, representing the effects of impurities. If the defects are randomly distributed, as in the case of oxygen vacancies, the quenched fluctuations in the impurity potential will obey

$$\overline{V_D(\mathbf{r}, z) V_D(\mathbf{r}', z')} = \Delta\delta(\mathbf{r} - \mathbf{r}')\delta(z - z'), \tag{8.19}$$

with $\overline{V_D(\mathbf{r}, z)} = 0$. The overbar represents an average over the impurity disorder. An approximate formula for the coefficient Δ is given by Fisher et al. [45], namely,

$$\Delta \approx \frac{1}{4}\gamma_I^2 v_c n_I \left(\frac{T_c}{T_c - T}\right)^2 \frac{\phi_0^4}{16\pi^2\lambda_{ab}^2}, \tag{8.20}$$

where v_c is the volume of the unit cell in the underlying crystal, n_I is the fractional occupation number of impurities in this cell and

$$\gamma_I = d[\ln T_c(n_I)]/dn_I \tag{8.21}$$

is a dimensionless impurity coupling constant (typically of order unity). Note that disorder due to a frozen distribution of grain or twin boundaries is qualitatively different. The randomness is *correlated* by virtue of the planar nature of the disorder in this case.

For an isotropic superconductor, the last term in Eq. (8.15) comes from the expansion of the total line energy, $E_i = \varepsilon_1 \int_0^L (1 + |d\mathbf{r}_i/dz|^2)^{1/2} dz$. In this case, $\tilde{\varepsilon}_1 = \varepsilon_1$. For the anisotropic layered compounds under consideration here, $\tilde{\varepsilon}_1$ is considerably smaller,

$$\tilde{\varepsilon}_1 = \frac{M_\perp}{M_z}\varepsilon_1, \tag{8.22}$$

where $M_\perp$ is the in-plane effective mass and $M_z \approx 10^2 M_\perp$ is the much larger effective mass describing the weak Josephson coupling between the planes. The formula (8.22) applies when the flux lines are dense, $n_0\lambda_{ab}^2 \gg 1$ [21, 67]. In the opposite limit $n_0\lambda_{ab}^2 \lesssim 1$, the electromagnetic coupling between CuO_2 planes is important and one has [45]

$$\tilde{\varepsilon}_1 \approx \varepsilon_1/\ln \kappa. \tag{8.23}$$

Abrikosov's mean-field treatment of the transition at H_{c1} neglects disorder and assumes that the vortices form a triangular lattice of rigid rods with density $n_0 = B/\phi_0$ parallel to the z axis, so that the last term of (8.15) vanishes. Flux lines begin to penetrate when the first term changes sign, i.e., when $H \geq H_{c1} = 4\pi\varepsilon_1/\phi_0$.

A *full* statistical treatment of the partition function associated with Eq. (8.15) entails integration of $\exp[-G/(k_B T)]$ over all vortex trajectories $\{\mathbf{r}_i(z)\}$. The grand canonical partition function, for example, is

$$Z_{gr} = \sum_{N=0}^{\infty} \frac{1}{N!} \int \mathcal{D}\mathbf{r}_1(z) \cdots \int \mathcal{D}\mathbf{r}_N(z)\, e^{-G/(k_B T)}. \tag{8.24}$$

If we also wish to average over the quenched random impurity disorder, we should calculate

$$
\overline{\ln Z} = \frac{\int \mathcal{D}V_{\mathrm{D}}(\mathbf{r}_\perp, z)\, \ln Z \exp\left(-\frac{1}{2\Delta}\int d^2 r_\perp \int dz\, V_{\mathrm{D}}^2(\mathbf{r}_\perp, z)\right)}{\int \mathcal{D}V_{\mathrm{D}}(\mathbf{r}_\perp, z)\, \exp\left(-\frac{1}{2\Delta}\int d^2 r_\perp \int dz\, V_{\mathrm{D}}^2(\mathbf{r}_\perp, z)\right)}. \tag{8.25}
$$

8.2.1.1 Thermal wandering of a single line

For now, we work with flux liquids at high enough densities and temperatures that weak, oxygen-vacancy, disorder can be neglected. One can then estimate when the Abrikosov theory breaks down from a simple random-walk argument [20]. We consider a *single* flux line $\mathbf{r}(z)$ and determine how far it wanders perpendicular to the z axis as it traverses the sample. The relevant path integral is

$$
\langle |\mathbf{r}(z) - \mathbf{r}(0)|^2 \rangle = \frac{\int \mathcal{D}\mathbf{r}(s)\, |\mathbf{r}(z) - \mathbf{r}(0)|^2 \exp\left[-\frac{\tilde{\varepsilon}_1}{2k_{\mathrm{B}}T}\int_0^L \left(\frac{d\mathbf{r}}{ds}\right)^2 ds\right]}{\int \mathcal{D}\mathbf{r}(s)\, \exp\left[-\frac{\tilde{\varepsilon}_1}{2k_{\mathrm{B}}T}\int_0^L \left(\frac{d\mathbf{r}}{ds}\right)^2 ds\right]}
$$

$$
= \frac{2k_{\mathrm{B}}T}{\tilde{\varepsilon}_1}|z|, \tag{8.26}
$$

which shows that the vortex "diffuses" as a function of the time-like variable z,

$$
\langle |\mathbf{r}(z) - \mathbf{r}(0)|^2 \rangle^{1/2} = (2Dz)^{1/2} \tag{8.27}
$$

with diffusion constant

$$
D = \frac{k_{\mathrm{B}}T}{\tilde{\varepsilon}_1} = \frac{M_\perp}{M_z} \frac{4\pi k_{\mathrm{B}}T}{\phi_0 H_{c1}}. \tag{8.28}
$$

At $T = 77$ K, we take $H_{c1} \approx 10^2$ G and $M_z/M_\perp \approx 10^2$ and find $D = 10^{-6}$ cm, so that vortex lines wander a distance of order 1 μm while traversing a sample of thickness 0.01 cm.

These close encounters will occur quite frequently in fields of order 1 T or more, in which vortices are separated by distances of order 500 Å or less. Collisions between neighboring vortices must now be taken into account. In terms of the "entanglement length" discussed in the introduction,

$$
\ell_z \equiv \frac{1}{2Dn_0} = \frac{\tilde{\varepsilon}_1}{2k_{\mathrm{B}}Tn_0}, \tag{8.29}
$$

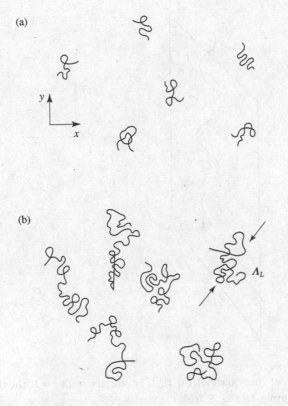

Fig. 8.7. Projections of flux lines onto the *xy*-plane in (a) the disentangled regime and (b) when the lines are on the verge of entanglement. The average projected area of one of the flux lines determines the spatial extent Λ_L of its wave function in the boson analogy.

we conclude that collisions and entanglement of vortex lines will alter the Abrikosov theory whenever

$$L > \ell_z, \tag{8.30}$$

i.e., for $B > B_{x1}$.

The above criterion was derived by neglecting intervortex interactions, which should reduce the asymptotic "diffusion constant" D measured at scales large relative to ℓ_z. Marchetti has studied vortex wandering in a dense flux liquid [68] and finds that this reduction is by about a factor of ten at most. Physical arguments leading to similar conclusions for quantum liquids were presented many years ago by Feynman [69].

In Fig. 8.7, we show six neighboring flux lines, projected down the z axis, for two different values of the diffusion constant D in Eq. (8.28). These line projections can be modelled as two-dimensional random walks. The mean-square projected area occupied by the lines is of order

$$\Lambda_L^2 \equiv \frac{2\pi k_B T}{\tilde{\varepsilon}_1} L, \tag{8.31}$$

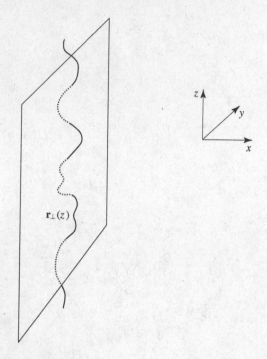

Fig. 8.8. A flux line trapped by a twin boundary that occupies the *yz*-plane.

i.e., of order the square of Eq. (8.27) evaluated with $z = L$, the thickness of the sample. In Fig. 8.7(a), the vortex projections do not overlap significantly as they traverse the sample. Overlap is beginning in Fig. 8.7(b), however, marking the onset of the new physics associated with HTC superconductors.

8.2.1.2 The binding energy for binding of a vortex line to planar and linear pins

Consider a single flux line with trajectory $\mathbf{r}_\perp(z)$ wandering along the z axis in a slab of thickness L near a twin boundary (see Fig. 8.8). This planar pin lies in the yz-plane and attracts the line with a binding energy U_0 per unit length at zero temperature. The physically relevant quantity for transport experiments at liquid-nitrogen temperatures, however, is the binding *free* energy $U(T) = U_0 - TS$. The binding free energy includes a contribution due to the reduced entropy S associated with localizing the line near the twin boundary. The binding free energy is given by the path integral

$$e^{U(T)L/(k_B T)} = \frac{\int \mathcal{D}\mathbf{r}(z) \exp\left[-\frac{\tilde{\varepsilon}_1}{k_B T} \int_0^L \left(\frac{d\mathbf{r}_\perp}{dz}\right)^2 dz - \frac{1}{k_B T} \int_0^L V[\mathbf{r}_\perp(z)]\, dz \right]}{\int \mathcal{D}\mathbf{r}(z) \exp\left[-\frac{\tilde{\varepsilon}_1}{k_B T} \int_0^L \left(\frac{d\mathbf{r}_\perp}{dz}\right)^2 dz \right]}.$$

(8.32)

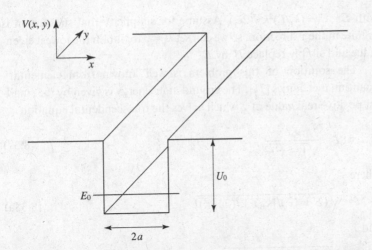

Fig. 8.9. The potential $V(x, y)$ presented to a vortex by the planar pin of Fig. 8.8.

Here, $V(x, y)$ is the binding potential due to the twin, which is defined to vanish as $x \to \pm\infty$. The denominator is included to subtract off the free energy of a line wandering far from the twin.

Equation (8.32) is just the Feynman path integral for a quantum-mechanical particle moving in imaginary time [70]. When this equation is viewed as a problem in classical statistical mechanics, the transfer matrix along z for this partition function is the exponential of the Hamiltonian of a fictitious quantum system. As $L \to \infty$, the binding free energy $U(T)$ is then given by the negative of the lowest eigenvalue of the associated "Schrödinger equation" [70],

$$\left(-\frac{(k_B T)^2}{2\tilde{\varepsilon}_1} \nabla_\perp^2 + V(\mathbf{r}_\perp) \right) \psi(\mathbf{r}_\perp) = E\psi(\mathbf{r}_\perp). \tag{8.33}$$

Note that $k_B T$ plays the role of $\hbar$, whereas $\tilde{\varepsilon}_1$ plays that of the mass of the fictitious particle. The density distribution of flux lines projected down the z axis is given by the modulus squared of the ground-state wave function. See Chapter 7 for a more detailed discussion. We model the twin by a potential $V(\mathbf{r}_\perp)$ that is independent of y, $V(\mathbf{r}_\perp) \equiv V(x)$, and take $V(x)$ to be a square well of depth U_0 and width $2a$ in the x direction (see Fig. 8.9),

$$V(x) = \begin{cases} -U_0, & -a < x < a, \\ 0, & \text{otherwise.} \end{cases} \tag{8.34}$$

Upon assuming a plane-wave state along y,

$$\psi(x, y) = e^{ik_y y}\psi(x), \tag{8.35}$$

we recover the Schrödinger equation for a one-dimensional particle in a box,

$$\left(-\frac{(k_B T)^2}{2\tilde{\varepsilon}_1} \frac{d^2}{dx^2} + V(x) \right) \psi(x) = E'\psi(x), \tag{8.36}$$

with $E = E' + (k_B T)^2 k_y^2/(2\tilde{\varepsilon}_1)$. Assume for simplicity that the system is infinite in the y direction, so we can set $k_y = 0$ to obtain the lowest eigenvalue and simply replace E' by E.

The solution of this problem is well known from elementary quantum mechanics [71]. The ground-state energy is given by the smallest positive real value of ξ which solves the transcendental equation

$$\cot \xi = \frac{\xi x}{\sqrt{1 - x^2 \xi^2}}, \tag{8.37}$$

where

$$\xi = \sqrt{(E + U_0)/[(k_B T)^2/(2\tilde{\varepsilon}_1 a^2)]} \tag{8.38a}$$

and

$$x = \sqrt{[(k_B T)^2/(2\tilde{\varepsilon}_1 a^2)]/U_0}. \tag{8.38b}$$

When $x \ll 1$, the well is very deep and we find the usual particle-in-a-box result for $U(T) \equiv -E$,

$$U(T) \approx U_0 - \frac{\pi^2}{4} \frac{(k_B T)^2}{2\tilde{\varepsilon}_1 a^2}. \tag{8.39}$$

The second term is due to quantum zero-point motion and represents for classical flux lines the loss of entropy due to confinement. For $x \gg 1$, on the other hand, the flux line is only weakly bound,

$$U(T) \approx U_0^2/[(k_B T)^2/(2\tilde{\varepsilon}_1 a^2)]. \tag{8.40}$$

Consider, for simplicity, a vortex line well below $H_{c2}^0(T)$ so that $\tilde{\varepsilon}_1$, U_0 and a are approximately temperature-independent. The binding free energy renormalized by thermal wandering then takes the form

$$U(T) = U_0 f(T/T^*), \tag{8.41}$$

where, according to Eqs. (8.39) and (8.40),

$$f(x) \approx 1 - \frac{\pi^2}{4} x^2, \quad x \ll 1, \tag{8.42}$$

$$f(x) \approx 1/x^2, \quad x \gg 1. \tag{8.43}$$

The crossover temperature in Eq. (8.41) is given by

$$k_B T^* = \sqrt{2\tilde{\varepsilon}_1 a^2 U_0}. \tag{8.44}$$

A schematic plot of $U(T)$ as a function of temperature is shown in Fig. 8.10. Upon taking $U_0 = \varepsilon_1/\ln \kappa_{ab}$ and $a = \xi_{ab}$ to model the HTC materials, we see that there is a large downward renormalization of the binding energy above a characteristic temperature,

$$k_B T^* = \sqrt{2 M_\perp/M_z} \, \varepsilon_1 \xi_{ab}/\ln^{1/2} \kappa_{ab}. \tag{8.45}$$

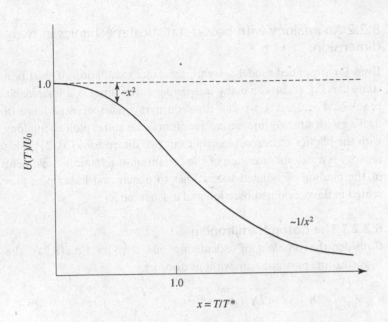

Fig. 8.10. Thermal renormalization of the free energy of binding for binding of a flux line to a twin boundary.

The short coherence length and large mass anisotropies typical of HTC materials conspire to make this temperature small, of order 10 K in BSCCO ($M_z/M_\perp = 3600$, $H_{c1} \approx 100$ G), for example. If melting occurs close to $H_{c2}^0(T)$, ε_1 and ξ_{ab} are strongly temperature-dependent in the liquid phase. One must now solve (8.45) self-consistently to determine the crossover temperature. Thermal renormalization always becomes important sufficiently close to H_{c2}^0.

A similar analysis applies to linear pins oriented along the z axis. Pins of this kind can be created by bombarding initially clean YBCO samples with tin ions [72]. The statistical mechanics of one flux line is now equivalent to the quantum mechanics of a particle in a cylindrically symmetric well. Particles are more weakly bound by an attractive potential in two dimensions than they are in the one-dimensional example discussed above when the "zero-point energy" $(k_B T)^2/(2\tilde{\varepsilon}_1 a^2)$ is large. The renormalized binding energy still takes the form (8.41). Although a result similar to (8.42) and (8.43) holds for small x, Eqs. (8.42) and (8.43) are now replaced by (see Chapter 7) [73]

$$f(x) \approx x^2 e^{-x^2} \tag{8.46}$$

when x is large. The loss of wandering entropy when a line is pinned by planar or linear pins makes an important contribution to a strongly temperature-dependent activation barrier in transport experiments [1].

8.2.2 An analogy with boson statistical mechanics in two dimensions

Even for simplified model systems, the exact calculation of partition sums like Eq. (8.24) for many *interacting* vortex lines is a formidable problem. Fortunately, we can draw on many years of experience in dealing with strongly interacting quantum fluids and exploit an analogy with the physics of boson superfluids in two dimensions [20, 21]. This analogy is a natural extension of the "quantum-mechanical" treatment of the binding of isolated vortex lines to planar and linear pins presented in the previous subsection and in Chapter 7.

8.2.2.1 The boson Hamiltonian

Consider the problem of calculating one term in Eq. (8.24), the N-vortex-line partition sum without disorder,

$$Z_N = \int \mathcal{D}\mathbf{r}_1(z) \cdots \mathcal{D}\mathbf{r}_N(z) e^{-G_N/(k_B T)}, \tag{8.47}$$

where

$$G_N = \frac{1}{2}\tilde{\varepsilon}_1 \sum_{j=1}^{N} \int_0^L \left(\frac{d\mathbf{r}_j(z)}{dz}\right)^2 dz + \sum_{i>j} \int_0^L V[\mathbf{r}_{ij}(z)]\, dz. \tag{8.48}$$

The grand partition function is then $Z_{gr} = \sum_{N=0}^{\infty} (1/N!) e^{\mu L N} Z_N$, where $\mu = H\phi_0/(4\pi) - \varepsilon_1$. The standard way of attacking such problems in classical statistical mechanics is to use the transfer matrix. It is not hard to show that the transfer matrix connecting neighboring constant-z slices of the partition function (8.47) is just the exponential of an N-particle Hamiltonian for quantum-mechanical particles propagating in imaginary time and interacting with a pair potential $V(\mathbf{r}_{ij}(z))$ [70]. All this means in practice is that, up to a multiplicative constant, Eq. (8.47) can be regarded as an integral over a quantum-mechanical matrix element,

$$Z_N = \int d\mathbf{r}_1' \cdots d\mathbf{r}_N' \int d\mathbf{r}_1 \cdots d\mathbf{r}_N \langle \mathbf{r}_1' \cdots \mathbf{r}_N' | e^{-\mathcal{H}_N L/(k_B T)} | \mathbf{r}_1 \cdots \mathbf{r}_N \rangle, \tag{8.49}$$

where the Hamiltonian is

$$\mathcal{H}_N = -\frac{(k_B T)^2}{2\tilde{\varepsilon}_1} \sum_{j=1}^{N} \nabla_j^2 + \sum_{i>j} V(|\mathbf{r}_1 - \mathbf{r}_j|). \tag{8.50}$$

The states $\langle \mathbf{r}_1, \ldots, \mathbf{r}_N \rangle$ and $\langle \mathbf{r}_1', \ldots, \mathbf{r}_N' |$ describe, respectively, the entry and exit points for the vortices in Fig. 8.4.

We now insert a complete set of many-particle energy eigenstates $|m\rangle$ of $\mathcal{H}_N$ into Eq. (8.49) whereupon we find that

$$Z_N = \sum_m |\langle m|\mathbf{p}_1 = 0, \ldots, \mathbf{p}_N = 0\rangle|^2 e^{-E_m L/(k_B T)}, \tag{8.51}$$

where $|\mathbf{p}_1 = 0, \ldots, \mathbf{p}_N = 0\rangle$ is the zero-momentum state,

$$|\mathbf{p}_1 = 0, \ldots, \mathbf{p}_N = 0\rangle = \int d\mathbf{r}_1 \cdots \int d\mathbf{r}_N |\mathbf{r}_1, \ldots, \mathbf{r}_N\rangle. \qquad (8.52)$$

The eigenstates of the permutation-symmetric Hamiltonian (8.50) may be classified in terms of their transformation properties under the permutation group. In principle, all types of states, bosonic, fermionic and those obeying "parastatistics," should enter the sum (8.51). Because the state (8.52) is itself *symmetric* under permutations, however, the matrix element in Eq. (8.51) vanishes for all but the permutation-symmetric boson eigenstates. We may thus rewrite Eq. (8.51) as a restricted sum,

$$Z_N = \sum_{\substack{\text{boson} \\ \text{states } m}} |\langle m | \mathbf{p}_1 = 0, \ldots, \mathbf{p}_N = 0 \rangle|^2 e^{-E_m L/(k_B T)}. \qquad (8.53)$$

The partition sum (8.54) should be contrasted with the sum appropriate for real bosons at a finite fictitious "temperature" β,

$$Z'_N = \sum_{\substack{\text{boson} \\ \text{states } m}} e^{-\beta E_m}, \qquad (8.54)$$

where $\beta = L/(k_B T)$. The difference between Z_N and Z'_N is in the weights involved in the projection onto the zero-momentum ground state in Eq. (8.53). As $L \to \infty$, the lowest-energy bosonic state dominates in both cases, however. The partition function Z'_N is in fact the *correct* quantity even for finite L for flux lines in a special torroidal geometry, which would be especially interesting to investigate experimentally [21, 22]. In this case, one expects a sharp Kosterlitz–Thouless transition from an entangled "superfluid" phase at large L to a disentangled "normal" phase at small L. This entanglement transition may become a more gradual crossover for the free boundary conditions embodied in Eq. (8.53), however [74].

The boson analogy is summarized in Table 8.1. The partition function (8.24) is just the grand canonical partition function for interacting bosons in two dimensions with chemical potential $\mu = H\phi_0/(4\pi) - \varepsilon_1$. As in Section 8.2.2, the trajectories of vortices across the sample are isomorphic to boson world lines. The thermal energy $k_B T$ plays the role of $\hbar$, with L corresponding to the distance $\beta\hbar$ traveled in the imaginary time direction. The parameter $\tilde{\varepsilon}_1$ plays the role of the boson mass. Table 8.1 shows clearly one reason why high-temperature superconductors are interesting: They allow us to explore a world of exceptionally light $(\tilde{\varepsilon}_1 \ll \varepsilon_1)$ bosons in which "Planck's constant" (i.e., $k_B T$) is ten times larger than it is in conventional materials.

Although the boson analogy is powerful, it does not extend in any simple way to *dynamic* phenomena. Real boson superfluids like ^{4}He flow without resistance through the finest capillary tubes. Entangled flux

Table 8.1. *Detailed correspondence of the parameters of a flux-line liquid with the mass, value of Planck's constant, reciprocal temperature* β *and pair potential of two-dimensional bosons*

Vortex lines	$\tilde{\varepsilon}_1$	$k_B T$	L	$H\phi_0/(4\pi)-\varepsilon_1$	$V[r_{ij}(z)]$
Two-dimensional bosons	m	$\hbar$	$\beta\hbar$	μ	Boson pair potential

liquids, on the other hand, may be sufficiently viscous to cause a polymeric glass transition!

8.2.2.2 The coherent-state path integral

As discussed in [4, 21, 59], practical calculations for flux lines in the boson representation are conveniently carried out using the coherent-state path-integral formulation of boson statistical mechanics in the grand canonical ensemble [75, 76]. The bosons are represented by a complex field $\psi(\mathbf{r}_\perp, z)$ and the grand partition function is

$$Z_{\text{gr}} = \int \mathcal{D}\psi(\mathbf{r}_\perp, z) \int \mathcal{D}\psi^*(\mathbf{r}_\perp, z)\, e^{-S(\psi,\psi^*)/(k_B T)}. \tag{8.55}$$

where the "action" is

$$S = \int_0^L dz \int d^2 r_\perp \left(k_B T \psi^* \partial_z \psi + \frac{(k_B T)^2}{2\tilde{\varepsilon}_1} |\nabla_\perp \psi|^2 - \mu |\psi|^2 \right.$$
$$\left. + \frac{1}{2} v_0 |\psi|^4 + V_D(\mathbf{r}_\perp, z) |\psi|^2 \right). \tag{8.56}$$

We have, for simplicity, used the cutoff delta-function pair potential (8.17); see [59] for arbitrary in-plane pair potentials. Note that we have now retained the disorder potential describing pinning by point impurities.

The amplitude of the field $\psi(\mathbf{r}_\perp, z)$ is related to the fluctuating local flux-line density

$$n(\mathbf{r}_\perp, z) = \sum_{j=1}^N \delta[\mathbf{r}_\perp - \mathbf{r}_j(z)]$$

$$= |\psi(\mathbf{r}_\perp, z)|^2. \tag{8.57}$$

"Superfluidity" of the bosons is equivalent to entanglement of the flux lines [70], so we expect that entanglement is accompanied by coherence in the *phase* of the order parameter

$$\psi(\mathbf{r}_\perp, z) = |\psi(\mathbf{r}_\perp, z)| e^{i\theta(\mathbf{r}_\perp, z)}, \tag{8.58}$$

i.e., long-range order in

$$G(\mathbf{r}_\perp, \mathbf{r}'_\perp; z, z') = \langle \psi(\mathbf{r}_\perp, z)\psi^*(\mathbf{r}'_\perp, z')\rangle. \tag{8.59}$$

Line crystals, on the other hand, are characterized by the *absence* of long-range order in $\psi(\mathbf{r}_\perp, z)$, even though the average flux-line density $|\psi(\mathbf{r}_\perp, z)|^2$ is nonzero, (see Section 8.2.2.3).

In the entangled phase, correlation functions can be calculated by expanding about the minimum of the action in Eq. (8.56),

$$\psi(\mathbf{r}_\perp, z) = \sqrt{n_0 + \pi(\mathbf{r}_\perp, z)}\, e^{i\theta(\mathbf{r}_\perp, z)}$$

$$\approx \sqrt{n_0}\left(1 + \frac{\pi}{2n_0}\right)e^{i\theta}, \tag{8.60}$$

where the average density of vortices is given by

$$n_0 = \mu/v_0 \tag{8.61}$$

and we keep only terms quadratic in $\theta(\mathbf{r}_\perp, z)$ and $\pi(\mathbf{r}_\perp, z)$. We can now easily calculate, for example, the structure function

$$S(q_\perp, q_z) = \overline{\langle |\hat{n}(\mathbf{q}_\perp, q_z)|^2\rangle}, \tag{8.62}$$

where $\hat{n}(\mathbf{q}_\perp, q_z)$ is the Fourier transform of the fluxon density. The angular brackets represent a thermal average, while the bar means a quenched average over the Gaussian pinning potential (8.19). The structure function takes the form [59]

$$S(q_\perp, q_z) = \frac{k_BT(n_0q_\perp^2/\tilde{\varepsilon}_1)}{q_z^2 + [\varepsilon(q_\perp)/(k_BT)]^2} + \Delta\left(\frac{(n_0q_\perp^2/\tilde{\varepsilon}_1)}{q_z^2 + [\varepsilon(q_\perp)/(k_BT)]^2}\right)^2 + \mathcal{O}(\Delta^2), \tag{8.63}$$

where $\varepsilon(q_\perp)/(k_BT)$ is the "Bogoliubov spectrum,"

$$\frac{\varepsilon(q_\perp)}{k_BT} = \sqrt{\left(\frac{k_BTq_\perp^2}{2\tilde{\varepsilon}_1}\right)^2 + \frac{n_0V_0}{\tilde{\varepsilon}_1}q_\perp^2}. \tag{8.64}$$

The meaning of this result (and its generalization to fluids with nonlocal interactions) will be discussed further in Section 8.3. For now, we simply note that Eq. (8.63) is the beginning of a systematic expansion in the strength of disorder. To leading order in Δ, disorder simply introduces "Lorentzian-squared" corrections into the correlations which characterize pure systems.

8.2.2.3 The Lindemann criterion

If the "superfluid" flux liquid loses its phase coherence even though the flux-line density $|\psi(r_\perp, z)|^2$ remains finite, this signals a transition to new phase without significant entanglement. If the disorder is negligible, this phase is presumably a line crystal [77].

To estimate when long-range order in $G(\mathbf{r}_\perp, \mathbf{r}'_\perp; z, z')$ disappears, we determine the parameter values such that the fluctuations in the phase exceed unity,

$$\overline{\langle \theta^2(\mathbf{r}_\perp, z)\rangle} \gtrsim 1. \tag{8.65}$$

To evaluate this "inverse Lindemann criterion," we first integrate out the amplitude fluctuation field $\pi(r_\perp, z)$ in Eq. (8.56), and obtain an effective action that depends on the phase only,

$$S_{\text{eff}} = \int d^2 r_\perp \int dz \left[\frac{(k_B T)^2 n_0^2}{2} \left(\frac{1}{K} |\nabla_\perp \theta|^2 + \frac{1}{B} (\partial_z \theta)^2 \right) \right.$$
$$\left. + i \frac{k_B T}{v_0} V_D(r_\perp, z)\, \partial_z \theta \right]. \tag{8.66}$$

Here,

$$K = n_0 \tilde{\varepsilon}_1 \tag{8.67}$$

and

$$B = n_0^2 v_0 \tag{8.68}$$

are the tilt and bulk modulus, respectively, of this simple model liquid of interacting flux lines [21].

The average in Eq. (8.65) is conveniently evaluated in Fourier space. In the absence of disorder, for example, we find

$$\langle \theta^2 \rangle = \frac{BK}{2\pi k_B T n_0} \int_0^\Lambda q_\perp \, dq_\perp \int_{-\infty}^\infty dq_z \frac{1}{Bq_\perp^2 + Kq_z^2}, \tag{8.69}$$

where the integral over $q_\perp$ is restricted by a cutoff Λ related to the inter-vortex spacing $d \approx n_0^{-1/2}$,

$$\Lambda = \sqrt{4\pi n_0}. \tag{8.70}$$

Upon approximating the Fourier integrals in this way (and including disorder), we find

$$\overline{\langle \theta^2 \rangle} \approx \frac{1}{\sqrt{\pi}} \bar{v}^{1/2} \left(1 - \frac{1}{16} \frac{\bar{\Delta}}{\bar{v}} \right), \tag{8.71}$$

where $\bar{v}$ and $\bar{\Delta}$ are important dimensionless coupling constants,

$$\bar{v} = v_0 \tilde{\varepsilon}_1 / (k_B T)^2 \tag{8.72a}$$

and

$$\bar{\Delta} = \Delta \tilde{\varepsilon}_1 / (k_B T)^3 \tag{8.72b}$$

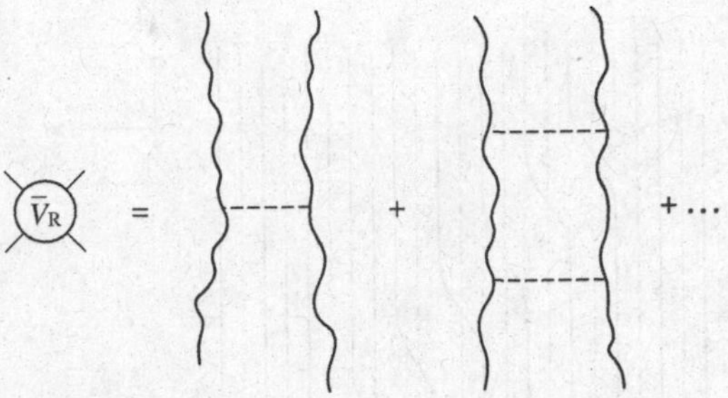

Fig. 8.11. Repeated interactions between two flux lines leading to the renormalized effective interaction discussed in the text.

It can be shown that thermal fluctuations in clean materials renormalize the coupling $\bar{v}$ in Eq. (8.71) down to very small values when the flux liquid is dilute [21]. Indeed, summing the ladder interactions between the two flux lines shown in Fig. 8.11 leads to the renormalized effective interaction,

$$\bar{v}_R = \frac{\bar{v}}{1 + [\bar{v}/(4\pi)] \ln[1/(n_0\lambda_{ab}^2)]}$$

$$\approx 4\pi/\ln[1/(n_0\lambda_{ab}^2)]. \tag{8.73}$$

Here, λ_{ab} represents the interaction range. Upon neglecting the term proportional to disorder in Eq. (8.71), we conclude using (8.65) that dilute line liquids in clean materials are unstable with respect to formation of a new phase whenever

$$\ln\left(\frac{1}{n_0\lambda^2}\right) \lesssim \mathcal{O}(1). \tag{8.74}$$

An equivalent formula for the reentrant part of the melting line $H_m(T)$ in Fig. 8.1(a) was derived in [22]. Note that inclusion of disorder in Eq. (8.71) *reduces* the phase fluctuations and hence *promotes* entanglement, as one would expect. Disorder-driven entanglement of flux lines (without thermal fluctuations) was suggested several years ago by Wordenweber and Kes and discussed by Brandt [78] on the basis of experimental observations of "dimensional crossover" in flux-pinning experiments.

A physical argument for the loss of phase coherence (and hence long-range order) in the correlation function (8.59) for line crystals (without disorder) can be constructed as follows: $\psi(r_\perp, z)$ and $\psi^*(r_\perp, z)$ are, respectively, creation operators for flux-line heads and tails, i.e., magnetic monopoles [79]. The composite operator in Eq. (8.59) creates an extra line at $(r'_\perp, z')$ (i.e., a column of interstitials in the solid) and destroys an existing line at $(r_\perp, z)$, creating a column of vacancies. The lowest-energy configuration is then a line of vacancies (for $z' > z$) or interstitials (for $z' < z$) connecting the two points with an energy σs

Fig. 8.12. The contribution of the lowest-energy solid phase to the correlation function (8.64), which inserts a flux head and tail into a crystalline vortex array. Dashed lines represent a row of vortices slightly behind the plane of the page. In (a), a vacancy is created at "time" z, which then propagates and is destroyed at time z'. Interstitial propagation from z' to z is shown in (b). The energy of the "string" defect connecting the head to the tail increases linearly with the separation in both cases and leads to the exponential decay of $G(\mathbf{r}_\perp, \mathbf{r}_\perp'; z, z')$.

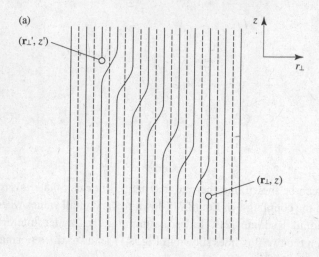

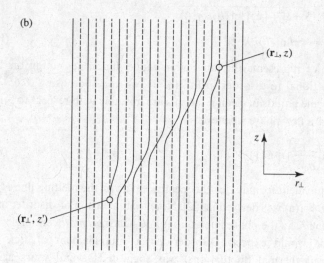

proportional to the length s of this "string." See Fig. 8.12. It follows that the correlation function (8.59) decays exponentially to zero (i.e., like $e^{-\sigma s/k_B T}$) for large separations in the crystalline phase. In a line *liquid*, on the other hand, it can be shown (using the hydrodynamic approach of Section 8.3) that the asymptotic string tension σ vanishes, implying that there is long-range order in $G(\mathbf{r}_\perp, \mathbf{r}_\perp'; z, z')$. Note that the magnitude of the string tension in the solid will depend on the orientation of the string and on the ordering of z and z'.

8.2.3 The importance of disorder in the dilute limit

Results such as Eq. (8.63) are correct when the flux liquid is dense, so that the effects of disorder are screened out by flux-line collisions and can be treated as a small perturbation. New physics arises, however, sufficiently close to H_{c1}, i.e., when the vortex lines are dilute and "screening" is no longer effective. In this limit, it can be shown that disorder must produce new physics whenever [59]

$$n_0 \lambda_{ab}^2 \lesssim e^{-4\pi(k_B T)^3/(\tilde{\varepsilon}_1 \Delta)}. \tag{8.75}$$

Upon combining equations (8.74) and (8.75), we see that a sliver of equilibrated vortex liquid stable against both crystallization and disorder *must* exist in the density range

$$e^{-4\pi(k_B T)^3/(\tilde{\varepsilon}_1 \Delta)} \lesssim n_0 \lambda_{ab}^2 \lesssim 1. \tag{8.76}$$

The constitutive relation between $B = n_0 \phi_0$ and H derived for pure systems in [21],

$$B(H) = \frac{\bar{v}}{4\pi}(H - H_{c1}) \ln\left(\frac{4\pi}{\bar{v}} \frac{\phi_0/\lambda_{ab}^2}{H_i - H_{c1}}\right), \tag{8.77}$$

can be combined with Eq. (8.75) to determine the line $H_i(T)$ in Fig. 8.1(b), below which disorder dominates. $H_i(T)$ is given by the solution of

$$(H_i - H_{c1}) \ln\left(\frac{4\pi}{\bar{v}} \frac{\phi_0/\lambda_{ab}^2}{H_i - H_{c1}}\right) = \frac{4\pi\phi_0}{\bar{v}\lambda_{ab}^2} e^{-4\pi(k_B T)^3/(\tilde{\varepsilon}_1 \Delta)}. \tag{8.78}$$

Note that $\phi_0/\lambda_{ab}^2 = \mathcal{O}(H_{c1})$, so that $(H_i - H_{c1})/H_{c1} \lesssim 4\pi/\bar{v}$. Because $\bar{v}$ is so large ($\bar{v} = \mathcal{O}(10^5)$), the disorder-dominated regime is typically much *smaller* than that shown in Fig. 8.1(b). Because $\tilde{\varepsilon}_1$ vanishes as $T \to T_c$, the region of Fig. 8.1(b) dominated by disorder rapidly shrinks to zero in this limit. It is not yet clear whether the region below this line is a vortex glass [45] or simply represents a crossover to new critical exponents at H_{c1}.

8.3 Correlations in flux liquids with weak disorder

8.3.1 Hydrodynamic treatment of correlations

The simplified boson model treated in Section 8.2 neglects nonlocal effects that are known to be quantitatively important at high fields in the crystalline phase [23, 39]. The long-wavelength behavior of static correlation functions can be rederived [59], however, via a simpler (and explicitly nonlocal) approach first proposed for the *dynamics* of flux liquids

[51]. Although the shear modulus of the Abrikosov flux lattice vanishes in a vortex liquid, the tilt and compressional moduli remain finite. Correlations in a dense vortex liquid with weak disorder can be described in terms of nonlocal tilt and compressional moduli, which can be estimated from their crystalline-phase values [23] over a wide range of fields and temperatures. In this section, we sketch this theory (which is analogous to the elastic description of a flux crystal) and discuss the physical interpretation of the correlation functions.

To describe a flux configuration such as that shown in Fig. 8.4, we use two basic hydrodynamic fields, a microscopic vortex density

$$n_{\mathrm{mic}}(\mathbf{r}_\perp, z) = \sum_{j=1}^{N} \delta^{(2)}[\mathbf{r}_\perp - \mathbf{r}_j(z)] \tag{8.79}$$

and a microscopic "tangent" field in the plane perpendicular to $\hat{z}$.

$$\mathbf{t}_{\mathrm{mic}}(\mathbf{r}_\perp, z) = \sum_{j=1}^{N} \frac{d\mathbf{r}_j(z)}{dz} \delta^{(2)}[\mathbf{r}_\perp - \mathbf{r}_j(z)]. \tag{8.80}$$

We now coarse grain these fields over several flux-line spacings to obtain smoothed density and tangent fields $n(\mathbf{r}_\perp, z)$ and $\mathbf{t}(\mathbf{r}_\perp, z)$.

The free energy of the liquid can be expanded to quadratic order in the density deviation $\delta n(\mathbf{r}_\perp, z) = n(\mathbf{r}_\perp, z) - n_0$ and in $\mathbf{t}(\mathbf{r}_\perp, z)$ as follows:

$$F = \frac{1}{2n_0^2} \int d^2 r_\perp \, dz \int d^2 r'_\perp \, dz' [K(\mathbf{r}_\perp - \mathbf{r}'_\perp, z - z') \mathbf{t}(\mathbf{r}_\perp, z) \cdot \mathbf{t}(\mathbf{r}'_\perp, z')$$

$$+ B(\mathbf{r}_\perp - \mathbf{r}'_\perp, z - z') \, \delta n(\mathbf{r}_\perp, z) \, \delta n(\mathbf{r}'_\perp, z')]$$

$$+ \int d^2 r_\perp \, dz \, V_{\mathrm{D}}(\mathbf{r}_\perp, z) \, \delta n(\mathbf{r}_\perp, z). \tag{8.81}$$

Here $V_{\mathrm{D}}(\mathbf{r}_\perp, z)$ is the quenched random Gaussian potential discussed in Section 8.2.1. The parameter B is a generalized bulk modulus for areal compressions and dilations perpendicular to the z axis, while K is the modulus for tilting the lines away from the direction of the applied field. Because we are dealing with *lines*, *not* simply oriented anisotropic particles, Eq. (8.81) must be supplemented with an "equation of continuity,"

$$\partial_z \delta n + \nabla_\perp \cdot \mathbf{t} = 0, \tag{8.82}$$

which reflects the fact that vortex lines cannot stop or start inside the medium. Correlation functions can be calculated by assuming that the probability of a particular line configuration is proportional to $\exp[-F/(k_{\mathrm{B}}T)]$ and imposing the constraint (8.82) on the statistical mechanics.

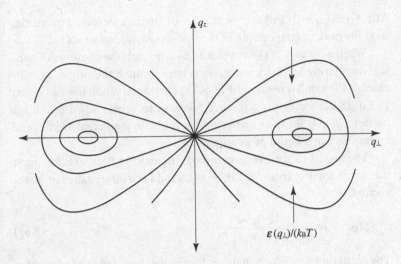

Fig. 8.13. Contours of constant scattering intensity for dense line liquids. The contours are linear near the origin and surround a maximum on the $q_\perp$ axis located approximately at the position of the first Bragg peak of the nearby triangular solid phase.

The condition (8.82) amounts to the requirement of "no magnetic monopoles" and can be implemented using a vector potential. Alternatively, we can set

$$\delta n = -n_0 \, \nabla_\perp \cdot \mathbf{u} \tag{8.83a}$$

and

$$\mathbf{t} = n_0 \frac{\partial \mathbf{u}}{\partial z}, \tag{8.83b}$$

where $\mathbf{u}(\mathbf{r}_\perp, z)$ is a two-component "displacement field." Equation (8.82) is now satisfied automatically and, with these substitutions, Eq. (8.81) differs from the standard elastic description of the Abrikosov flux lattice [80] only by the absence of a shear modulus. The structure function of the liquid is now easily found to be

$$\overline{S(\mathbf{q}_\perp, q_z)} \equiv \overline{\langle |\delta n(\mathbf{q})|^2 \rangle} = k_{\mathrm{B}}T \frac{n_0^2 q_\perp^2}{\hat{K}(\mathbf{q})q_z^2 + \hat{B}(\mathbf{q})q_\perp^2}$$

$$+ \Delta \left(\frac{n_0^2 q_\perp^2}{\hat{K}(\mathbf{q})q_z^2 + \hat{B}(\mathbf{q})q_\perp^2} \right)^2, \tag{8.84}$$

where $\hat{K}(\mathbf{q})$ and $\hat{B}(\mathbf{q})$ are the Fourier transforms of the functions in Eq. (8.81). This simple hydrodynamic result is identical in form to the hydro-dynamic limit of Eq. (8.63), with wavevector-dependent functions replacing the constant parameters of the boson model. Weak disorder again merely leads to a small correction to the results for pure systems. To simplify the remaining discussion, we shall henceforth neglect disor-der and, for the most part, simply replace $\hat{B}(\mathbf{q})$ and $\hat{K}(\mathbf{q})$ by constants B and K. We note, however, that $K(q_\perp \approx 0, q_z \approx 0)$ and $B(q_\perp \approx 0, q_z \approx 0)$ will determine the behavior near the origin of Fig. 8.13 whereas

$K(q_\perp \approx (\pi/d, q_z \approx 0)$ and $B(q_\perp \approx \pi/d, q_z \approx 0)$ are the controlling quantities near the peak corresponding to the first reciprocal-lattice vector.

When $\hat{K}(\mathbf{q})$ and $\hat{B}(\mathbf{q})$ are replaced by constants the contours of constant scattering intensity are straight lines through the origin, as indicated in the small-$\mathbf{q}$ region of Fig. 8.13. The conservation law embodied in Eq. (8.82) forces the structure function to vanish when $q_\perp = 0$ for nonzero q_z and leads to scattering contours very different from those appropriate to a liquid of point particles.

The results of the boson mapping discussed in Section 8.2 suggest that an alternative form for Eq. (8.84), valid for a larger range of wavevectors, is

$$S(q_\perp, q_z) = \frac{n_0^2 k_B T q_\perp^2 / K}{q_z^2 + \varepsilon^2(q_\perp)/(k_B T)^2}. \tag{8.85}$$

The scattering function remains a Lorentzian in q_z for fixed $q_\perp$, with a width given by the function $\varepsilon(q_\perp)/(k_B T)$ (see Fig. 8.13). In the Bogoliubov approximation of Section 8.2, this function can be expressed in terms of the elastic constants B and K as

$$\frac{\varepsilon(q_\perp)}{k_B T} = \left[\left(\frac{k_B T n_0 q_\perp^2}{2K} \right)^2 + \frac{B}{K} q_\perp^2 \right]^{1/2}, \tag{8.86}$$

which reduces to the prediction $\varepsilon(q_\perp)/(k_B T) = \sqrt{B/K} q_\perp$ for small $q_\perp$. A better approximation for $\varepsilon(q_\perp)$ when the flux liquid is dense is the Feynman formula [70],

$$\varepsilon(q_\perp) = \frac{(k_B T)^2 n_0 q_\perp^2}{2K S_2(q_\perp)}, \tag{8.87}$$

which expresses the "phonon–roton spectrum" in terms of the two-dimensional structure function of a constant-z cross section, $S_2(q_\perp)$. If we insist that the three-dimensional structure function has the Lorentzian form (8.85), Eq. (8.87) is the only result consistent with the sum rule

$$\int_{-\infty}^{\infty} \frac{dq_z}{2\pi} S(q_\perp, q_z) = n_0 S_2(q_\perp). \tag{8.88}$$

The physical meaning of the function $\varepsilon(q_\perp)$ for classical line liquids is illustrated in Fig. 8.14. Upon using (8.85) and (8.87) to determine the decay of in-plane density fluctuations $\delta\hat{n}(q_\perp, z)$ along the z axis, we find

$$\bar{S}(q_\perp, z) \equiv \langle \delta\hat{n}(\mathbf{q}_\perp, z)\, \delta n^*(\mathbf{q}_\perp, 0) \rangle$$

$$= S_2(q_\perp) e^{-\varepsilon(q_\perp)|z|/(k_B T)}, \tag{8.89}$$

which is the behavior summarized in Eq. (8.5) of the introduction. The qualitative behavior of $\varepsilon(q_\perp)$ obtained by using a realistic

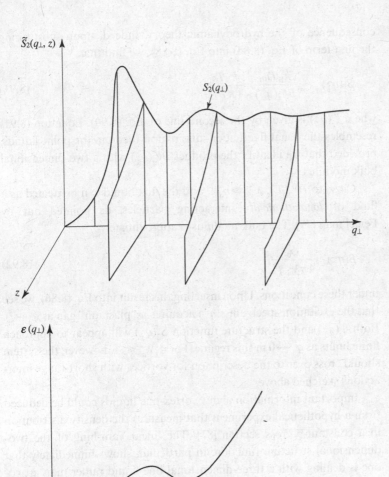

Fig. 8.14. The mixed structure function $\tilde{S}(q_\perp, z)$ showing the decay of the in-plane Fourier components of the density in a line liquid. For fixed $q_\perp$, the decay is exponential in $|z|$.

Fig. 8.15. A schematic diagram of the function $\varepsilon(q_\perp)$ which controls the decay of density fluctuations along the z axis. A minimum in this "phonon–roton spectrum" appears when the two-dimensional structure function of a dense liquid is inserted into Eq. (8.87).

two-dimensional structure function in Eq. (8.87) is shown in Fig. 8.15. Density fluctuations with wavelength $q_\perp$ evidently decay due to line wandering along the z axis over a length scale given by

$$\ell_z(q_\perp) = k_B T / \varepsilon(q_\perp)$$

$$\underset{q_\perp \to 0}{\approx} (K/B)^{1/2} / q_\perp. \tag{8.90}$$

The two-dimensional structure function $S_2(q_\perp)$ shown in Fig. 8.14 is similar to that of a dense liquid of points in two dimensions. It is unusual only in that it goes to *zero* (linearly) as $q_\perp \to 0$. This incompressibility of a constant-z cross section of a line liquid is a direct consequence of the hydrodynamic theory. Indeed, upon substituting the first term of Eq. (8.84) into Eq. (8.88), we find that

$$S_2(q_\perp)_{q_\perp \to 0} \approx \frac{k_B T n_0}{B \ell_z(q_\perp)} = \frac{k_B T n_0}{(KB)^{1/2}} q_\perp, \tag{8.91}$$

where $\ell_z(q_\perp)$ is given by the second line of Eq. (8.90). Equation (8.91) resembles the usual fluctuation–dissipation theorem for point liquids, provided that we identify the product $B\ell_z(q_\perp)$ with a two-dimensional bulk modulus.

Close to $H_{c2}^0(T)$, $\lambda_{ab} \gg n_0^{-1/2}$ and the flux liquid can be treated as a fluid of *logarithmically* interacting particles, as pointed out by Feigel'man [77]. The bulk modulus is approximately

$$B(q_\perp) \approx \frac{\phi_0^2}{4\pi \lambda_{ab}^2 q_\perp^2}, \tag{8.92}$$

under these conditions. Upon inserting this result into Eq. (8.86), we see that the excitation spectrum $\varepsilon(q_\perp)$ acquires a "plasmon" gap as $q_\perp \to 0$. Both $\ell_z(q_\perp)$ and the structure function $S_2(q_\perp)$ will appear to approach finite limits as $q_\perp \to 0$ in this regime. For $q_\perp \lambda_{ab} \ll 1$, however, the system should cross over to the description for vortices with short-range interactions sketched above.

Important information about vortex-line liquids could be deduced from a hypothetical experiment that measures the density distribution in a constant-z cross section [81]. The linear vanishing of the two-dimensional structure function, in particular, shows immediately that one is dealing with a three-dimensional line liquid rather than a two-dimensional liquid of points [82]. We see from Eq. (8.91) that the slope determines the product of elastic constants KB. For cross sections in the middle of a slab of finite thickness L, the two-dimensional structure function will finally level out at very small $q_\perp$. The approach to a constant begins for $q_\perp \lesssim q_\perp^*$, where $\ell_z(q_\perp^*) = L/2$, i.e., for wavelengths less than

$$q_\perp^* = 2(K/B)^{1/2}/L. \tag{8.93}$$

An experimental determination of the crossover wavevector $q_\perp^*$ would fix the *ratio* of elastic constants K/B.

8.3.2 Hexatic order in vortex liquids

Thus far, we have said relatively little about hexatic order in vortex liquids. As pointed out by Chudnovsky [52], the Larkin–Ovchinnikov model [19] of impurity disorder acting on an Abrikosov flux lattice destroys translational order but leaves the orientational order intact.

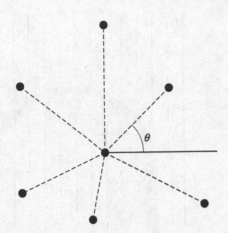

Fig. 8.16. Definition of the bond angles entering $\psi_6(\mathbf{r}_\perp, z)$. The bonds run between nearest-neighbor flux lines in a constant-z cross section. The angles are defined with respect to an external axis.

A high-temperature equilibrium hexatic liquid was proposed on the basis of a dislocation-loop model at about the same time [42]. The definition of the orientational order parameter is reviewed in Fig. 8.16, for a constant-z cross section with seven neighboring flux lines. The local orientational order parameter associated with the central site is obtained by averaging over its six bond angles,

$$\psi_6(\mathbf{r}_\perp, z) = \frac{1}{6} \sum_{j=1}^{6} e^{6i\theta(\mathbf{r}_\perp, z)}. \tag{8.94}$$

The recent observations of a low-temperature hexatic vortex glass in BSCCO [27, 28] suggest the possibility of an equilibrated hexatic vortex liquid at higher temperatures, and are consistent with a suggestion by Worthington *et al.* [32] regarding YBCO. The potential for hexatic order in vortex liquids is important because a nonzero hexatic stiffness constant amplifies the intervortex shear viscosity, as discussed in [51]. A broken orientational symmetry at low temperatures would, morcover, make some sort of phase transition with increasing temperature *inevitable* when this symmetry was restored.

In the absence of disorder, the hexatic order should appear just above the melting line in Fig. 8.1(a) [42]. The hexatic phase results if melting of the Abrikosov flux lattice is driven by a proliferation of thermally activated dislocations. A heavily defected vortex crystal with a finite concentration of unbound dislocation loops is one possible description of an entangled flux liquid. Screw dislocations are especially effective at producing entanglement [78]. A dislocation loop in a flux-line lattice with $\hat{\mathbf{H}} \parallel \mathbf{z}$ is shown in Fig. 8.17. All such loops are highly constrained, in the sense that they must lie in the plane defined by the z axis and their Burgers vector [83, 84]. Note that these loops have a mixed edge-and-screw character. As Fig. 8.17 indicates, the average distance

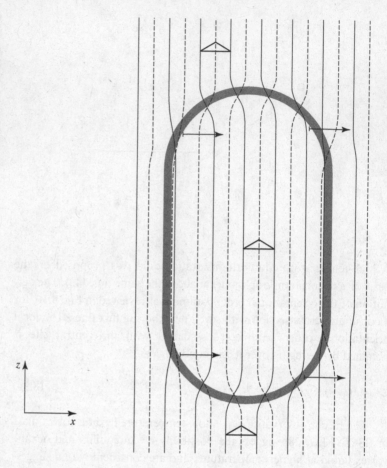

Fig. 8.17. Dislocation loop in a flux-line solid. Dashed lines represent vortices just behind the plane of the figure. Such loops always lie in the plane spanned by their Burgers vector and the z axis. The orientation of the three triangles is the same, showing that the loop has only a small effect on orientational order.

between the edge and screw components of this loop gas determines the translational correlation lengths parallel and perpendicular to the z axis in the hexatic liquid.

Two basic results about this model of the flux liquid can be proved [42]. First, the presence of unbound dislocation loops is sufficient to melt the lattice, in the sense that the equilibrium shear modulus vanishes. Although this conclusion may seem obvious, we are unaware of an analogous result for the Larkin–Ovchinnikov model of disorder [19], which neglects dislocations entirely. The second result is that dislocations do *not* destroy the long-range order in the orientational correlation function, $G_6(\mathbf{r}_\perp) = \langle \psi_6(\mathbf{r}_\perp, z)\psi_6^*(0, 0) \rangle$, i.e.,

$$\lim_{\substack{r_\perp \to \infty \\ z \to \infty}} G_6(\mathbf{r}_\perp, z) \neq 0. \tag{8.95}$$

This result is suggested by the geometrical construction in Fig. 8.17: The three triangles lie in different constant-z planes, but all have the same orientation, suggesting that dislocation loops have only a minor effect on bond-orientational order. This broken rotational symmetry means that the flux liquid resists deformations in the bond-angle field, as in the hexatic phase discussed in Chapter 2 [85]. As a result, the hydrodynamic free energy in Eq. (8.81) must be augmented by a hexatic contribution,

$$\delta F_H = \frac{1}{2} \int d^2 r_\perp \, dz \, (K_\perp |\nabla_\perp \theta|^2 + K_z |\partial_z \theta|^2). \tag{8.96}$$

The Frank constants $K_\perp$ and K_z are related to the energies of edge and screw-dislocation cores [42].

Note that the loop in Fig. 8.17 may be viewed as a virtual pair of edge dislocations propagating along the z axis in the boson picture of Section 8.2. Loop unbinding thus provides a mechanism by which crystalline two-dimensional quantum crystals can melt into quantum hexatics, entangled in imaginary time. Little is known about the hexatic superfluid phase predicted by this approach.

When disorder is included, the hexatic-to-crystal transition will be washed out, with hexatic order remaining down to $T=0$ [52] within the Larkin–Ovchinnikov model. However, as pointed out in [42], Chudnovsky's analysis of hexatic order within the Larkin-Ovchinnikov models does not allow for a random field that acts directly on the bond-angle field. A small random field of this kind will be generated in realistic samples whenever two impurities in close proximity single out a preferred direction for the local crystallographic axes. This field will, in principle, eventually destroy long-range orientational order at sufficiently large length scales. The orientational correlation length can, however, greatly exceed the translational correlation length, in qualitative agreement with Chudnovsky's ideas. Toner has developed this point further [86] and finds that a coupling to an underlying lattice with a 4-fold symmetry can in fact stabilize hexatic order out to *arbitrarily* large distances, because the broken orientational symmetry then becomes Ising-like.

Hexatic order could arise at low temperatures in another way. Suppose that entanglement in an isotropic vortex liquid creates long relaxation times sufficient to prohibit the formation of *any* significant translational order upon cooling below $H_m(T)$. Hexatics may appear upon undercooling simply because this is the most ordered state compatible with entanglement.

8.3.3 Planar pins, linear pins and boson localization

Thus far, we have discussed only the effects of point disorder, such as oxygen vacancies, on correlation functions. Since pinning by oxygen vacancies at liquid-nitrogen temperatures is weak [34], one can imagine regimes in which the physics on large scales is dominated instead by strongly pinning *correlated* structures, such as planar twin boundaries [87] or linear defects produced by bombardment with heavy ions [72]. It is straightforward to determine how the addition of correlated disorder modifies the structure function in dense flux liquids far from the irreversibility line by the methods sketched above. One can go somewhat further in understanding flux-line configurations, however, if point pins are ignored entirely and one considers only disorder that is perfectly correlated along the direction of the field. The behavior of flux-line configurations with increasing magnetic field is then closely related to the physics of boson localization [1, 88–91].

We have already seen how an isolated planar or linear defect interacts with one flux line in Section 8.2.1.2. There is always at least one bound state, although the binding weakens at high temperatures. The squared modulus of the "wave function" $\psi(\mathbf{r}_\perp)$ associated with such states determines the spatial distribution of a typical flux-line configuration projected down the z axis. The spatial extent of such wave functions will increase as the binding becomes weak.

Suppose now that *many* planar or linear defects oriented along the z axis (and passing completely through the sample) are present and let us imagine that the magnetic field is slowly increased above H_{c1} at a fixed temperature. The first few flux lines which enter will be trapped in the bound states discussed above. Repulsive interactions between lines will eventually limit the capacity of the defect array to absorb more flux quanta, however. At some point, the localization length $\ell_\perp$ which describes the spatial extent (perpendicular to the z axis) of the most energetic flux lines will diverge at a two-dimensional localization transition [91],

$$\ell_\perp \sim \frac{1}{|H_\ell(T) - H|^{\nu_\perp}}, \tag{8.97}$$

where $\nu_\perp$ is a localization exponent and $H_\ell(T)$ is a critical field above which localization takes place.

Above $H_\ell(T)$, a finite fraction of the vortex lines will be part of a relatively mobile flux liquid, the dynamics of which will be discussed in Section 8.4. Flux flow of this viscous liquid near a twin boundary or around a cylindrical linear pin is discussed in [51]. Below $H_\ell(T)$, all flux lines will be localized in a "Bose-glass" phase with qualitatively different (and more sluggish) transport properties. The dynamic consequences of the transition to the Bose glass were worked out in [1]. Linear defects may be particularly helpful in pinning the flux liquid above $H_\ell(T)$: Mobile flux lines will presumably entangle, not only with each other but also with the lines which remain anchored to linear defects. This may be one reason why linear defects are so effective at raising the temperature of the irreversibility line [72]. An underlying "Bose-glass" transition due to twinning disorder provides an interesting alternative to "vortex-glass" explanations of resistivity experiments [46], which assume that microscopic point-like disorder dominates at large length scales. The correct physics for many samples almost certainly involves a *combination* of point-like and correlated disorder, with an enhanced concentration of point disorder concentrated within the twin planes [87]. See the articles cited in [1] for details.

8.4 Dynamics near the irreversibility line

In this section we discuss the dynamics of flux liquids. This can be done by adapting extensive studies of the hydrodynamics of point vortices in superfluid and superconducting films [92, 93]. The hydrodynamic normal modes both of isotropic and of hexatic flux liquids have now been worked out [51]. The characteristic frequencies of these modes depend *both* on the vortex friction $\gamma(T)$ and on the intervortex viscosity $\eta(T)$ discussed in the introduction. The tilt mode, in particular, may be responsible for the peaks at the irreversibility line observed in [29, 30]. We shall restrict our attention here, however, to flux-flow-transport experiments, focusing on four different possible explanations for the irreversibility line in Fig. 8.1(b).

There are two broad categories of pins that can keep flux lines in place. *Point-like microscopic disorder* includes microscopic defects such as oxygen vacancies and can be represented by a weak potential that fluctuates on scales much shorter than the intervortex spacing. Intervortex distances typically exceed 100 Å even in the most intense magnetic fields. Point-like disorder affects vortices much as a sheet of sandpaper influences the motion of poker chips sliding on its surface. Although oxygen-vacancy disorder may be effective at pinning flux lines at low temperatures (typical pinning energies are 4 K or less), it is unlikely that this sort of pinning, by itself, could keep flux lines in place at liquid-nitrogen temperatures. Even if there are many (weakly pinning) oxygen vacancies per flux line, vortices can move in response to an external current by displacing only a small line segment at a time. The energetic barrier to motion is not substantial at high temperatures, provided that the vortex lines are dense enough to screen out the randomness at large length scales [59].

Correlated or *mesoscopic disorder* appears to be necessary to pin flux lines at the high fields and temperatures envisioned for many practical applications. Examples of inhomogeneous disorder are inclusions of other phases, twin and grain boundaries and the linear defects of Civale *et al.* [72]. The pinning energies are typically much stronger than those provided by oxygen vacancies and the spacing between these pins is usually much *larger* than the intervortex spacing. The elastic constants of the underlying Abrikosov flux lattice play an important role when a few strong pins are present because they allow flux lines to be pinned collectively. The strong pins play the role of nails holding a carpet to a wooden floor. The entire carpet can be fixed in place with just a few nails because it has a shear modulus.

8.4.1 Explanations of the irreversibility line

Whether the irreversibility line in disordered systems (see Fig. 8.1(b)) is related to melting in pure systems depends on the degree and type of disorder in a particular sample. Here, we shall simply define the irreversibility line to be the locus in the temperature–field phase diagram where the magnetization for field-cooled samples disagrees with the result obtained for samples cooled initially at zero field [64]. The temperature-dependent resistivity in a magnetic field drops to zero at or near the irreversibility line. The irreversibility temperature of the bismuth-based superconductors (BSCCO) at $H = 1$ T is only 30 K, even though the temperature corresponding to the upper critical field is 85 K. Understanding the physical origin of this line is of considerable practical importance, since dissipationless supercurrents will not flow above it. Four very different explanations of the irreversibility line have been proposed. It is difficult to distinguish among the alternatives using measurements of the resistivity alone. Here, we briefly describe each of the hypotheses and then propose a transport experiment that would distinguish unambiguously among them.

8.4.1.1 Thermally assisted flux flow

In its simplest form, the model of thermally assisted flux flow (TAFF) ignores both the connectivity of individual flux lines and their mutual interactions. The system is then regarded as an ideal gas of disconnected flux bits subject to a Lorentz force (induced by externally imposed currents) and pinning. The combined effects of pinning and a uniform Lorentz force are often idealized in a tilted-washboard potential. In this oversimplified form, the TAFF model is more a set of assumptions useful for parameterizing data than a real theory. The model leads to an Arrhénius temperature dependence for the flux-flow resistivity [66],

$$\rho_p = \rho_0 e^{-U_p/(k_B T)} \tag{8.98}$$

where U_p is interpreted as a pinning energy.

8.4.1.2 Melting

Because the shear modulus of the Abrikosov flux lattice drops at a melting transition, flux flow in the presence of a few strong pins should change dramatically at this point. The melting curve in the (H, T)-plane determined from a simple Lindemann criterion [23] is in fact often quite close to the observed "melting" or irreversibility line determined by vibrating-reed experiments [29]. Intervortex interactions are taken into account automatically in this approach. An appropriate theory should take into account that melting of vortex lines in three dimensions is probably a first-order transition. The experimental changes observed in the vicinity of the irreversibility line are often smooth and continuous, suggesting that disorder has smeared out any underlying melting transition. First-order transitions are a striking feature of clean samples in the appropriate experimental geometry, however [16].

8.4.1.3 The vortex-glass hypothesis

Disorder acting on the complex condensate order parameter of superconductors is reminiscent of disordered XY spin systems. This suggests the possibility of a finite temperature spin-glass-like transition [45]. Unlike the TAFF model, which leads to a nonzero (but possibly quite small) flux-flow resistivity at all nonzero temperatures, the vortex-glass hypothesis predicts that the linear resistivity vanishes below a well-defined temperature T_g. The way in which it vanishes is determined by a universal critical exponent. The onset of this transition is signaled by a diverging disorder correlation length, which measures the range over which a (disordered) condensate-order-parameter configuration responds to, say, a change in boundary conditions. Although the theory is almost entirely phenomenological and relies on experimental and numerical fitting procedures to determine the relevant critical exponents, good fits to transport experiments have been obtained with this approach [46].

8.4.1.4 The "polymer"-glass hypothesis

The vortex-glass hypothesis assumes that the system undergoes a glass transition due to extrinsic, impurity-induced pinning constraints. An alternative source of slow relaxation times is the *intrinsic* constraints associated with entanglement in a liquid of vortex lines [21, 61]. Even in the absence of point-like (or correlated) disorder, one might expect a polymer-like glass transition with decreasing temperature, i.e., at a temperature at which the intervortex viscosity becomes effectively infinite. Although the system may remain disordered upon cooling, it would behave as if it had a shear modulus because of this large viscosity. The effects of a few strong pins could then propagate, just as in an Abrikosov flux lattice with a shear modulus. As discussed in the introduction and in Chapter 7, the validity of this picture requires a large (several times $k_B T$ [51]) barrier to flux cutting.

8.4.2 "Viscous electricity"

In the liquid phase, all the above ideas are consistent with the flow-velocity field Eq. (8.10), which describes flux liquids on scales large relative to the intervortex spacing. Weak microscopic pinning centers are assumed to be incorporated into a renormalized Bardeen–Stephen friction constant γ. Once the vortex velocity field is known, we also know the distribution of the electric field perpendicular to $\hat{z}$ inside the superconductor [36],

$$\mathbf{E}_\perp(\mathbf{r}) = \frac{n_v^0 \phi_0}{c} \hat{z} \times \mathbf{v}(\mathbf{r}). \tag{8.99}$$

The steady-state flux-flow velocity is thus linearly related to the in-plane electric field. The Lorentz force is linear in the external current $\mathbf{j}_T$, see Eq. (8.9). Using these well-known relations, we can rewrite Eq. (8.10) in the suggestive form

$$-\delta^2 \nabla_\perp^2 \mathbf{E}_\perp + \mathbf{E}_\perp = \rho_0 \mathbf{j}_T, \tag{8.100}$$

where $\rho_0 = (n_0 \phi_0/c)^2/\gamma$ is the flux-flow resistivity of a sample with only homogeneous disorder and $\mathbf{j}$ is the current. We refer to this result as the equation of "viscous electricity," because it is just Ohm's law modified by a viscous term proportional to the Laplacian of the electric field and the viscous length scale:

$$\delta(T, H) = \sqrt{\eta(T,H)/\gamma(T,H)}. \tag{8.101}$$

The four competing theories of flux-flow resistivity near the irreversibility line all give different predictions for the parameters γ and η, as summarized in Table 8.2.

If there is no finite-temperature phase transition, as predicted by the TAFF theory, then $\gamma(T) \sim e^{U_p/(k_B T)}$ at low temperatures. The TAFF hypothesis makes no prediction for $\eta(T)$. If the irreversibility line is associated with freezing of the vortex liquid into a translationally ordered Abrikosov flux lattice, $\gamma(T)$ should remain *finite* at the melting temperature. If the melting transition is weakly first order, $\eta(T)$ will increase above the transition and then jump discontinuously to infinity. If a polymer-like glass transition is responsible for the irreversibility line, one again expects a finite $\gamma(T)$ near a glass transition T_g and it may be possible to fit $\eta(T)$ to the standard Vogel–Fulcher form [94]

$$\eta(T) \approx \eta_0 \exp[c/(T - T_g)]. \tag{8.102}$$

In contrast to the above three possibilities, $\gamma(T)$ is expected to *diverge* at the vortex-glass transition, provided that the effect of weak pins is incorporated into a renormalized friction coefficient. If we identify the length δ in Eq. (8.101) with the diverging vortex-glass correlation length ξ_G of Fisher *et al.* [45], we see that the viscosity $\eta = \delta^2 \gamma$ should also diverge. The vortex-glass theory predicts that, at the transition, both γ and ξ_G diverge according to [45]

$$\gamma(T) \sim \frac{1}{\rho(T)} \sim \frac{1}{|T - T_{vg}|^{\nu(z-1)}},$$

$$\xi_G(T) \sim \frac{1}{|T - T_{vg}|^{\nu}}, \tag{8.103}$$

where $\rho(T)$ is the resistivity, T_{vg} is the vortex-glass transition temperature and the exponents were obtained from fits to the experiments of [46]. Assuming that $\eta \sim \xi_G^2 \gamma$, we find

$$\eta(T) \sim \frac{1}{|T - T_{vg}|^{\nu(z+1)}}. \tag{8.104}$$

The exponents ν and z must, unfortunately, be fit to simulation or experiments at the moment, since there is no analytical theory. One set of experimental estimates [46] lies in the ranges $\nu \sim 1.7$–1.8 and $z \sim 4.7$–5.0.

The predictions of these different theories of the irreversibility line are summarized in Table 8.2. An experiment capable of measuring $\eta(T)$ and $\gamma(T)$ and thus distinguishing among the four competing explanations for the irreversibility line is shown in Fig. 8.18. The left-hand portion of an initially clean (twin-free!) superconducting sample with homogeneous disorder only is irradiated by protons or neutrons to produce a high density of much stronger pins, so that the linear resistivity is effectively zero in this region. The right-hand portion remains clean. With the magnetic field normal to the plane of the figure, a current flows through the interface between the clean and dirty regions. It is desirable for this interface to be very sharp; interfacial widths of a few hundred or thousand ångström units may be possible. A series of voltage taps bisecting the interface can then be used to obtain the electric-field distribution inside the sample. The electric field $E_x(x)$

Table 8.2 *Predictions of four competing explanations of the physics of the irreversibility line (the TAFF theory does not make an explicit prediction for the parameter η)*

Theory	$\gamma(T)$	$\eta(T)$
TAFF (traditional)	$\gamma \sim e^{U_p/(k_B T)}$	Undefined
Vortex glass	$\gamma \sim (1/|T - T_{vg}|^{6.6})$	$\eta \sim (1/|T - T_{vg}|^{10})$
"Polymer" glass transition	γ finite	$\eta \sim \exp(c/|T - T_g|)$
Freezing/melting	γ finite	η *large*; grows and jumps discontinuously to infinity at T_m if the transition is weakly first order

determines via Eq. (8.99) the hydrodynamic flux-line-velocity $v_y(x)$ so one can obtain the flux-flow-velocity profile inside the sample. As shown in Fig. 8.18, the solution of Eq. (8.10) for $\mathbf{v}(x, y)$ rises from zero to its limiting homogeneous-disorder value over a distance δ. By measuring both the asymptotic velocity v_∞ and the width δ (which we assume is greater than the width of the defect profile), one can determine both $\gamma(T)$ and $\eta(T)$. A discussion of experiments of this kind with a finite-frequency driving term has been given by Marchetti [95]. For a discussion of related ideas applied to the Bose glass transition, see [15].

Acknowledgements

Many of the results described here were obtained in collaboration with H. S. Seung, M. C. Marchetti and P. Le Doussal. I am grateful for these collaborations, as well as for stimulating discussions with D. J. Bishop, D. S. Fisher, P. L. Gammel, A. Kapitulnik, P. H. Kes, C. A. Murray, M. Tinkham and T. K. Worthington. This chapter is adapted with permission from *Phenomenology and Applications of High Temperature Superconductors*, edited by K. S. Bedell *et al.* (Addison-Wesley, New York 1992), pp. 187–239.

References

[1] D. R. Nelson and V. M. Vinokur, *Phys. Rev.* B **48**, 13 060 (1993); T. Hwa, D. R. Nelson and V. M. Vinokur, *Phys. Rev.* B **48**, 1167 (1993); D. R. Nelson and L. Radzihovsky, *Phys. Rev.* B **54**, R6845 (1996).

[2] J. Lidmar and M. Wallin, *Europhys. Lett.* **47**, 494 (1999); see also D. R. Nelson and V. M. Vinokur, *Phys. Rev.* B **61**, 5917 (2000).

[3] T. Hwa, P. Le Doussal, D. R. Nelson and V. Vinokur, *Phys. Rev. Lett.* **71**, 3545 (1993); P. Le Doussal and D. R. Nelson, *Physica* C **232**, 69 (1994).

[4] U. Tauber and D. R. Nelson, *Phys. Rep.* **289**, 157 (1997).

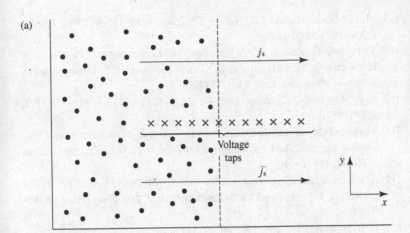

(a)

j_s

× × × × × × × × × × × × ×

Voltage
taps

$\bar{j}_s$

y

x

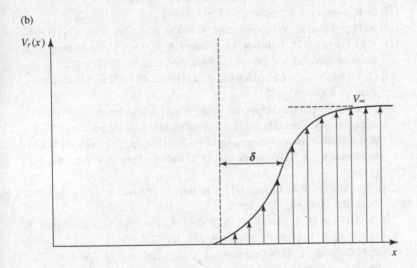

(b)

$V_r(x)$

V_∞

δ

x

Fig. 8.18. An idealized experiment allowing an independent determination of the key hydrodynamic parameters γ and η [51]. The left-hand half of the sample in (a) has a high density of pinning centers. A series of voltage taps indicated by crosses determines the velocity profile shown in (b): the voltage drop between any two crosses is proportional to the *area* under the corresponding portion of the velocity-profile curve.

[5] N. Hatano and D. R. Nelson, *Phys. Rev.* B **56**, 8651 (1997).

[6] L. Balents and D. R. Nelson, *Phys. Rev.* B **52**, 12 951 (1995).

[7] W. K. Kwok, J. Fendrich, U. Welp, S. Fleshler, J. Downey and G. W. Crabtree, *Phys. Rev. Lett.* **72**, 1088 (1994).

[8] T. Giamarchi and P. Le Doussal, *Phys. Rev.* B **52**, 1242 (1995).

[9] J. Kierfeld, T. Nattermann and T. Hwa, *Phys. Rev.* B **55**, 626 (1997).

[10] D. S. Fisher, *Phys. Rev. Lett.* **78**, 1964 (1997).

[11] M. Gringas and D. Huse, *Phys. Rev.* B **53**, 15193 (1996).

[12] T. Nattermann and S. Scheidl, *Adv. Phys.* **49**, 607 (2000); see also D. S. Fisher, *Phys. Rep.* **301**, 113 (1998).

[13] M. H. Theunissen, E. Van der Drift and P. H. Kes, *Phys. Rev. Lett.* **77**, 159 (1996); J. Pastoriza and P. H. Kes, *Phys. Rev. Lett.* **75**, 3525 (1995).

[14] D. Lopez, W. K. Kwok, H. Safar, R. J. Olsson, A. M. Petrean, L. Paulius and G. W. Crabtree, *Phys. Rev. Lett.* **82**, 1277 (1999); G. W. Crabtree, D. Lopez, W. K. Kwok, H. Safar, L. M. Paulius, in *Proceedings of the MOS 99*, *J. Low. Temp. Phys.* **117** (1999).

[15] M. C. Marchetti and D. R. Nelson, *Phys. Rev.* B **59**, 13624 (1999); *Physica* C **330**, 105 (2000).

[16] Y. Paltiel, E. Zeldov, Y. Myasoedov, M. L. Rappaport, G. Gung, S. Bhattacharya, M. J. Higgins, Z. L. Xiao, E. Y. Andrei, P. L. Gammel and D. J. Bishop, *Phys. Rev. Lett.* **85**, 3712 (2000).

[17] A. A. Abrikosov, *Zh. Éksp. Teor. Fiz.* **32**, 1442 (1957) [*Sov. Phys. JETP* **5**, 1174 (1957)].

[18] The possibility of a melting transition *very* close to H_{c2} (triggered by a vanishing shear modulus) was suggested long ago in R. Labusch, *Phys. Status Solidi* **32**, 439 (1969).

[19] A. I. Larkin, *Zh. Éksp. Teor. Fiz.* **58**, 1466 (1970) [*Sov. Phys. JETP* **31**, 784 (1970)]; A. I. Larkin and Yu. N. Ovchinnikov, *J. Low Temp. Phys.* **34**, 409 (1979).

[20] D. R. Nelson, *Phys. Rev. Lett.* **60**, 1973 (1988).

[21] D. R. Nelson and S. Seung, *Phys. Rev.* B **39**, 9158 (1989).

[22] D. R. Nelson, *J. Statist. Phys.* **57**, 511 (1989).

[23] A. Houghton, R. A. Pelcovits and A. Sudbo, *Phys. Rev.* B **40**, 6763 (1989).

[24] P. L. Gammel, D. J. Bishop, G. J. Dolan, J. R. Kwo, C. A. Murray, L. F. Schneemeyer and J. V. Waszczak, *Phys. Rev. Lett.* **59**, 2592 (1987).

[25] G. T. Dolan, G. V. Chandrasekar, T. R. Dinger, C. Feild and F. Holtzberg, *Phys. Rev. Lett.* **62**, 827 (1989).

[26] Even disordered flux arrays can "melt" if vortex lines begin to move appreciably on experimental time scales at sufficiently high temperatures. See the striking decoration photographs in R. N. Kleiman, P. L. Gammel, L. F. Schneemeyer, J. V. Waszczak and D. J. Bishop, *Phys. Rev. Lett.* **62**, 2331 (1989).

[27] C. A. Murray, P. L. Gammel, D. J. Bishop, D. B. Mitzi and A. Kapitulnik, *Phys. Rev. Lett.* **64**, 2312 (1990).

[28] D. G. Grier, C. A. Murray, C. A. Bolle, P. L. Gammel, D. J. Bishop, D. B. Mitzi and A. Kapitulnik, *Phys. Rev. Lett.* **66**, 2270 (1991).

[29] P. L. Gammel, L. F. Schneemeyer, J. V. Waszczak and D. J. Bishop, *Phys. Rev. Lett.* **61**, 1666 (1988).

[30] D. E. Farrell, J. P. Rice and D. M. Ginzberg, *Phys. Rev. Lett.* **67**, 1165 (1991).

[31] E. H. Brandt, P. Esquinazi and G. Weiss, *Phys. Rev. Lett.* **62**, 2330 (1989).

[32] T. K. Worthington, F. H. Holtzberg and C. A. Field, *Cryogenics* **30**, 417 (1990).

[33] A similar interpretation of *low*-field resistivity data was proposed earlier by R. B. van Dover, L. F. Schneemeyer, E. M. Gyorgy and J. V. Waszczak, *Phys. Rev.* B **39**, 4800 (1989).

[34] M. Tinkham, *Helv. Phys. Acta* **61**, 443 (1988).

[35] Melting near H_{c1} has now been seen in computer simulations by L. Xing and Z. Tesanovic, *Phys. Rev. Lett.* **65**, 794 (1990). As predicted in [20], this transition occurs at very small reduced fields, $(H_m - H_{c1})/H_{c1} \ll 1$. There are, however, large demagnetizing factors in the usual slab-like single-crystalline geometries. As a result, $B = H$ to an excellent approximation and it is the vortex density which is held fixed experimentally. The melting transition

then occurs for $B \lesssim B_m = \phi_0/\lambda^2 \approx 100$ G, a region that is relatively easy to probe experimentally. For a detailed discussion of melting near H_{c1} from this point of view see [45]. For an interesting simulation of a lattice model of flux lines at *high* fields, see Y.-H. Li and S. Teitel, *Phys. Rev. Lett.* **66**, 3301 (1991).

[36] See, e.g., the articles in *Superconductivity*, edited by R. D. Parks (Dekker, New York, 1969) Vol. II.

[37] I am indebted to M. Feigel'man for discussions on this point. A similar argument for reentrant melting in a real quantum-mechanical system of two-dimensional fermions has been applied to electrons (and their image charges) on a thin substrate of liquid helium by P. Platzman, *Phys. Rev. Lett.* **50**, 2021 (1983).

[38] See comment (4) on p. 9157 of [21].

[39] E. H. Brandt, *Phys. Rev.* B **34**, 6514 (1986); *Phys. Rev. Lett.* **63**, 1106 (1989).

[40] S. Sengupta, C. Dasgupta, H. R. Krishnamurthy, G. I. Menon and T. V. Ramakrishnan, Bangalore preprint.

[41] S.-T. Chui, *Europhys. Lett.* **26**, 197 (1994).

[42] M. C. Marchetti and D. R. Nelson, *Phys. Rev.* B **42**, 9938 (1990).

[43] E. Brezin, D. R. Nelson and A. Thiaville, *Phys. Rev.* B **31**, 7124 (1985).

[44] *The Collected Papers of L. D. Landau*, ed. D. ter Haar (Gordon and Breach–Pergamon, New York, 1965) p. 193.

[45] M. P. A. Fisher, *Phys. Rev. Lett.* **62**, 1415 (1989); D. S. Fisher, M. P. A. Fisher and D. Huse, *Phys. Rev.* B **43**, 130 (1991).

[46] R. H. Koch, V. Foglietti, W. J. Gallagher, G. Koren, A. Gupta and M. P. A. Fisher, *Phys. Rev. Lett.* **63**, 1151 (1989).

[47] D. R. Nelson, in *Phase Transitions and Critical Phenomena*, Vol. 7, edited by C. Domb and J. L. Lebowitz (Academic Press, New York, 1983), and references therein. See pp. 69–71 for a formula relating the shear viscosity to the density of free dislocations.

[48] S. Doniach and B. Huberman, *Phys. Rev. Lett.* **17**, 1169 (1979).

[49] D. S. Fisher, *Phys. Rev.* B **22**, 1190 (1980).

[50] In this review, we reserve the word "viscosity" to apply to shear forces in a plane perpendicular to the magnetic field due to *inter*vortex interactions. The Bardeen–Stephen coupling of vortex motion to the underlying ionic lattice will be called a "friction," not a "viscosity."

[51] C. M. Marchetti and D. R. Nelson, *Phys. Rev.* B **42**, 9938 (1990); *Physica* C **174**, 40 (1991).

[52] E. M. Chudnovsky, *Phys. Rev.* B **40**, 11355 (1989); *Phys. Rev. Lett.* **65**, 3060 (1990); *Phys. Rev.* B **43**, 7831 (1991).

[53] R. Seshadri and R. M. Westervelt, *Phys. Rev. Lett.* **66**, 2774 (1991).

[54] This length has been denoted ξ_z elsewhere. We use the notation ℓ_z here to avoid confusion with the superconducting coherence length.

[55] As will be discussed in Section 8.2, one should make the replacement $(M_\perp/M_s) \rightarrow 1/\ln \kappa$ in this formula when $n_0 \lambda_{ab}^2 \lesssim 1$, where n_0 is the areal vortex density.

[56] W. R. White, A. Kapitulnik and M. R. Beasley, *Phys. Rev. Lett.* **66**, 2826 (1991).

[57] L. I. Glazman and A. E. Koshelev, *Phys. Rev.* B **43**, 2835 (1991). These authors suggest that there is a sharp phase transition at B_{x2}, instead of the gradual crossover discussed here.

[58] E. M. Forgan, D. McK. Paul, H. A. Mook, P. A. Timmins, H. Keller, S. Sutton and J. S. Abell, *Nature* **343**, 735 (1990).

[59] D. R. Nelson and P. Le Doussal, *Phys. Rev.* B **42**, 10113 (1990).

[60] The line-crossing energy in these models can, in fact, be made as large or small as desired. See the cutoff potential used, for example, in Eq. (5.3) of [21], which leads to Eq. (8.18) in this review.

[61] S. P. Obukhov and M. Rubinstein, *Phys. Rev. Lett.* **65**, 1279 (1990).

[62] J. R. Clem, private communication 1989; E. H. Brandt and A. Sudbo, *Phys. Rev. Lett.* **66**, 2378 (1991) and references therein.

[63] S. P. Obukhov and M. Rubinstein, *Phys. Rev. Lett.* **66**, 2279 (1991).

[64] A. P. Malozemoff, T. K. Worthington, Y. Yeshurun and F. Holtzberg, *Phys. Rev.* B **38**, 7203 (1988) and references therein.

[65] This may still be an appropriate description below the melting line, however. See M. B. Feigel'man and V. M. Vinokur, *Phys. Rev.* B **41**, 8986 (1990); M. V. Feigel'man, V. B. Geshkenbein, A. I. Larkin and V. M. Vinokur, *Phys. Rev. Lett.* **63**, 2303 (1989).

[66] For a review, see P. H. Kes, J. Aarts, J. van den Berg and J. A. Mydosh, *Supercond. Sci. Technol.* **1**, 241 (1989).

[67] V. G. Kogan, *Phys. Rev.* B **24**, 1572 (1981).

[68] M. C. Marchetti, *Phys. Rev.* B **43**, 8012 (1991).

[69] R. P. Feynman, *Phys. Rev.* **91**, 1291 (1953).

[70] R. P. Feynman and A. R. Hibbs, *Quantum Mechanics and Path Integrals* (McGraw-Hill, New York, 1965); R. P. Feynman, *Statistical Mechanics* (Benjamin, Reading, MA, 1972).

[71] See, e.g., E. Merzbacher, *Quantum Mechanics*, 2nd Edition (Wiley, New York, 1970) pp. 105–108.

[72] L. Civale, A. D. Marwick, T. K. Worthington, M. A. Kirk, J. R. Thompson, L. Krusin-Elbaum, Y. Sun, J. R. Clem and F. Holtzberg, *Phys. Rev. Lett.* **67**, 648 (1991).

[73] L. D. Landau and E. M. Lifshitz *Quantum Mechanics* (Pergamon, New York, 1965) Section 45.

[74] M. P. A. Fisher and D. H. Lee, *Phys. Rev.* B **39**, 2756 (1989).

[75] N. V. Popov, *Functional Integrals and Collective Excitations* (Cambridge University Press, New York, 1981).

[76] J. W. Negele and J. Orland, *Quantum Many-Particle Systems* (Addison-Wesley, New York, 1988) Chapters 1 and 2.

[77] Feigel'man has made the intriguing alternative suggestion that the loss of phase coherence leads to a *disentangled* flux liquid even in the limit $L \to \infty$. Such a phase would correspond to a normal liquid at $T = 0$ in the boson language. See M. Feigel'man, *Physica* A **168**, 319 (1990).

[78] R. Wordenweber and P. H. Kes, *Phys. Rev.* B **34**, 494 (1986); E. H. Brandt, *Phys. Rev.* B **34**, 6514 (1986); *Jap. J. Appl. Phys.* **26**, 151 (1987).

[79] For a related application of this idea to directed polymer melts, see P. Le Doussal and D. R. Nelson, *Europhys. Lett.* **15**, 161 (1991). See also J. Selinger and R. Bruinsma, *J. Physique* II **2**, 1215 (1992).

[80] P. G. de Gennes and J. Matricon, *Rev. Mod. Phys.* **36**, 45 (1964).

[81] Flux-decoration measurements probing the distribution of vortex lines that enter or leave the surface of a superconducting sample are close to this idealized experiment. Here, however, one must also account for the special nature of a bounding surface and the magnetic field outside the sample. See M. C. Marchetti and D. R. Nelson, *Phys. Rev.* B **47**, 12 214 (1993) and D. Huse, *Phys. Rev.* B **46**, 8621 (1992).

[82] The only caveat is that a two-dimensional liquid of points could mimic the correlations in a cross section of a line liquid if its interactions were of sufficiently long range. A $1/r$ pair potential, in particular, leads to a two-dimensional structure function which vanishes linearly with $q_\perp$.

[83] F. R. N. Nabarro and A. T. Quintanilha, in *Dislocations in Solids*, edited by F. R. N. Nabarro (North-Holland, Amsterdam, 1980) Vol. 5.

[84] For important early work on this subject, see R. Labusch, *Phys. Rev. Lett.* **22**, 9 (1966).

[85] D. R. Nelson and B. I. Halperin, *Phys. Rev.* B **19**, 2457 (1979).

[86] J. Toner, *Phys. Rev. Lett.* **66**, 2523 (1991).

[87] W. K. Kwok, U. Welp, G. W. Crabtree, K. G. Vandervoort, R. Hulschere and J. Z. Liu, *Phys. Rev. Lett.* **64**, 966 (1990).

[88] J. A. Hertz, L. Fleischman and P. W. Anderson, *Phys. Rev. Lett.* **43**, 942 (1979).

[89] M. Ma, B. I. Halperin and P. A. Lee, *Phys. Rev.* B **34**, 3136 (1986).

[90] T. Giamarchi and H. J. Schulz, *Europhys. Lett.* **3**, 1287 (1987).

[91] D. S. Fisher and M. P. A. Fisher, *Phys. Rev. Lett.* **61**, 1847 (1988); M. P. A. Fisher, P. B. Weichman, G. Grinstein and D. S. Fisher, *Phys. Rev.* B **40**, 546 (1989).

[92] V. Ambegaokar, B. I. Halperin, D. R. Nelson and E. D. Siggia, *Phys. Rev.* B **21**, 1806 (1980).

[93] B. I. Halperin and D. R. Nelson, *J. Low Temp. Phys.* **36**, 599 (1979).

[94] See, e.g., R. Zallen, *The Physics of Amorphous Solids* (Wiley, New York, 1983).

[95] M. C. Marchetti, *J. Appl. Phys.* **69**, 5185 (1991).

Chapter 9
Statistical mechanics of directed polymers

Preface

In this chapter, we discuss applications of the boson mapping and other ideas to the statistical mechanics of columnar materials. Physical realizations of these materials include condensed, oriented phases of DNA, various types of polymer nematics and the hexagonal columnar phase of discotic liquid crystals. In the absence of external electric and magnetic fields, the constraint of rotational invariance requires a "softer" version of the effective boson theory appropriate for vortex lines in superconductors. The resulting theory allows a statistical treatment (in a grand canonical ensemble) of many interacting directed polymers with a broken rotational symmetry and leads to interesting results for the X-ray structure function.

Some of the results here have implications for the field of "stereography," in which information about three-dimensional density distributions is to be inferred from information contained in one or more cross sections through the material [1]. In the context of equilibrated arrays of directed polymers oriented along the z axis, one can infer the average length of the polymers from the behavior of the structure function of a two-dimensional xy cross section at small wavevectors. See [2] for a more detailed account of these ideas.

One of the more intriguing predictions reviewed here is the idea that two-dimensional polymers in a nematic medium can have a *continuously variable* Flory exponent [3]. Maier and Radler recently carried out measurements of the equilibrium properties of single DNA molecules electrostatically bound to isotropic fluid lipid membranes and checked the famous Flory result for the radius-of-gyration exponent (see Chapter 5) $\nu = \frac{3}{4}$ [4]. By repeating this experiment on a substrate with tilted lipid molecules (or, alternatively, with ordered "Gemini lipids" or triblock copolymers) one might be able to test the prediction of a continuously variable exponent $\nu \geq \frac{8}{9}$ in the presence of nematic-like order.

This chapter concludes with an account of defects in hexagonal columnar phases of directed polymers, i.e., the analogue for polymers of the Abrikosov flux lattice. Despite the softer underlying elasticity theory, ideas about intermediate hexatic [5] and supersolid [6] phases for thermally excited vortex arrays in superconductors are applicable here as well. The concept of a "supersolid" phase with a finite concentration of infinitely long vacancy and interstitial *lines* is intimately connected with the line-like nature of the molecules themselves. The sharp transition from a hexagonal columnar solid of lines to a supersolid with unbound vacancies and interstitials is blurred by the finite concentration of polymer "heads" and "tails" present when the polymers are of finite length. See [7] for a more detailed account of the statistical mechanics of vacancy and interstitial strings in hexagonal columnar crystals. Two interesting predictions of this study are that interstitials are preferred over vacancies for a wide range of parameters and that both vacancies and interstitials often have a *lower* symmetry than that of the lattice in which they are embedded. The latter prediction (which also applies to lines of vacancies and interstitials in the Abrikosov flux lattice [6]) has been tested by experiments in which vacancies and interstitials are created artificially with laser tweezers in two-dimensional crystals of colloidal particles [8].

Related theories allow an analysis of modifications of the structure of hexagonal columnar crystals when the chirality is strong [9]. Chirality competes with the tendency for perfect crystalline order and one result of this competition can be a fascinating braided "moiré" state. Related ideas about braiding may apply to vortex arrays in superconductors with supercurrents running *parallel* to the direction of the magnetic field [10]. For a discussion of the effects of chirality on hexatic order in chiral polymer nematics, see the articles cited in [11]. For experiments on the "line-hexatic" phase of DNA, see the recent work by Strey *et al.* [12].

9.1 Introduction

Much effort was expended in the twentieth century toward understanding correlations and dynamics of dense liquids of point-like atoms and molecules [13]. Particularly noteworthy is the pioneering work by Ornstein and Zernike [14], who argued that the density fluctuations are Gaussian and determined the form of the static correlation functions at long wavelengths. More recently, interest has focused on assemblies of *extended* objects, with the topology of lines [15], surfaces [16], or even more complicated architectures arising in solutions with surface-active molecules [17]. In this review, we focus on dense, interacting arrays of linear polymers *oriented* along one preferred direction in space. Although they are less common than isotropic polymer melts [15], there are nevertheless many important physical examples of such directed polymers – see below. Because of the directed nature, one can probably go further in understanding the statics and dynamics than one can for isotropic polymers. Important simplifications arise both in the microscopic and in the long-wavelength theories because self-avoidance of an individual line is unimportant: The significant physics lies in entanglement and interline interactions.

This distinction is illustrated by the various field theories which allow a detailed statistical description of isotropic and directed polymer systems. As noted by de Gennes [18], summing over configurations of isotropic polymers in a good solvent is equivalent to understanding the $n \to 0$ limit of an interacting n-component spin system. The theory becomes applicable to dense polymer melts upon introduction of an external magnetic field for the spins, which acts like a polymer chemical potential [19]. As will be explained in this chapter, *directed* polymers are instead equivalent to the many-body quantum mechanics of bosons in two dimensions at zero temperature [20, 21]. For an understanding of isotropic polymers that goes beyond simple Flory theory (e.g., to obtain equations of state, universal scaling functions and precise dynamical exponents), the $n \to 0$ limit of the equivalent spin system must usually be solved perturbatively in an expansion in $\varepsilon = 4 - d$, where d is the dimensionality of space [22]. The boson field theory for directed polymers, on the other hand, allows many results to be obtained directly in the physical dimension $d = 3$.

The mapping onto quantum mechanics (see also Chapters 7 and 8) uses a formal correspondence between polymer trajectories and Feynman path integrals, a relationship that had already been exploited for conventional polymers in the 1960s [23]. Applications in this case, however, are usually limited, for example, to mean-field descriptions of average density variations near a boundary, because self-avoidance appears as an awkward nonlocal interaction between different slices of imaginary time in the path integral. The correspondence to quantum mechanics is much more precise for directed polymers: This problem is *identical* in many respects to the low-temperature physics of a film of real bosons such as ^{4}He atoms with a slightly different pair potential.

There is a wide variety of interesting physical systems consisting of extended one-dimensional objects with a preferred direction [24–26]. Biological macromolecules such as DNA [27], filamentous bacteriophages [28], helical synthetic polypeptides such as poly (γ-benzyl glutamate) (PBG) [28, 29], liquid crystals composed of stacks of disk-shaped molecules [30, 31] and micelles of amphiphillic molecules [32] can all form crystalline columnar phases with in-plane order, as well as nematic phases with fluid-like in-plane order. Although stiffer chains align more easily, some nematic polymer liquid crystals can also be formed with chains of relatively low rigidity, by alternating a nematogenic unit with a flexible hydrocarbon spacer [33–35]. The transition from isotropic melt to nematic can be achieved experimentally by lowering the temperature [36] or, more frequently, by increasing the concentration. Steric repulsion is sufficient to produce alignment at high enough concentration, although many other interactions can be present depending on the material: Van der Waals attraction, electrostatic forces [37, 38], hydration forces, etc. Ferrofluids [39] and electrorheological fluids [40] are also composed of chains of particles, in this case aligned by external magnetic or electric fields.

As discussed in the two preceeding chapters, another physical realization of directed "polymers" arises in very clean type-II superconductors in a strong magnetic field. Quantized vortex filaments play the role of "polymers" in this case. The familiar Abrikosov flux lattice is similar to the columnar hexagonal phase of real polymeric systems. In the presence of thermal fluctuations, which are particularly pronounced in the new high-temperature cuprate superconductors, the Abrikosov crystal can melt into a liquid of directed lines interacting with a repulsive potential (see Chapter 7). The average line direction is determined by the external magnetic field. In the absence of strong vortex pinning, the boson analogy is particularly appropriate for the vortex liquid [41–43].

There are, however, significant differences between the directed polymer-like systems discussed above and flux lines. Although oriented nematic polymers in a solvent wander along a preferred axis, just as thermally excited flux lines do, the average polymer direction represents a spontaneous, rather than externally imposed, broken symmetry. Even if the monomer chains are aligned by an external electric or magnetic field, the lines are typically of variable length and need not span the system, unlike flux lines. Nematic polymers can, moreover, make relatively low-energy hairpin turns, because of the symmetry of the director field under $\mathbf{n} \to -\mathbf{n}$. Such "back-tracking" can usually be ignored for flux lines [41–43].

Variable line lengths can be incorporated into the boson field theory by adding a source [20], in analogy with the des Cloiseaux treatment of isotropic polymer melts [19]. A related source term accounts for hairpin excitations and the soft broken symmetry of polymer nematics can be treated by coupling the bosons to a "gauge" field representing the nematic degrees of freedom [21]. The resulting statistical description goes well beyond conventional many-body boson quantum mechanics [44] and illustrates the deep and challenging issues which arise in soft condensed matter physics, even in the simplified context of directed polymer melts.

As discussed above, directed polymer melts crystallize at high concentrations or low temperatures, often into columnar phases. Columnar hexagonal phases are crystalline in the two directions perpendicular to the average line axis, but remain unordered along the remaining direction [45]. The nature of *defects* in columnar hexagonal phases consisting of very long polymers is quite different from that of defects in conventional three-dimensional crystals of point particles. Isolated dislocation loops, for example, are more constrained: Dislocations must lie in a *plane* spanned by their Burgers vectors and the average column direction [46]. Vacancies and interstitials differ even more dramatically from defects in point crystals. As will be discussed later, each polymer end marks the terminus of a *line* of vacancy or interstitial defects [47, 48], in contrast to the point-like vacancy and interstitial defects in conventional crystals. These vacancy and interstitial lines are themselves a directed polymer "melt" of excitations relative to a perfect columnar crystal. Their statistical mechanics can be also treated by the techniques of this chapter.

One potentially important aspect of directed polymer melts will be neglected here, namely, the effects of chirality. The helical molecular architecture of such common biopolymers as DNA, PBG and xanthan (a polysaccharide) can lead to twisted cholesteric order in the local direction [49, 50]. The treatment presented here is restricted to scales smaller than the cholesteric pitch. A number of remarkable effects due to chirality are described in [9, 51].

In Section 9.2 we discuss the alignment of a single polymer by a nematic solvent in two and three dimensions. The theoretical description of directed polymer melts is reviewed in Section 9.3. The mapping onto a boson field theory is explained in Section 9.4 and Section 9.5 describes results special to polymer nematics with a "soft" broken symmetry. A simpler hydrodynamic approach due originally to de Gennes [52] and developed further in [20, 21] and by Selinger and Bruinsma [53] is illustrated in Section 9.6 [54]. The statistical mechanics of vacancy and interstitial lines in columnar crystals is described in Section 9.7. We discuss in some detail a possible entropic transition to a crystal with infinitely long defect lines [7, 48]. See [55] for a closely related transition for the Abrikosov flux lattice in high-temperature superconductors.

9.2 One polymer in a nematic solvent

9.2.1 Three dimensions

Issues connected with the soft broken symmetry in a polymer nematic are already evident for a single polymer embedded in a nematic solvent [24]. If we characterize the polymer trajectory by a single-valued

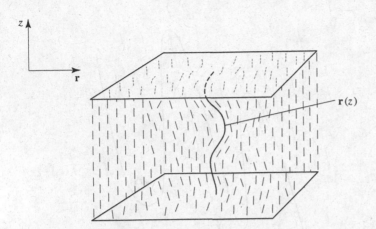

Fig. 9.1. An isolated polymer in a short-chain nematic solvent. The Goldstone modes of the nematic matrix produce anomalous wandering of the polymer transverse to the z axis.

function of the z coordinate $\mathbf{r}(z) = [x(z), y(z)]$, as in Fig. 9.1, the free energy in the one-Frank-constant approximation reads

$$F = \frac{1}{2}\kappa \int_0^L dz \left(\frac{d^2\mathbf{r}}{dz^2}\right)^2 + \frac{1}{2}g \int_0^L dz \left|\frac{d\mathbf{r}(z)}{dz} - \delta\mathbf{n}(\mathbf{r}(z), z)\right|^2$$

$$+ \frac{1}{2}K \int_0^L dz \int d^2r (\nabla\delta\mathbf{n})^2. \tag{9.1}$$

Here, the nematic degrees of freedom are represented by a deviation $\delta\mathbf{n}(\mathbf{r})$ of a unit vector $\mathbf{n}(\mathbf{r})$ away from the z axis,

$$\mathbf{n}(\mathbf{r}) = \frac{1}{\sqrt{1 + (\delta\mathbf{n})^2}} \begin{pmatrix} \delta n_x \\ \delta n_y \\ 1 \end{pmatrix}$$

$$\approx \begin{pmatrix} \delta\mathbf{n} \\ 1 \end{pmatrix}, \tag{9.2}$$

and K is the Frank constant [45]. The probability $\Pr(\{\mathbf{r}(z)\})$ of a particular polymer trajectory is given by a functional integral over the director field and a path integral over $\mathbf{r}(z)$ (with $k_B \equiv 1$),

$$\Pr(\{\mathbf{r}(z)\}) = \frac{\int \mathcal{D}\delta\mathbf{n}(\mathbf{r}) \exp(-F/T)}{\int \mathcal{D}\mathbf{r}(z) \int \mathcal{D}\delta\mathbf{n}(\mathbf{r}) \exp(-F/T)}. \tag{9.3}$$

The coupling κ is the rigidity of the polymer against bending, which is related to the polymer persistence length ℓ_p in the isotropic phase of the solvent by [15]

$$\ell_p = \kappa/T. \tag{9.4}$$

Fig. 9.2. Large-scale crumpling of a polymer in a nematic medium due to hairpin excitations of radius r_h and spacing ℓ_h.

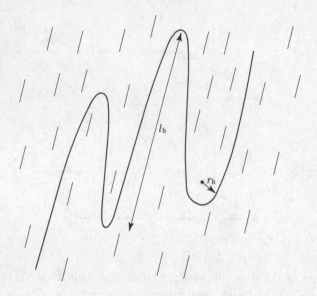

In an isotropic solvent, the polymer will always crumple on scales large relative to ℓ_p, where the description in terms of small bending fluctuations about a straight line breaks down. An isolated polymer in a nematic solvent will also crumple, but only on much larger scales, due to the hairpin excitations shown in Fig. 9.2. The description (9.1) in terms of a single-valued function $\mathbf{r}(z)$ is correct only for distances less than the hairpin spacing ℓ_h ($\ell_h \gg \ell_p$), which is estimated below.

The coupling g describes interactions between the polymer and the aligned nematogens which tend to straighten its trajectory. If we represent a point on the polymer by the *three*-dimensional position vector $\mathbf{R}(z) = [\mathbf{r}(z), z]$, then this coupling arises from the small-tipping-angle expansion of a rotationally invariant coupling of the form

$$-g\left(\frac{d\mathbf{R}}{dz} \cdot \mathbf{n}\right)^2. \tag{9.5}$$

Together, the couplings κ and g determine a length scale

$$r_h = \sqrt{\kappa/g}, \tag{9.6}$$

which is the radius of the hairpins shown in Fig. 9.2. To see this, we estimate the energy of a hairpin, assuming that the nematogens themselves do not follow the abrupt rotation of the polymer. The contribution from the first two terms of Eq. (9.1) to the energy ε_h of a hairpin with radius of curvature r is then approximately [21, 24]

$$\varepsilon_h \sim \frac{\kappa}{r} + gr. \tag{9.7}$$

Minimizing over r leads to $r_h = \sqrt{\kappa/g}$ and a hairpin energy

$$\varepsilon_h \approx \sqrt{\kappa g}. \tag{9.8}$$

The density of hairpins along the chain is $n_h = a^{-1} \exp(-\varepsilon_h/T)$, where a is the size of a monomer, so the hairpin spacing is $\ell_h = n_h^{-1}$, i.e.,

$$\ell_h \approx a \exp(\sqrt{g\kappa}/T). \tag{9.9}$$

Note that $\ell_h \gg \ell_p$ at low temperatures or large κ, showing that the polymer is stretched out considerably below the nematic ordering transition of the solvent.

On scales λ such that $\sqrt{\kappa/g} \ll \lambda \ll \ell_h$, the coupling κ can be neglected relative to g and we can ignore hairpins [24]. If $K = \infty$, director fluctuations are frozen out completely and it is then easy to show that the polymer executes a simple random walk perpendicular to the z axis as $Z \to \infty$,

$$\langle |\mathbf{r}(z) - \mathbf{r}(0)|^2 \rangle \approx \frac{2T}{g} |z|, \tag{9.10}$$

where the angle brackets represent a path integral over the probability distribution (9.3). In a sample of thickness L, the polymer will thus have size $\sim L$ parallel to $\hat{z}$ and dimension $\sim L^{1/2}$ perpendicular to it. The situation is more complicated for finite K, however, because of the appearance of $\mathbf{r}(z)$ in the argument of $\delta\mathbf{n}(\mathbf{r}(z), z)$ in the coupling term. In an approximation with the replacement

$$\delta\mathbf{n}(\mathbf{r}(z), z) \to \delta\mathbf{n}(0, z) \tag{9.11}$$

for a polymer wandering near the z axis, de Gennes found a result of the form [24]

$$\langle |\mathbf{r}(z) - \mathbf{r}(0)|^2 \rangle = \frac{2T}{g} |z| + \frac{T}{2\pi K} |z| \ln\left(\frac{|z|}{a}\right). \tag{9.12}$$

Long-range director correlations along the z axis thus lead to a logarithmically divergent transverse wandering. Although the polymer remains stretched out, the "softer" broken symmetry associated with the confining nematogens when $K < \infty$ produces a singular correction to Eq. (9.10). A more systematic renormalization-group treatment of this problem (using the field theory discussed in the next section and allowing for different Frank constants) leads to the same result with, however, a factor of two difference in the coefficient of the logarithm [21].

Fig. 9.3. A polymer interacting with direction fluctuations in a two-dimensional nematic medium and described by its complex coordinate z as a function of arclength s. Arrows indicate the sense of tilt if the "nematic" is a Langmuir film of amphiphillic molecules with tilted hydrocarbon chains.

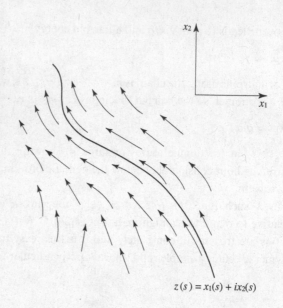

$$z(s) = x_1(s) + ix_2(s)$$

9.2.2 Two dimensions

We now consider the configurations of a chain in a two-dimensional nematic matrix [21]. Although it would be more difficult than its three-dimensional counterpart, an experiment at a two-dimensional interface might be possible: Imagine a long polymer chain with N monomers adsorbed at an air–water interface that is also covered with a tilted monolayer Langmuir–Blodgett film. The projection of the tilted hydro-carbon chains onto the plane of the interface plays the role of a direc-tor $\mathbf{n}$, *without*, however, the inversion symmetry $\mathbf{n} \rightarrow -\mathbf{n}$. See Fig. 9.3. We parameterize the local tilt by putting $\mathbf{n} = (\cos \theta, \sin \theta)$, Thermal fluctu-ations destroy the tilt order above T_c via a Kosterlitz–Thouless [56] vortex-unbinding transition. Even below T_c, however, there is no privi-leged direction in the tilt field since a broken continuous symmetry is impossible in two dimensions for systems with short-range interactions. Thus, one expects the polymer to crumple even for $T < T_c$ and one can ask for the wandering exponent ν governing the mean end-to-end dis-tance $R \sim N^\nu$, where N is the number of monomers. Note that $\nu = 1$ in the *three*-dimensional problem discussed above. The problem for $d = 2$ is complicated because, below T_c, the nematic field has long-range cor-relations decaying with a continuously varying exponent:

$$\langle \mathbf{n}(\mathbf{x}) \cdot \mathbf{n}(\mathbf{x}') \rangle = \langle e^{i\theta(\mathbf{x})} e^{-i\theta(\mathbf{x}')} \rangle$$
$$\sim |\mathbf{x} - \mathbf{x}'|^{-\eta(T)}. \tag{9.13}$$

The exponent $\eta(T)$ varies continuously from $\eta = 0$ at $T = 0$ to $\eta = \frac{1}{4}$ at the transition [56]. One may assume that the tilt fluctuations are described at long wavelengths by the usual XY-model free energy $\frac{1}{2}K \int d^2x \, (\nabla\theta)^2$ with a single Frank constant K [21]; for $T<T_c$ one has $\eta = \eta(T) = T/(2\pi K)$ [56]. Above T_c the correlations decay exponentially and one then expects that the wandering exponent of the polymer is the two-dimensional self-avoiding-random-walk value $\nu=\frac{3}{4}$ (or $\nu=\frac{1}{2}$ for a phantom chain).

The rotationally invariant two-dimensional version of the free energy (9.1) reads

$$F=\frac{1}{2}\kappa \int ds \left(\frac{d^2\mathbf{r}}{ds^2}\right)^2 - g\int ds \left(\frac{d\mathbf{r}(s)}{ds}\cdot\mathbf{n}[\mathbf{r}(s)]\right)^2 + \frac{1}{2}K\int d^2r \, (\nabla\mathbf{n})^2, \qquad (9.14)$$

where $\mathbf{r}(s) = [x_1(s), x_2(s)]$ is the position of the polymer parameterized as a function of arclength s. The g coupling appears squared because the free energy must remain unchanged under $d\mathbf{r}/ds \to -d\mathbf{r}/ds$ for polymers with local inversion symmetry along their backbone. A simple random-walk argument gives an interesting prediction for the exponent ν below T_c, at least for $g\to\infty$ [21]. In the infinite-g limit one expects that the polymer is totally aligned with the local field, so we equate the two unit vectors $d\mathbf{r}/ds$ and $\mathbf{n}$. In a convenient complex notation this equality reads

$$\frac{dz}{ds} = e^{i\theta(z(s))}, \qquad (9.15)$$

where we denote the position of the polymer by $z(s) = x_1(s) + ix_2(s)$. Upon integrating over s we find

$$\langle|z(s)-z(0)|^2\rangle_{z(t),\theta} = \int_0^s\int_0^s du \, du' \, \langle e^{i[\theta(z(u))-\theta(z(u'))]}\rangle_{z(t),\theta}, \qquad (9.16)$$

where the average is over both polymer configurations $z(t)$ and over the smoothly varying director angle $\theta(z)$. We now neglect the reaction of the polymer on the nematic field and replace the correlation on the right-hand side of (9.16) by the correlation of the unperturbed nematic field given by Eq. (9.13). This procedure amounts to replacing the problem by that of determining the wandering of the tangent curves of a pure nematic polymer, as indicated in Fig. 9.3. We are led to an integral equation, namely,

$$\langle|z(s)-z(0)|^2\rangle_{z(t)} = \int_0^s\int_0^s du \, du' \left\langle\frac{\text{constant}}{|z(u)-z(u')|^\eta}\right\rangle_{z(t)}. \qquad (9.17)$$

Upon assuming that $|z(s)-z(0)|$ scales as s^ν, we find that the self-consistent value of ν entering a typical polymer size $R\sim N^{\nu(T)}$ is [21]

$$\nu(T) = \frac{2}{2+\eta(T)}. \qquad (9.18)$$

Although we derived this result only in the limit $g \to \infty$, it is plausible that this relation for the continuously variable Flory exponent holds in the limit $N \to \infty$ for arbitrary g.

According to Eq. (9.18) ν is always close to $\nu = 1$, continuously decreasing with temperature from $\nu(T=0) = 1$ to $\nu(T_c^-) = \frac{8}{9}$. Note that ν always exceeds $\frac{3}{4}$, the value for a self-avoiding random walk in two dimensions. Our neglect of self-avoidance for $T < T_c$ is thus self-consistent because of the relatively small number of self-intersections that occur for $\nu > \frac{3}{4}$. The exponent $\nu(T)$ should jump discontinuously from $\nu(T_c^-) = \frac{8}{9}$ to the constant value $\nu(T) = \frac{3}{4}$ above the Kosterlitz–Thouless transition.

9.3 A model of polymer nematics

It is straightforward to generalize (9.1) to describe multiple flux lines wandering through a nematic medium. For now, we neglect free ends and hairpins, which will be included later when the problem is recast as a boson field theory. As before we describe the position of the ith polymer as it traverses the nematic medium by a function $R_j(z) = (\mathbf{r}_j(z), z)$. The N polymer lines interact with each other and the nematic field via a free energy adapted from [24],

$$
F = \frac{1}{2} g \sum_{j=1}^{N} \int dz \left(\frac{d\mathbf{r}_j}{dz} - \delta \mathbf{n}(\mathbf{r}_j(z), z) \right)^2
$$

$$
+ \frac{1}{2} \sum_{\substack{i,j=1 \\ i \neq j}}^{N} \int dz \, V(|\mathbf{r}_i(z) - \mathbf{r}_j(z)|) + F_N(\delta \mathbf{n}). \tag{9.19}
$$

The parameter g controls the coupling between the local polymer direction and a "background" nematic field $\delta \mathbf{n}(\mathbf{r}, z)$. This coupling is the only one allowed by rotational invariance, to lowest order in $d\mathbf{r}_j/dz$ and $\delta \mathbf{n}$. The potential $V(\mathbf{r})$ describes interactions between different polymer strands. The probability of a particular field configuration is proportional to $\exp(-F/T)$ and averages are calculated by integrating over both $\delta \mathbf{n}(\mathbf{r})$ and polymer configurations $\{\mathbf{r}_j(z)\}$.

The nematic free energy reads

$$
F_n = \frac{1}{2} \int d^2 r_\perp \int dz \left(K_1 (\nabla_\perp \cdot \delta \mathbf{n})^2 + K_2 (\nabla_\perp \times \delta \mathbf{n})^2 + K_3 (\partial_z \delta \mathbf{n})^2 \right.
$$

$$
\left. + \frac{1}{2} \chi_a H^2 |\delta \mathbf{n}|^2 \right), \tag{9.20}
$$

where the $\{K_i\}$ are the usual Frank constants for splay, twist and bend and $\delta \mathbf{n}(\mathbf{r}) = (\delta n_x(\mathbf{r}_\perp, z), \delta n_y(\mathbf{r}_\perp, z))$ is a vector representing a small deviation of the director field $\mathbf{n}(\mathbf{r})$ from its average orientation along the

z axis, $\mathbf{n(r)} \approx (\delta\mathbf{n}, 1)$. We have allowed for an external field $\mathbf{H} \| \mathbf{z}$ that aligns the sample via the anisotropy in magnetic susceptibility χ_a. The pair potential $V(r)$ could, for example, represent a hard-core delta-function interaction between polymer strands. For charged polyelectrolytes like DNA, when the mean polymer spacing is much larger than the hard-core radius, it is more appropriate to take $V(r)$ to be the interaction energy per unit length which arises from an electrolyte with Debye screening [27],

$$V(\mathbf{r}) = 2u_0 K_0(\kappa_D r). \tag{9.21}$$

Here κ_D^{-1} is the Debye or hydration screening length and u_0 is related to the linear charge density λ along the backbone of the polymer and the dielectric constant ε of the solvent by

$$u_0 \approx \lambda^2 / \varepsilon. \tag{9.22}$$

We have assumed that $\kappa_D^{-1} > d$, the hard-core polymer diameter.

The simplest physical interpretation of the free energy F is of polymers aligned by a nematic solvent with Frank constants $\{K_i\}$. In this case one should add a term (unimportant at long wavelengths)

$$\frac{1}{2}\kappa' \sum_{j=1}^{N} \int dz \left(\frac{d^2\mathbf{r}}{dz^2} \right)^2 \tag{9.23}$$

to Eq. (9.19) to allow for the bending rigidity of the polymer that augments the bending Frank constant K_3 of the solvent. As discussed in [21], F also describes *dense* nematic polymers in an *isotropic* solvent. The director $\mathbf{n(r)}$ then represents a coarse-grained nematic field obtained by averaging over the polymer tangents in a hydrodynamic averaging volume (see Fig. 9.4). Deviations of the orientation of any individual polymer from this average direction are described by the coupling g. This interpretation of F is especially appropriate for polymers made of nematic molecules connected by flexible hydrocarbon spacers [31]. In this case $\delta\mathbf{n(r)}$ describes fluctuations in the orientations of individual nematogens, while $\mathbf{r}_j(z)$ describes how the nematogens are threaded together by the hydrocarbon spacers. Note that the potential $V(\mathbf{r})$ represents a *scalar* interpolymer interaction within a constant-z plane. The coupling g, when the nematic modes are integrated out, leads both to a scalar interaction, due to the longitudinal modes of the nematic phase, and to a *vectorial* interaction between polymer tangents, due to the transverse nematic modes. Since $\delta\mathbf{n} \approx \delta\mathbf{r}_j / \partial z$ the Frank coupling $K_3(\partial_z \delta\mathbf{n})^2$, in particular, describes the bending rigidity of the polymers, with $\kappa = K_3 a_0^2$, where a_0 is the interpolymer spacing.

As discussed in Section 9.7, Eq. (9.19) leads immediately to the standard columnar hexagonal elastic free energy [45] when the polymers

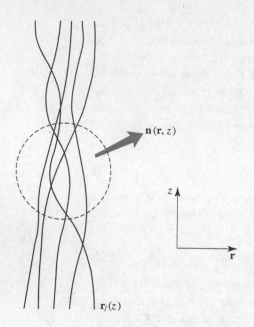

Fig. 9.4. The hydrodynamic averaging volume surrounding a small region of a nematic polymer containing many strands of polymer. The averaging over the polymer tangents in this volume defines a coarse-grained director field **n**(r, z), which then tends to align the individual polymers that pass through the region.

crystallize. Another check of our assumption that polymers interacting with a "nematic background" field are equivalent to dense nematic polymers in an isotropic solvent is provided by the hydrodynamic approach to correlation functions [52, 53], which gives identical results for these two systems in the limit of long wavelengths [21].

The parameter g can be estimated from the standard description of order in the nematic state of the polymer [57]. Assume that the distribution of tipping angles of the polymer tangents away from the z axis in the nematic phase is described by a probability

$$\mathcal{P}(\theta) \propto e^{-\alpha\theta^2/2}, \tag{9.24}$$

so that $\langle\theta^2\rangle = \langle|d\mathbf{r}_j/dz|^2\rangle = 1/\alpha$. The first term in Eq. (9.19), on the other hand, leads to

$$\left\langle\left|\frac{d\mathbf{r}_j}{dz}\right|^2\right\rangle \approx \frac{T}{g\ell^*}, \tag{9.25}$$

where ℓ^* is the short-distance cutoff appropriate to polymers interacting with, e.g., the pair potential (9.21): ℓ^* is the distance the polymer wanders along **z** before it feels the confining effects of its neighbors. Upon comparing these results for $\langle|d\mathbf{r}_j/dz|^2\rangle$, we find an expression for g,

$$g = T\alpha/\ell^*. \tag{9.26}$$

For polymers with only a hard-core excluded-volume interaction, ℓ^* would be the Odijk "deflection length" [26]. The length ℓ^* for polymers with an *arbitrary* in-plane interaction potential is estimated in Section 9.7.

9.4 Mapping onto boson quantum mechanics

9.4.1 The transfer-matrix approach

We first consider a polymer nematic with very large background Frank constants $\{K_i\}$, or, equivalently, one subjected to a very intense magnetic field $\mathbf{H}$ so that the director fluctuations are effectively frozen out. The system is then closely related to the vortex lines described in Chapter 8. The discussion here is simply a prelude for treating polymers with *free ends*, which have no analogue in vortex systems. When free ends are included, the model also describes ferro- and electrorheological fluids. Upon setting $\delta\mathbf{n}(\mathbf{r}, z) = 0$ in Eq. (9.19), the partition function for N lines in a sample of thickness L is a multidimensional path integral,

$$Z_N = \int \mathcal{D}\mathbf{r}_1(z) \cdots \mathcal{D}\mathbf{r}_N(z) e^{-F_N/T}, \tag{9.27}$$

where

$$F_N = \frac{1}{2} g \sum_{j=1}^{N} \int_0^L \left(\frac{d\mathbf{r}_j(z)}{dz} \right)^2 dz + \sum_{i>j} \int_0^L V[\mathbf{r}_{ij}(z)] dz. \tag{9.28}$$

The standard way of attacking such problems in classical statistical mechanics is to use the transfer matrix. It is not hard to show that the transfer matrix connecting neighboring constant-z slices of the partition function (9.27) is just the exponential of an N-particle Hamiltonian for quantum-mechanical particles propagating in imaginary time and interacting with a pair potential $V(\mathbf{r}_{ij}(z))$ [58]. Equation (9.27) can be reexpressed as an integral over a quantum-mechanical matrix element,

$$Z_N = \int d\mathbf{r}_1' \cdots d\mathbf{r}_N' \int d\mathbf{r}_1 \cdots d\mathbf{r}_N \langle \mathbf{r}_1' \cdots \mathbf{r}_N' | e^{-\mathcal{H}_N L/T} | \mathbf{r}_1 \cdots \mathbf{r}_N \rangle, \tag{9.29}$$

where the Hamiltonian is

$$\mathcal{H}_N = -\frac{T^2}{2g} \sum_{j=1}^{N} \nabla_j^2 + \sum_{i>j} V(|\mathbf{r}_i - \mathbf{r}_j|). \tag{9.30}$$

The states $|\mathbf{r}_1, \ldots, \mathbf{r}_N\rangle$ and $\langle \mathbf{r}_1', \ldots, \mathbf{r}_N'|$ describe, respectively, the entry and exit points for the directed polymers in Fig. 9.5.

We now insert a complete set of many-particle energy eigenstates $|m\rangle$ of $\mathcal{H}_N$ into Eq. (9.29) and find that

$$Z_N = \sum_m |\langle m | \mathbf{P}_1 = 0, \ldots, \mathbf{P}_N = 0 \rangle|^2 e^{-E_m L/T}, \tag{9.31}$$

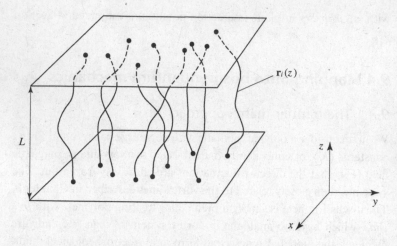

where $|\mathbf{P}_1 = 0, \ldots, \mathbf{P}_N = 0\rangle$ is the zero-momentum state,

$$|\mathbf{P}_1 = 0, \ldots, \mathbf{P}_N = 0\rangle = \int d\mathbf{r}_1 \cdots \int d\mathbf{r}_N |\mathbf{r}_1, \ldots, \mathbf{r}_N\rangle. \tag{9.32}$$

As discussed in Chapter 8, the eigenstates of the permutation-symmetric Hamiltonian (9.30) may be classified in terms of their transformation properties under the permutation group. In principle, all types of states, bosonic, fermionic and those obeying "parastatistics," should enter the sum (9.31). Because the state (9.32) is itself *symmetric* under permutations, however, the matrix element in Eq. (9.31) vanishes for all but the permutation-symmetric boson eigenstates. We may thus rewrite Eq. (9.31) as a restricted sum,

$$Z_N = \sum_{\substack{\text{boson} \\ \text{states } m}} |\langle m | \mathbf{P}_1 = 0, \ldots, \mathbf{P}_N = 0\rangle|^2 e^{-E_m L/T}. \tag{9.33}$$

The partition sum (9.33) should be contrasted with the sum appropriate for real bosons at a finite β,

$$Z'_N = \sum_{\substack{\text{boson} \\ \text{states } m}} e^{-\beta E_m}. \tag{9.34}$$

The difference between Z_N and Z'_N is in the weights involved in the projection onto the zero-momentum ground state in Eq. (9.33). As $L \to \infty$, the lowest-energy bosonic state dominates in both cases, however.

The trajectories of polymers across the sample are isomorphic to boson world lines. The thermal energy T plays the role of $\hbar$, with L corresponding to the distance $\beta\hbar$ traveled in the imaginary time direction. The parameter g plays the role of the boson mass m (see Table 9.1).

Table 9.1 *The detailed correspondence of the*
parameters for a polymer nematic with the
mass, value of Planck's constant, reciprocal
temperature β and pair potential of two-
dimensional bosons

Directed polymers	Two-dimensional bosons
g	m
T	$\hbar$
L	$\beta\hbar$
$V(\mathbf{r})$	Boson pair potential

9.4.2 The coherent-state path-integral representation

Results for correlation functions of interacting bosons near $T=0$ (i.e.,
polymer nematics in the thermodynamic limit $L\rightarrow\infty$) are conveniently
derived in a coherent-state path-integral representation of the grand
canonical partition function (see also Chapter 8)

$$Z_{gr} = \sum_{N=0}^{\infty} \frac{1}{N!} e^{L\mu N/T} Z_N. \tag{9.35}$$

Here μ is the chemical potential per unit length of the polymer, which
can be controlled in a nematic solution of a polymer by varying the
osmotic pressure. The bosons are represented by a complex field $\psi(\mathbf{r}_\perp,
z)$ and the grand partition function is

$$Z_{gr} = \int \mathcal{D}\psi(\mathbf{r}_\perp, z) \int \mathcal{D}\psi^*(\mathbf{r}_\perp, z)\, e^{-S(\psi,\psi^*)/T}, \tag{9.36}$$

where the "action" is [44]

$$S = \int_0^L dz \int d^2r_\perp \left(T\psi^* \partial_z \psi + \frac{T^2}{2g} |\nabla_\perp^2 \psi|^2 - \mu |\psi|^2 \right)$$

$$+ \int_0^L dz \int d^2r \int d^2r'\, V(|\mathbf{r}-\mathbf{r}'|) |\psi(\mathbf{r}, z)|^2 |\psi(\mathbf{r}', z)|^2. \tag{9.37}$$

To allow for polymers of finite length, which start and stop inside
the solvent, we add a source to Eq. (9.37),

$$\frac{S}{T} \rightarrow \frac{S}{T} - h \int dz \int d^2r\, (\psi + \psi^*). \tag{9.38}$$

As illustrated in Fig. 9.6, the source coupling h is proportional to the
probability per unit area and per unit "time" along z of starting or

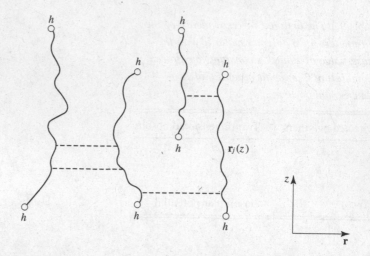

terminating a polymer. If the areal polymer density is $\rho_0 \approx a_0^{-2}$, the typical length of polymer associated with this Poisson-like process is [20, 21]

$$\ell = \sqrt{\rho_0}/h. \tag{9.39}$$

As discussed in [21], a graphic proof of the equivalence of Eqs. (9.36) and (9.37) to a system of directed polymers results from expanding Z_{gr} in the potential $V(r)$ and in h, the fugacity of a polymer end. The Green function propagator in this expansion is

$$G_0(\mathbf{r}, z) = \langle \psi(\mathbf{r}, z)\psi^*(0, 0)\rangle_0$$
$$= \theta(z)\frac{1}{4\pi Dz}e^{-r^2/(4Dz)}, \tag{9.40}$$

where $\theta(z)$ is the unit step function, the "diffusion constant" D is

$$D = T/(2g) \tag{9.41}$$

and the average $\langle\ \rangle_0$ is evaluated with respect to the Gaussian action

$$S_0 = \int_0^L dz \int d^2r\, [\psi^*(\partial_z - D\nabla_\perp^2)\psi]. \tag{9.42}$$

Equation (9.40) is identical to the propagator used in the more conventional virial expansion for the equilibrium problem of directed flux lines in [59]. In terms of the coherent-state approach, we can think of $\psi^*(0, 0)$ as creating a polymer at $(0, 0)$ and $\psi^*(\mathbf{r}, z)$ as destroying a polymer at $(\mathbf{r}, z)$. Although this produces the usual diffusive-random-walk propagator for $z > 0$, $G_0(\mathbf{r}, z) = 0$ when $z < 0$, showing that propagation backward in the time-like variable z is impossible.

Figure 9.6 may be viewed as a Feynman graph for the expansion of Z_{gr} in $V(r)$ and h, where the solid lines represent the polymer propagators and the dashed lines represent interactions. A total of n interacting polymers is represented by the sum of all terms proportional to h^{2n} in the expansion. Because propagation backward in z is forbidden, closed polymer loops are impossible.

The amplitude of the complex field $\psi(\mathbf{r}_\perp, z)$ is related to the fluctuating local line density of polymer according to

$$\rho(\mathbf{r}_\perp, z) = \sum_{j=1}^{N} \delta[\mathbf{r}_\perp - \mathbf{r}_j(z)]$$
$$= |\psi(\mathbf{r}_\perp, z)|^2. \tag{9.43}$$

"Superfluidity" of the bosons is equivalent to entanglement of the polymers (see Chapter 8), so we expect that entanglement of oriented polymers is accompanied by coherence in the *phase* of the order parameter

$$\psi(\mathbf{r}_\perp, z) = |\psi(\mathbf{r}_\perp, z)| e^{i\theta(\mathbf{r}_\perp, z)}, \tag{9.44}$$

i.e., long-range order in

$$G(\mathbf{r}_\perp, \mathbf{r}_\perp'; z, z') = \langle \psi(\mathbf{r}_\perp, z)\psi^*(\mathbf{r}_\perp', z') \rangle. \tag{9.45}$$

Line crystals, on the other hand, are usually characterized by the *absence* of long-range order in $\psi(\mathbf{r}_\perp, z)$, even though the average flux-line density $|\psi(\mathbf{r}_\perp, z)|^2$ is nonzero (see, e.g., [55]).

Correlation functions in the entangled melt can be calculated by expanding about the minimum of the action of Eq. (9.37)

$$\psi(\mathbf{r}_\perp, z) = \sqrt{\rho_0 + \pi(\mathbf{r}_\perp, z)} e^{i\theta(\mathbf{r}_\perp, z)}$$

$$\approx \sqrt{\rho_0} \left(1 + \frac{\pi}{2\rho_0} \right) e^{i\theta}, \tag{9.46}$$

where the average density of vortices is given by

$$\rho_0 = \mu/V_0, \tag{9.47}$$

with

$$V_0 = \int d^2r \, V(r)$$
$$= 4\pi u_0/\kappa_D^2 \tag{9.48}$$

for the potential of Eq. (9.21).

Upon keeping only terms quadratic in $\theta(\mathbf{r}_\perp, z)$ and $\pi(\mathbf{r}_\perp, z)$, one can easily calculate, for example, the structure function

$$S(q_\perp, qz) = \langle |\hat{\rho}(\mathbf{q}_\perp, q_z)|^2 \rangle, \tag{9.49}$$

where $\hat{\rho}(\mathbf{q}_\perp, q_z)$ is the Fourier transform of the fluxon density. The structure function takes the form [20, 21]

$$S(q_\perp, q_z) = \rho_0 \frac{2(\ell^{-1} + Dq_\perp^2)}{q_z^2 + \bar{\varepsilon}(q_\perp)^2}, \tag{9.50}$$

where $D = T/(2g)$ and

$$\bar{\varepsilon}(q_\perp) = (\ell^{-1} + Dq_\perp^2)\left(\ell^{-1} + Dq_\perp^2 + \frac{2\rho_0 \hat{V}(q_\perp)}{k_B T}\right). \tag{9.51}$$

Here $\hat{V}(q_\perp)$ is the Fourier transform of the pair potential of the polymer,

$$\hat{V}(q_\perp) = \frac{4\pi u_0}{\kappa_D^2 + q_\perp^2}. \tag{9.52}$$

If $h = 0$ (i.e., a typical length of polymer is $\ell = \infty$), and we assume that we have a delta-function interaction potential

$$V(\mathbf{r}) = V_0 \delta^{(2)}(\mathbf{r}), \tag{9.53}$$

$\bar{\varepsilon}(q_\perp)$ reduces to the well-known Bogoliubov superfluid excitation spectrum [44],

$$\bar{\varepsilon}(q_\perp) = \sqrt{\left(\frac{T}{2g} q_\perp^2\right)^2 + \frac{\rho_0 V_0}{g} q_\perp^2}. \tag{9.54}$$

Note that the effect of finite chain lengths is to produce a gap in the more general excitation spectrum (9.51) at $q_\perp = 0$. Although it is correct at long wavelengths, the Bogoliubov form for $\bar{\varepsilon}(q_\perp)$ is only qualitatively correct for all wavevectors $q_\perp$ unless the interactions are weak. A better approximation for $\bar{\varepsilon}(q_\perp)$ when the directed polymers are dense is the Feynman formula [58]

$$\bar{\varepsilon}(q_\perp) = \frac{Tq_\perp^2}{2gS_2(q_\perp)}, \tag{9.55}$$

which expresses the "phonon–roton spectrum" in terms of the *two-dimensional* structure function of a constant-z cross section, $S_2(q_\perp)$. If we insist that the three-dimensional structure function has the Lorentzian form (9.50), Eq. (9.55) is the only result consistent with the sum rule

$$\int_{-\infty}^{\infty} \frac{dq_z}{2\pi} S(q_\perp, q_z) = n_0 S_2(q_\perp). \tag{9.56}$$

The physical meaning of $\bar{\varepsilon}(q_\perp)$ for polymer nematics is illustrated by the contours of constant scattering shown in Fig. 9.7. The function $\bar{\varepsilon}(q_\perp)$ determines the shape of this characteristic "butterfly" diffraction pattern of directed polymers. We now use Eqs. (9.50) and (9.55) to determine the decay of in-plane density fluctuations $\delta\hat{\rho}(\mathbf{q}_\perp, z)$ along the z axis, with the result (see Fig. 9.8)

$$\tilde{S}(\mathbf{q}_\perp, z) \equiv \langle \delta\hat{\rho}(\mathbf{q}_\perp, z)\delta\hat{\rho}^*(\mathbf{q}_\perp, 0)\rangle$$
$$= S_2(q_\perp)e^{-\bar{\varepsilon}(q_\perp)|z|}, \tag{9.57}$$

thus generalizing results for vortices to polymers of finite length.

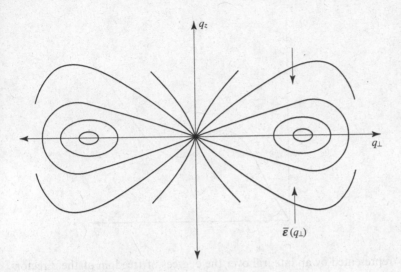

Fig. 9.7. Contours of constant scattering intensity for a dense directed polymer melt. The contours are linear near the origin and surround a maximum on the $q_\perp$ axis located approximately at the position of the first Bragg peak of the nearby triangular solid phase.

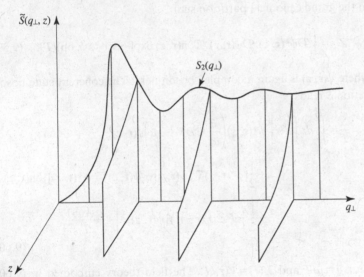

Fig. 9.8. The mixed structure function $\tilde{S}(q_\perp, z)$ showing the decay of the in-plane Fourier components of the density in a directed polymer melt. For fixed $q_\perp$, the decay is exponential in $|z|$.

The qualitative behavior of $\bar{\varepsilon}(q_\perp)$ obtained by using a realistic two-dimensional structure function in Eq. (9.55) is shown in Fig. 9.9. Density fluctuations with wavelength $q_\perp$ evidently decay due to line wandering along the z axis over a length scale given by

$$\ell_z(q_\perp) = 1/\bar{\varepsilon}(q_\perp). \tag{9.58}$$

9.5 Correlations in polymer nematics with soft broken symmetry

The same methods lead to correlation functions in polymer nematics with $H=0$ in Eq. (9.20). The "soft" spontaneous broken symmetry is

Fig. 9.9. A schematic diagram of the function $\varepsilon(q_\perp)$ which controls the decay of the density fluctuations along the z axis. A minimum in this "phonon–roton spectrum" appears when the two-dimensional structure function of a dense liquid is inserted into Eq. (9.55).

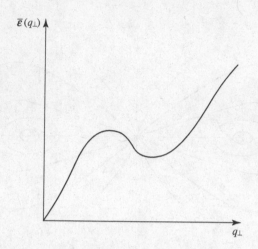

represented by an integral over the degrees of freedom of the director in the grand canonical partition sum

$$Z_{gr} = \int \mathcal{D}\psi^*(\mathbf{r}, z)\, \mathcal{D}\psi(\mathbf{r}, z)\, \mathcal{D}\delta\mathbf{n}(\mathbf{r}, z)\, \exp[-S(\psi^*, \psi, \delta\mathbf{n})/T], \quad (9.59)$$

where $\psi(\mathbf{r}, z)$ is again a complex boson field. The coherent state boson action S reads

$$\frac{S}{T} = \int_0^L dz \int d^2r \left[\psi^*(\mathbf{r}, z)\left(\frac{\partial}{\partial z} - D\nabla_\perp^2 - \bar{\mu}\right)\psi(\mathbf{r}, z) \right.$$

$$+ \frac{1}{2}[\psi^*(\mathbf{r}, z)\nabla_\perp\psi(\mathbf{r}, z) - \psi(\mathbf{r}, z)\nabla_\perp\psi^*(\mathbf{r}, z)] \cdot \delta\mathbf{n}(\mathbf{r}, z)$$

$$\left. + \frac{1}{2}\int d^2r'\; \bar{V}(\mathbf{r} - \mathbf{r}')|\psi(\mathbf{r}, z)|^2 \right] + \frac{F_N(\delta\mathbf{n})}{T},$$

$$(9.60)$$

with $\bar{\mu} = \mu/T$ and $\bar{V}(\mathbf{r}) = V(\mathbf{r})/T$. The field theory embodied in (9.60) differs from the action (9.37) for directed polymers with a strong aligning field in the coupling of the boson current $\psi^* \nabla_\perp\psi - \psi \nabla_\perp\psi^*$ to the director field. As discussed in [21], rotational invariance forces the coefficient of $(\psi^* \nabla_\perp\psi - \psi \nabla_\perp\psi^*) \cdot \delta\mathbf{n}$ to be exactly half that of $\psi^* \partial_z\psi$ in the second-quantized coherent-state representation. Finite lengths of polymer may be incorporated as in Eq. (9.38).

Correlations may be constructed as before, by using Eq. (9.46) and evaluating Gaussian integrals in $\pi(\mathbf{r}, z)$, $\theta(\mathbf{r}, z)$ and $\delta\mathbf{n}(\mathbf{r}, z)$. The structure function takes the form [21]

$$S(\mathbf{q}_\perp, q_z) = \rho_0 \frac{\rho_0 q_\perp^2 + 2\tilde{K}(\mathbf{q})(\ell^{-1} + Dq_\perp^2)}{\tilde{K}(\mathbf{q})[q_z^2 + \bar{\varepsilon}^2(q_\perp)] + \frac{1}{2}\rho_0 q_\perp^2 \bar{\varepsilon}^2(q_\perp)/(\ell^{-1} + Dq_\perp^2)}, \quad (9.61)$$

where

$$\tilde{K}(\mathbf{q}) = \frac{K_1 q_\perp^2 + K_3 q_z^2}{T} \tag{9.62}$$

and $\tilde{\varepsilon}(q_\perp)$ is given by Eq. (9.51). Similar manipulations show that K_2 and K_3 are unrenormalized by the interpolymer interactions, while the renormalized Frank constant K_1^R is [21]

$$K_1^R(\mathbf{q}_\perp, q_z) = K_1 + \frac{1}{2}\rho_0 T \frac{\tilde{\varepsilon}(q_\perp)}{q_z^2 + \tilde{\varepsilon}^2(q_\perp)} \frac{1}{(\ell^{-1} + Dq_\perp^2)}. \tag{9.63}$$

Upon taking the limit $\tilde{K}(\mathbf{q}) \to \infty$, we recover from Eq. (9.61) the results for polymers with an externally imposed direction discussed in Section 9.4.

The small-wavevector limit of (9.61) when the Frank constants are finite differs from the corresponding limit of Eq. (9.50),

$$S(\mathbf{q}_\perp, q_z) = k_B T \frac{\rho_0 q_\perp^2 + (K_1 q_\perp^2 + K_3 q_z^2)/G}{B q_\perp^2 + [B/(G\rho_0^2) + q_z^2](K_1 q_\perp^2 + K_3 q_z^2)}, \tag{9.64}$$

where $B = \rho_0^2 V_0$ and $G = \ell T/(2\rho_0)$.

When $G \to \infty$, (9.64) reduces to the prediction [53] of a hydro-dynamic theory originally due to de Gennes [52]. Finite-length effects distort the contours near the origin (see Fig. 9.10), however, and may be important in fitting real scattering data. Fits to (9.64) should lead to a direct experimental determination of the Frank constants, the bulk modulus B and the mean polymer length. Upon taking the limit $\mathbf{q} \to 0$ in (9.63) we find the dependence of the renormalized splay elastic constant K_1^R on the polymer length ℓ,

$$K_1^R = K_1 + \frac{1}{2}T\ell\rho_0, \tag{9.65}$$

in agreement with a prediction of Meyer [25], but in disagreement with a suggestion by de Gennes [24].

Overall, we expect a "butterfly" diffraction pattern qualitatively similar to that displayed in Fig. 9.7. The peaks in these diffraction contours along the $\mathbf{q}_\perp$ axis should appear at approximately the same position as the delta-function Bragg peak of the columnar hexagonal crystal which presumably exists at higher concentrations. The main difference for polymers with a soft broken symmetry appears near the origin, as indicated in Fig. 9.10. Ao et al. [60] have, in fact, presented X-ray diffraction data on PBG that is quite similar to these predictions. The data, however, are insufficient for a quantitative determination of the elastic moduli.

The above discussion focused on polymer melts without hairpin

Fig. 9.10. Contours of constant scattering intensity near the origin for nematic polymers. The characteristic square-root contours are rounded off as indicated by the dashed lines unless the polymers are very long.

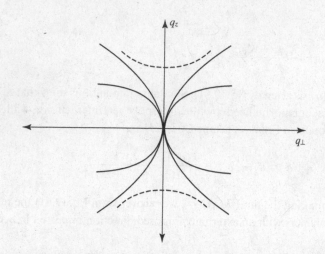

Fig. 9.10. Contours of constant scattering intensity near the origin for nematic polymers. The characteristic square-root contours are rounded off as indicated by the dashed lines unless the polymers are very long.

configurations like those in Fig. 9.2. We can account for hairpins by adding a term to the action (9.61), namely

$$S \to S - \frac{w}{2} \int dz \, d^2 r \, [\psi^2 + (\psi^*)^2]. \tag{9.66}$$

The coupling $w \propto \exp[-\varepsilon_h/(k_B T)]$, where the hairpin energy ε_h is $\varepsilon_h = O(\sqrt{g\kappa})$ as in Eq. (9.8). These terms create and destroy *pairs* of polymer lines at a single point and so add hairpins to the theory with probability proportional to w. The effect of hairpins on the structure function (9.61) is similar to the effect of the finite polymer length ℓ. Now, however, the $n \to 0$ trick [18] is required in order to exclude a number of unphysical Feynman graphs [21].

9.6 The hydrodynamic treatment of line liquids

Readers unfamiliar with the quantum mechanics of bosons may be relieved to know that most of the basic results for directed polymers at long wavelengths can be derived using a simpler "hydrodynamic" approach, similar to that derived by Ornstein and Zernike [14] over 80 years ago. By "hydrodynamic," we mean that the theory focuses on slowly varying spatial variations, without necessarily implying an explicit treatment of low-frequency time-dependent phenomena. Although it was originally developed for polymer nematics [52], "hydrodynamics" is also applicable to the vortex-liquid state of high-temperature superconductors [61] and to ferrorheological and electrorheological fluids. The theory leads immediately to unusual behavior of the three-dimensional structure function near the origin of reciprocal space. Density fluctuations in a constant-z cross section also provide a precise signature that one is dealing with a liquid of

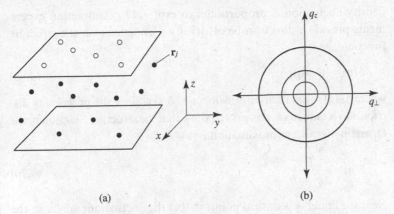

Fig. 9.11. (a) A schematic diagram of a fluid of point-like atoms or molecules at positions $\{r_j\}$ confined between two plates. (b) Contours of constant scattering intensity with a maximum near the origin for the situation shown in (a). These familiar results should be contrasted with the behavior of line liquids shown in Figs. 9.7 and 9.10.

lines rather than a liquid of points. Under favorable circumstances, it is even possible to deduce the average length of the lines purely from information contained in the cross section! Freeze–fracture experiments on polymer nematics would be an interesting way to implement this suggestion. Freeze–fracture studies of *two* different constant-z cross sections would allow a test of Eq. (9.57). Here, we present an elementary derivation of the "Bogoliubov spectrum" using this approach. See [54] for a more extensive review.

To provide a point of reference and establish notation, we first briefly review the Ornstein–Zernike theory for long-wavelength fluctuations in liquids of N isolated atoms or molecules. With the particles at positions $\{r_j\}$ shown in Fig. 9.11(a) we associate the microscopic density

$$\rho_{mic}(\mathbf{r}) = \sum_{j=1}^{N} \delta^{(3)}(\mathbf{r} - \mathbf{r}_j). \tag{9.67}$$

From this microscopic density, we construct a smoothed-out density field $\rho(\mathbf{r})$ by averaging (9.67) over a hydrodynamic averaging volume centered at $\mathbf{r}$. As in Landau's theory of phase transitions, we expand the free energy in deviations $\delta\rho(\mathbf{r}) = \rho(\mathbf{r}) - \rho_0$ of the density from its average value ρ_0. The Ornstein–Zernike approximation consists of keeping only quadratic terms and leading-order gradients in this expansion,

$$F(\delta\rho) = \frac{1}{2} \int d^3r \, [A|\nabla\delta\rho|^2 + B(\delta\rho)^2]. \tag{9.68}$$

The coefficient of $(\delta\rho)^2$ is proportional to the bulk modulus. The Landau approach to the liquid–gas phase transition consists of replacing this term by a polynomial in $\delta\rho$ that has minima at two densities below the critical temperature and pressure at which the bulk modulus vanishes.

Away from such phase transitions, density correlations are easily calculated from Eq. (9.68) upon assuming that the probability of a density fluctuation is proportional to $\exp(-F/T)$. Diffraction experiments provide a direct probe of density correlations via the structure function

$$S(\mathbf{q}) = \langle |\delta\hat{\rho}(\mathbf{q})|^2 \rangle, \qquad (9.69)$$

where $\delta\hat{\rho}(\mathbf{q})$ is the Fourier transform of $\delta\rho(\mathbf{r})$. Since the probability distribution is Gaussian, it is easy to show that the structure function in the Ornstein–Zernike approximation is just

$$S(\mathbf{q}) = \frac{T}{Aq^2 + B}. \qquad (9.70)$$

Near the liquid–gas critical point, so that the fluctuations are large, the structure function has a local maximum at the origin and the contours of constant scattering in, say, the $(q_\perp, q_z)$-plane are circles as shown in Fig. 9.11(b). In fluids of small prolate particles oriented along the z axis, there would be anisotropic gradient couplings in (9.68), leading to elliptical contours flattened along the z axis.

Now consider a configuration of very long directed polymers in an external field, as shown in Fig. 9.5. The basic hydrodynamic fields are now a microscopic vortex density

$$\rho_{\mathrm{mic}}(\mathbf{r}_\perp, z) = \sum_{j=1}^{N} \delta^{(2)}[\mathbf{r}_\perp - \mathbf{r}_j(z)] \qquad (9.71)$$

and a microscopic "tangent" field in the plane perpendicular to $\hat{z}$,

$$\mathbf{t}_{\mathrm{mic}}(\mathbf{r}_\perp, z) = \sum_{j=1}^{N} \frac{d\mathbf{r}_j(z)}{dz} \delta^{(2)}[\mathbf{r}_\perp - \mathbf{r}_j(z)]. \qquad (9.72)$$

As before, we coarse grain these fields to obtain smoothed density and tangent fields $\rho(\mathbf{r}_\perp, z)$ and $\mathbf{t}(\mathbf{r}_\perp, z)$.

We now expand the free energy of the liquid to quadratic order in the density deviation $\delta\rho(\mathbf{r}_\perp, z) = \rho(\mathbf{r}_\perp, z) - \rho_0$ and in $\mathbf{t}(\mathbf{r}_\perp, z)$,

$$F = \frac{1}{2\rho_0^2} \int d^2r_\perp \int dz [A_\perp |\nabla_\perp \delta\rho|^2 + A_z(\partial_z \delta\rho)^2 + B(\delta\rho)^2 + K|\mathbf{t}|^2]. \qquad (9.73)$$

The parameter B is a bulk modulus for areal compressions and dilations perpendicular to the z axis, while K is the modulus for tilting the lines away from the direction of the applied field. Terms involving gradients of $\mathbf{t}$, or more generally *nonlocal* elastic moduli, could also be introduced. Because we are dealing with *lines*, rather than simply oriented anisotropic particles, Eq. (9.73) must be supplemented with an "equation of continuity,"

$$\partial_z \delta\rho + \nabla_\perp \cdot \mathbf{t} = 0, \qquad (9.74)$$

which reflects the fact that vortex lines cannot stop or start inside the medium. Correlation functions can be calculated by assuming that the probability of a particular line configuration is proportional to $\exp(-F/T)$ and imposing the constraint (9.74) on the statistical mechanics.

To implement the constraint, we can set

$$\delta\rho = -\rho_0 \nabla_\perp \cdot \mathbf{u} \quad \text{and} \quad \mathbf{t} = \rho_0 \frac{\partial \mathbf{u}}{\partial z}, \tag{9.75}$$

where $\mathbf{u}(\mathbf{r}_\perp, z)$ is a two-component "displacement field." Equation (9.74) is now satisfied automatically. The structure function for the liquid is easily found to be

$$S(\mathbf{q}_\perp, q_z) = \langle |\delta\hat{\rho}(\mathbf{q}_\perp, q_z)|^2 \rangle = \frac{\rho_0^2 T q_z^2}{K q_z^2 + B q_\perp^2 + A_\perp q_\perp^4 + A_z q_z^2 q_\perp^2}. \tag{9.76}$$

The contours of constant scattering intensity are straight lines near the origin, as in the small-$\mathbf{q}$ region of Fig. 9.7. The conservation law embodied in Eq. (9.74) forces the structure function to vanish when $\mathbf{q}_\perp = 0$ for *any* nonzero q_z and leads to scattering contours very different from those appropriate to point particles (compare with Fig. 9.11(b)).

If we assume that $A_z \ll A_\perp$, Eq. (9.76) has precisely the form of the Bogoliubov spectrum (9.54), with

$$\bar{\varepsilon}(q_\perp) = \sqrt{\frac{A_\perp}{K} q_\perp^4 + \frac{B}{K} q_\perp^2}. \tag{9.77}$$

Allowing for $A_z \neq 0$ simply shifts the location of the q_z-pole in the denominator of (9.76) which determines the large-$|z|$ behavior of the correlation function (9.57). "Hydrodynamics" is thus both more general and less cumbersome than the traditional Bogoliubov approach for real bosons, which usually proceeds via tedious manipulations of creation and destruction operators in a second-quantized many-body Hamiltonian [62].

9.7 Defects in hexagonal columnar crystals

9.7.1 The Lindemann melting criterion for polymer crystals

Before discussing defects in columnar crystals, it is important to estimate the parameter values for which crystalline order is expected for the model of polymer nematics displayed in Eq. (9.19). By "crystalline order," we mean the usual columnar hexagonal phase [45], with a translational broken symmetry only in the two directions perpendicular to alignment. In general, a triangular "line lattice" is expected at low temperatures or high densities. A more precise estimate arises from the Lindemann criterion, which states that the lattice will melt (presumably via a first-order phase transition) whenever the root-mean-square thermal phonon displacements exceed a fixed fraction of the lattice constant. We set the external field H to zero, so the direction of polymer alignment is a soft, spontaneously broken symmetry, and restrict our attention to an isotropic solvent.

Before deriving the continuum elastic description of the line lattice, note first that Eq. (9.19) is invariant under rotations about the z axis and also invariant under a small rotation about an axis perpendicular to z. The latter symmetry is realized because the system is invariant under

$$\mathbf{r}_i(z) \rightarrow \mathbf{r}_i(z) + \boldsymbol{\theta}_0 z,$$
$$\delta\mathbf{n}(\mathbf{r}, z) \rightarrow \delta\mathbf{n}(\mathbf{r}, z) + \boldsymbol{\theta}_0, \tag{9.78}$$

which has the form of a Galilean transformation with "velocity" $\boldsymbol{\theta}_0$ if z is interpreted as "time." These symmetries will constrain the form of the elastic Hamiltonian.

Upon substituting

$$\mathbf{r}_i(z) \equiv \mathbf{R}_i + \mathbf{u}_i(z) \tag{9.79}$$

in Eq. (9.19), where the $\{\mathbf{R}_i\}$ are the fixed sites of an equilibrium triangular lattice, the free energy (9.19) can be expanded in the displacements $\mathbf{u}_i(z)$. We assume that an external chemical potential or osmotic pressure has fixed the concentration of polymer and hence a_0, the equilibrium lattice constant. The quadratic part of the free energy then becomes, in terms of the *continuum* displacement field $\mathbf{u}(\mathbf{r}, z)$ obtained from $\mathbf{u}_i(z)$,

$$F = \int d^2 r\, dz \left(\frac{1}{2} g \left| \frac{\partial \mathbf{u}}{\partial z} - \delta\mathbf{n} \right|^2 + \mu u_{ij}^2 + \frac{1}{2}\lambda u_{kk}^2 + \frac{1}{2}K_1(\nabla_\perp \cdot \delta\mathbf{n})^2 \right.$$
$$\left. + \frac{1}{2}K_2 |\nabla_\perp \cdot \delta\mathbf{n}|^2 + \frac{1}{2}K_3(\partial_z \delta\mathbf{n})^2 \right), \tag{9.80}$$

where $u_{ij}(\mathbf{r}, z) = \frac{1}{2}(\partial_i u_j + \partial_j u_i)$ and μ and λ are Lamé coefficients determined by the pair potential $V(\mathbf{r})$. For the pair potential (9.21), μ and λ can be extracted from closely related results for flux lattices in superconductors [63, 64]. Approximate values valid for all a_0 greater than the hard-core radius of the polymer are

$$\mu \approx \frac{1}{4} u_0 \kappa_D^2 K_0''(\kappa_D a_0) \tag{9.81}$$

and

$$\mu + \lambda \approx 4\pi u_0 \kappa_D^2 K_0''(\kappa_D a_0)/(\kappa_D a_0)^2. \tag{9.82}$$

These approximations are chosen to reduce to the correct results for $\kappa_D a_0 \ll 1$ and to exhibit a qualitatively correct dependence on a_0 ($\sim \exp(-\kappa_D a_0)$) for $\kappa_D a_0 \gg 1$. More accurate approximations in the limit $\kappa_D a_0 \gg 1$ would lead to $\mu \approx \lambda \sim u_0 \kappa_D^2 K_0''(\kappa_D a_0)$.

The next step is to integrate out the director field. The g coupling locks $\delta \mathbf{n}$ to $\partial \mathbf{u}/\partial z$,

$$\delta \mathbf{n} \approx \frac{\partial \mathbf{u}}{\partial z}, \tag{9.83}$$

for length scales larger than $(K/g)^{1/2}$, where K is a typical local Frank constant. To leading order in the gradients we then find the standard continuum description of a columnar hexagonal crystal [45], namely

$$F = \int d^2r \, dz \left(\mu u_{ij}^2 + \frac{1}{2}\lambda u_{kk}^2 + \frac{1}{2}K_3(\partial_z^2 \mathbf{u})^2 \right). \tag{9.84}$$

The rotational symmetries discussed above insure that only the *symmetrized* strain matrix u_{ij} and $\partial_z^2 \mathbf{u}$ (not $\partial_z \mathbf{u}$) enter this elastic energy. Using Eq. (9.84), we can easily estimate the mean-square phonon displacement,

$$\langle |\mathbf{u}(\mathbf{r}, z)|^2 \rangle = T \int \frac{d^2 q_\perp}{(2\pi)^2} \int \frac{dq_z}{2\pi} \left(\frac{1}{\mu q_\perp^2 + K_3 q_z^4} + \frac{1}{(2\mu + \lambda)q_\perp^2 + K_3 q_z^4} \right). \tag{9.85}$$

The first term of (9.85) (corresponding to shear deformations) is typically much larger than the second longitudinal term, which we henceforth neglect [65]. The next step depends on the magnitude of the dimensionless ratio [66]

$$\gamma = \left(\frac{K_3 \Lambda_z^4}{\mu \Lambda_\perp^2} \right)^{1/4}, \tag{9.86}$$

where Λ_z and $\Lambda_\perp$ are wavevector cutoffs in the directions along and perpendicular to $\hat{z}$, respectively. A convenient choice to cut off the perpendicular integration is a circle with radius $\Lambda_\perp = 4\pi/(\sqrt{3}a_0)$, which conserves the area of the original hexagonal Brillouin zone. For the z cutoff we assume $\Lambda_z = \xi_z^{-1}$, where ξ_z is the hard-core size of a monomer. For an isotropic solvent, K_3 is related to the bending rigidity κ by [24]

$$K_3 \approx \kappa/a_0^2. \tag{9.87}$$

Up to constants of order unity, we then have

$$\gamma \approx \left(\frac{\kappa}{\mu} \right)^{1/4} \Big/ \xi_z, \tag{9.88}$$

where $\kappa = T\ell_p$ according to Eq. (9.4). Using typical experimental values (e.g., $\ell_p = 600$ Å) for DNA polymers [28], we find that $\gamma \gg 1$. Although we shall consider only this situation in detail, we also expect that $\gamma \approx a_0/\xi_z > 1$ in discotic liquid-crystal columnar phases, where ξ_z is now the average spacing of the disks in a columnar stack.

Since $\gamma \gg 1$, we can take the limit $\Lambda_z \to \infty$ in (9.85), and find

$$\langle u^2 \rangle = \frac{\sqrt{2}}{4\pi} \frac{T\Lambda_\perp^{1/2}}{\mu^{3/4} K_3^{1/4}}. \tag{9.89}$$

The condition $\langle u^2 \rangle < c_L^2 a_0^2$ determines the region of stable hexagonal columnar crystal according to the Lindemann criterion [67]. The Lindemann constant is $c_L \approx 0.1$ for melting of crystals of point-like particles in three dimensions. For the experimentally observed first-order melting transition of flux lines in high-temperature superconductors, $c_L \approx 0.15$–0.30 provides the best fit to the data [43]. The *equality* $\langle u^2 \rangle = c_L^2 a_0^2$ determines the critical melting density for a hexagonal columnar phase as a function of experimental parameters such as temperature, bending rigidity and the Debye length using Eqs. (9.81) and (9.87). For an alternative approach to this problem, see [68].

The same criterion emerges from considering thermal fluctuations of polymers in real space. Consider one representative trajectory $\mathbf{r}(z)$ in the confining potential "cage" provided by its surrounding polymers in a triangular lattice. The free energy of this single strand may be written

$$F_1 = \frac{1}{2}\kappa \int_0^L \left(\frac{d^2\mathbf{r}}{dz^2}\right)^2 dz + \int_0^L V_1[\mathbf{r}(z)]\, dz, \tag{9.90}$$

where κ is the bending rigidity and $V_1(\mathbf{r})$ is chosen to mimic the interactions in Eq. (9.19). We now expand $V_1(\mathbf{r})$ about its minimum at $\mathbf{r} = 0$ to find

$$F_1 = \frac{1}{2}\kappa \int_0^L \left(\frac{d^2\mathbf{r}}{dz^2}\right)^2 dz + \frac{1}{2}k \int_0^L r^2(z)\, dz. \tag{9.91}$$

Up to constants of order unity, we have, using Eq. (9.21),

$$k \approx \frac{d^2 V(\mathbf{r})}{dr^2}\bigg|_{r=a_0}$$

$$= 2u_0 \kappa_D^2 K_0''(\kappa_D a_0), \tag{9.92}$$

i.e., k is of order the shear modulus given by Eq. (9.81).

Let us assume that a typical polymer wanders a perpendicular distance r, from one side of the parabolic confining potential to the other, and that it takes a "time" ℓ along the z axis to do this. From (9.90), the free energy of the segment length ℓ is then approximately

$$F_1 \approx \frac{L}{\ell}\left(\frac{\kappa r^2}{\ell^3} + kr^2\ell\right). \tag{9.93}$$

Upon optimizing ℓ, we find that the preferred segment length is

$$\ell^* = (\kappa/k)^{1/4}$$
$$\approx (\kappa/\mu)^{1/4}, \tag{9.94}$$

while the corresponding free energy of the segment is

$$f_1^* = \kappa^{1/4}k^{3/4}r^2$$
$$\approx \kappa^{1/4}\mu^{3/4}r^2. \tag{9.95}$$

We now assume that successive segments of length ℓ^* along the polymer fluctuate independently and apply the equipartition theorem to obtain

$$\langle r^2\rangle \approx T/\kappa^{1/4}\mu^{3/4}. \tag{9.96}$$

Upon setting $\langle r^2\rangle \equiv c_L^2 a_0^2$, we find that the melting point is given by

$$T_m = c_L^2 \kappa^{1/4}\mu^{3/4}a_0^2, \tag{9.97}$$

which agrees with the result of using Eq. (9.89), up to numerical constants, since $\kappa \approx K^3 a_0^2$. Note that Eq. (9.94) is the generalization of the Odijk length [26] to directed polymers with interactions not limited to a simple hard-core repulsion. A similar analysis with a tilt-modulus term proportional to $\int_0^L (d\mathbf{r}/dz)^2\,dz$ replacing the bending rigidity in Eq. (9.90) is presented by Podgornik and Parsegian [27].

9.7.2 Defects in columnar hexagonal crystals

Consider the set of all possible thermal excitations about the ground state of a columnar hexagonal crystal. The *phonons* embodied in Eq. (9.84) lead to a finite reduction of the translational order parameter

$$\rho_G(T) = \langle e^{i\mathbf{G}\cdot\mathbf{u}(\mathbf{r}_\perp, z)}\rangle$$
$$= \exp[-\tfrac{1}{2}G_iG_j\langle u_i(\mathbf{r}_\perp, z)u_j(\mathbf{r}_\perp, z)\rangle]. \tag{9.98}$$

See [45] for a more detailed discussion. Here, $\mathbf{G}$ is a reciprocal-lattice vector and $\langle u_i(\mathbf{r}_\perp, z)u_j(\mathbf{r}_\perp, z)\rangle$ can be calculated as in Eq. (9.85). *Dislocation loops* are topologically distinct excitations, which, when they proliferate at a melting transition, presumably drive $\rho_G(T)$ to zero at a first-order melting transition. Dislocation loops are more constrained than their counterparts in crystals of point particles: As discussed in Chapter 8, dislocation loops lie in a plane spanned by their Burgers vector and the average direction of the field (see Fig. 9.12) [69].

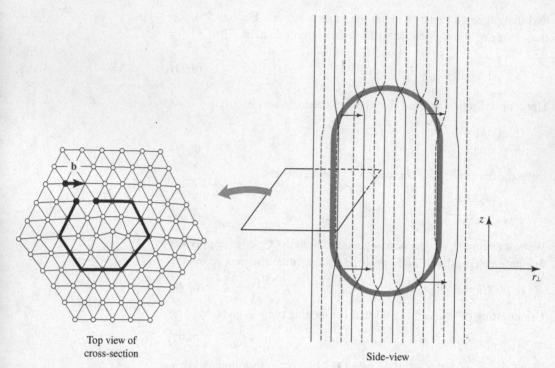

Top view of
cross-section

Side-view

Fig. 9.12. A dislocation loop in a polymer crystal. Dashed lines represent polymers just behind the plane of the figure. Such loops lie in the plane spanned by their Burgers vector (indicated by the horizontal arrows) and the z axis. The top view of the indicated cross section shows the conventional Burgers construction in two dimensions.

Vacancies and interstitials differ even more dramatically from the analogous defects in crystals of point particles. Assume for the moment that the polymers are longer than all other relevant length scales. Then, the number of strands in a constant-z cross section is conserved, which means that vacancies and interstitials must be *lines* instead of points. The point-like nature of vacancies and interstitials in conventional crystals ensures that they are present in equilibrium at all finite temperatures for entropic reasons [70]. However, because such imperfections have an energy proportional to their *length* in polymer crystals, they cannot extend completely across an equilibrated macroscopic sample at low temperatures. A typical fluctuation at low temperatures might consist of the vacancy–interstitial pair shown in Fig. 9.13. Unlike the dislocation loop in Fig. 9.12, this loop is not constrained to lie in a single plane. Vacancy and interstitial lines can be absorbed or emitted by dislocation loops, thus allowing them to become nonplanar, at the expense of becoming attached to a string of vacancies or interstitials. If we think of the z axis as "time," Fig. 9.13 represents excitation of a virtual particle–antiparticle pair. The planar dislocation loop in Fig. 9.12 represents "motion" by "gliding" of a dislocation parallel to the Burgers vector. "Climbing" of a dislocation, or "motion" perpendicular to the Burgers vector, requires vacancies or interstitials.

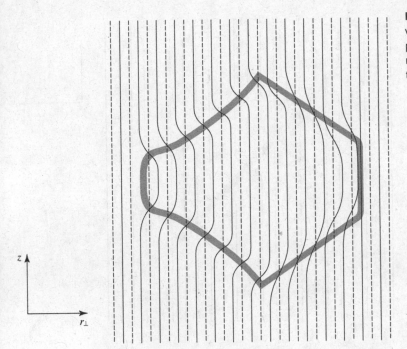

Fig. 9.13. A vacancy–interstitial pair in a polymer crystal. Dashed lines represent polymers just below the plane of the figure.

We can now discuss effects of polymer free ends in hexagonal columnar crystals. As discussed by Predecki and Statton [47], polymer heads and tails must be connected by vacancy or interstitial lines. See Fig. 9.14 [71]. If the head were in the upper-right-hand corner and the tail in the lower-left-hand corner, the connecting string would consist of interstitials instead of vacancies. At low temperatures or high densities, heads and tails will be tightly bound together, because of the linear potential associated with the strings joining them. A positive string tension then ensures that the boson correlation function (9.45) decays to zero as $|\mathbf{r}_\perp - \mathbf{r}_\perp'| \to \infty$. See [21] and Chapter 8.

As the temperature increases, the strings will get longer. The string tension is anisotropic, so it is natural to ask whether the separation between a head and tail will be predominantly parallel or perpendicular to the direction of the polymer: See Fig. 9.15. We estimate these tensions here for vacancies; estimates for vertical interstitial strings (which usually have a higher energy [7]) would differ only by constants of order unity. There is no distinction between vacancies and interstitials for horizontal strings. The energy per unit length of the vertical string of defects in Fig. 9.15(a) is determined by the shear modulus,

$$\varepsilon_d^z = \mu a_0^2. \tag{9.99}$$

The horizontal string tension in Fig. 9.15(b) follows by noting that the distance along z over which each polymer distorts as it crosses the defect

Fig. 9.14. A polymer "head" at $(\mathbf{r}_\perp, z)$ connected by a string of vacancies to a polymer tail at $(\mathbf{r}'_\perp, z')$.

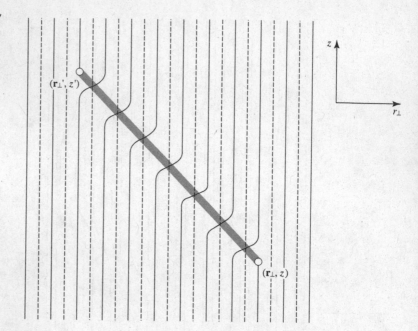

Fig. 9.15. A schematic diagram of polymer head and tail defects separated by (a) a vertical line of vacancies and (b) an analogous horizontal defect line, denoted a "lock-in fault line" in [48].

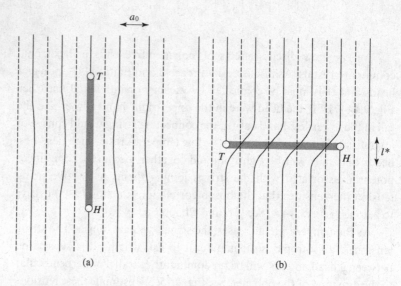

(a)   (b)

line should be of order ℓ^*. We know from Eq. (9.95) that the energy of each distorted segment is $\kappa^{1/4}\mu^{3/4}a_0^2$. The energy per unit length is thus

$$\varepsilon_d^\perp = \kappa^{1/4}\mu^{3/4}a_0. \tag{9.100}$$

The ratio of these energies is

$$\frac{\varepsilon_d^z}{\varepsilon_d^\perp} \approx \left(\frac{\mu}{\kappa}\right)^{1/4} a_0 \approx \frac{a_0}{\ell^*}. \tag{9.101}$$

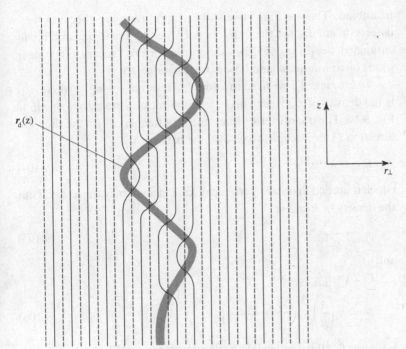

Fig. 9.16. A vacancy line $\mathbf{r}_d(z)$ (thick curve) meandering through a vortex crystal. The full lines show the flux lines which are in the same plane as the meandering vacancy. The dashed lines represent the flux lines in the neighboring plane.

If the polymers are relatively stiff, so that $\ell^* > a_0$, the separation will be primarily in the z direction.

Although the energies of vacancy and interstitial lines are proportional to their lengths, it is nevertheless possible for these defects to "proliferate" (i.e., to become infinitely long) at high temperatures (or low densities) for entropic reasons. Consider, for example, a vacancy wandering across a macroscopic sample of thickness L, as in Fig. 9.16. We assume that we have predominantly vertical trajectories here. A related entropy argument for horizontal strings has been given by Prost [48], who called them "lock-in fault lines." Let ℓ_z be the spacing between vacancy jumps along the z axis in Fig. 9.16. Since these jumps can be in any of six directions, the *free energy* of the defect line in a sample of thickness L is

$$F_d \approx \varepsilon_d^z L - T\left(\frac{L}{\ell_z}\right)\ln 6,\tag{9.102}$$

which becomes negative for $T > T_d$, where

$$T_d = \varepsilon_d^z \ell_z / \ln 6.\tag{9.103}$$

Above this temperature (provided that the crystal does not melt first), vacancies (or interstitials) will proliferate in a crystalline phase. At the same temperature, it will become very easy for polymer heads and tails to unbind. The resulting state will be both *crystalline* (because these defects do *not* destroy the translational order parameter ρ_G) and highly entangled. See [7] and [55] for a description of this "supersolid" state in terms of translational and boson order parameters.

To estimate T_d, we need to know ℓ_z. However, $\ell_z = n_{kink}^{-1}$, where n_{kink} is the density of polymer kinks along the vacancy trajectory shown in Fig. 9.16. The quantity $1/\ell^*$ is an "attempt frequency" (in the time-like direction $\hat{z}$) for attempts to go over the barrier,

$$E_{kink} \approx \kappa^{1/4} \mu^{3/4} a_0^2, \tag{9.104}$$

which is needed in order to produce a kink in a vacancy trajectory. Thus the density of kinks is

$$n_{kink} \approx \frac{1}{\ell^*} e^{-E_{kink}/T} \tag{9.105}$$

and

$$\ell_z(T) = n_{kink}^{-1}$$

$$\approx \left(\frac{\kappa}{\mu}\right)^{1/4} \exp\left(\frac{\kappa^{1/4} \mu^{3/4} a_0^2}{T}\right). \tag{9.106}$$

Equation (9.103) now leads to a self-consistent equation that determines the temperature (or concentration of polymer) at which the proliferation of defects occurs,

$$T_d \approx c_1 \kappa^{1/4} \mu^{3/4} a_0^2 \exp(-c_2 \kappa^{1/4} \mu^{3/4} a_0^2 / T_d), \tag{9.107}$$

where we have used Eq. (9.94) and have inserted constants c_1 and c_2 of order unity. Equation (9.107) is solved by writing

$$T_d \approx c_3 \kappa^{1/4} \mu^{3/4} a_0^2, \tag{9.108}$$

where c_3 is another numerical constant. This has the same form as the Lindemann condition (9.97).

9.7.3 The head–tail unbinding transition

Equations (9.97) and (9.108) specify phase boundaries in parameter space that differ only by a numerical constant. If $c_3 > c_L^2$, melting occurs before defects become plentiful and heads and tails will actually separate only in the melted polymer nematic. As discussed in [21], the strings connecting heads to tails "evaporate" once melting has occurred, producing a gas of unbound heads and tails. For $c_3 < c_L^2$, an intermediate phase with proliferating vacancies and interstitials is possible. The onset of such a phase would have interesting consequences. As argued by Frey *et al.* for vortex lines in superconductors [55], the lattice constant a_0 in a constant-z cross section will no longer be exactly locked to the size G of the first reciprocal-lattice vector according to the relation for a perfect crystal,

$$G = \frac{4\pi}{\sqrt{3}} \frac{1}{a_0}. \tag{9.109}$$

The discrepancy measures the density of vacancies or interstitials. Prost has argued that the elastic bending constant for polymer nematics should be greater by several orders of magnitude than nematic values, provided that horizontal strings proliferate [48]. Whether the intermediate phase actually occurs in a given directed-polymer crystal depends on the nonuniversal details of the solvent, interparticle interactions, etc.

The set of vacancy "strings" connecting heads and tails in a crystal may be viewed as a gas of directed living polymers. Methods developed for electrorheological fluids and ferrofluids [21] can be adapted to study this gas in more detail. The appropriate mathematical tool is the coherent-state path integral studied in Section 9.7.2.

The partition function for the one vacancy displayed in Fig. 9.16 is given by

$$Z_1 = \int \mathcal{D}r_\mathrm{d} \exp\left[-\frac{1}{2D} \int_0^L \left(\frac{dr_\mathrm{d}}{dz}\right)^2 dz \right], \tag{9.110}$$

with diffusion constant

$$\begin{aligned} D &\approx a_0^2/\ell_z \\ &\approx a_0^2 \left(\frac{\mu}{\kappa}\right)^{1/4} \exp\left(-\frac{c_2 \kappa^{1/4} \mu^{3/4} a_0^2}{T} \right) \end{aligned} \tag{9.111}$$

since the vacancy takes a "time" ℓ_z to hop one lattice constant a_0. The grand canonical partition function for many such vacancy strings is

$$Z_{gr} = \int \mathcal{D}\psi_\mathrm{d} \int \mathcal{D}\psi_\mathrm{d}^* e^{-\bar{S}_\mathrm{d}} \tag{9.112}$$

with

$$\bar{S}_\mathrm{d} = \int_0^L dz \int d^2 r_\perp \left[\psi_\mathrm{d}^*(\partial_z - D\,\nabla_\perp^2 - \bar{\mu})\psi_\mathrm{d} + \bar{V}_\mathrm{d}|\psi_\mathrm{d}|^4 - h(\psi_\mathrm{d}^* + \psi_\mathrm{d}) \right]. \tag{9.113}$$

$\psi_\mathrm{d}^*(\mathbf{r}_\perp, z)$ and $\psi_\mathrm{d}(\mathbf{r}_\perp, z)$ create heads and tails, respectively. The areal density of strings in a cross section is given by $\langle |\psi_\mathrm{d}|^2 \rangle$. The chemical potential $\bar{\mu} \equiv \mu/T$ is the negative of Eq. (9.102) divided by LT,

$$\bar{\mu} = n_{kink} \ln 6 - \varepsilon_\mathrm{d}^z/T, \tag{9.114}$$

and is positive once the defects have proliferated. We have assumed for simplicity that there is a repulsive delta-function potential between the strings,

$$V_\mathrm{d}(\mathbf{r}) \equiv T\bar{V}_\mathrm{d}\delta^{(2)}(\mathbf{r}). \tag{9.115}$$

The coupling $h>0$, which allows string heads and tails to be created in the medium, will be fixed below by the requirement that the three-dimensional density of heads or tails is

$$n_H = n_T \approx 1/(a_0^2 \ell_p), \tag{9.116}$$

where ℓ_p is the average length of a polymer. We assume that the polymers are very long, so the heads and tails are dilute. The strings connecting them can be dense in an areal cross section, however. *Long* polymers are necessary so we can neglect the fact that heads and tails are not only linked together by vacancy/interstitial strings, but also by a physical polymer chain.

As outlined in Section 9.7.2, we now expand Z_{gr} as a power series in h and $\bar{V}_d$. A typical term in this expansion bears a one-to-one relation between configurations of interacting strings, as discussed earlier (see Fig. 9.6). The perturbation series may be reorganized into the form

$$Z_{gr} = \sum_{N_m=0}^{\infty} \sum_{P=0}^{\infty} Z_{N_m}^P(D, \bar{V}_d) e^{\bar{\mu} L_m} h^{2P}, \tag{9.117}$$

where $Z_{N_m}^P(D, \bar{V}_d)$ is the partition function for N_m "monomers" of string $(N_m \approx L_m/\ell^*)$ distributed among P strings and $L_m \propto N_m$ is the *total* length along $\hat{z}$ occupied by the N_m monomers. Note that h^2 plays the role of a fugacity for pairs of heads/tails, while $e^{\bar{\mu}}$ controls the density of monomers.

The total number of strings obtained from (9.117) is

$$\langle P \rangle = h^2 \frac{\partial}{\partial (h^2)} \ln Z_{gr}$$

$$= \frac{\sum_{N_m=0}^{\infty} \sum_{P=0}^{\infty} P Z_{N_m}^P(D, \bar{V}_d, K_i) e^{\bar{\mu} L_m} h^{2P}}{Z_{gr}}$$

$$= \tfrac{1}{2} h(\langle \psi \rangle + \langle \psi^* \rangle) \Omega, \tag{9.118}$$

where Ω is the three-dimensional volume. The total string length is given by

$$\langle L_m \rangle = \frac{\partial}{\partial \bar{\mu}} \ln Z_{gr}$$

$$= \frac{\sum_{N_m=0}^{\infty} \sum_{P=0}^{\infty} L_m Z_{N_m}^P(D, \bar{V}_d) e^{\bar{\mu} L_m} h^{2P}}{Z_{gr}}$$

$$= \Omega \langle |\psi|^2 \rangle. \tag{9.119}$$

Note that the average length per string is

$$\ell_{av} = \frac{\langle L_m \rangle}{\langle P \rangle} = \frac{2 \langle |\psi|^2 \rangle}{h(\langle \psi \rangle + \langle \psi^* \rangle)}. \tag{9.120}$$

The partition function and its derivatives follow from evaluating (9.112) in the mean-field approximation. Upon setting $\psi(\mathbf{r}, z)$ to a constant value $\psi = \psi^* = \psi_0 = \sqrt{\rho_0}$, we have

$$\ln Z_{\mathrm{gr}} = -\Omega \min_{\psi_0}\left(-\bar{\mu}\psi_0^2 + \frac{1}{2}\bar{V}\psi_0^4 - 2h\psi_0\right),\qquad(9.121)$$

There are two limiting cases to consider. When $\bar{\mu} > 0$, the ordered state is only slightly perturbed by the small source field h. The strings are dense in this regime. The minimum of (9.121) is then given to lowest order in h by

$$\psi_0 \approx \sqrt{\frac{\bar{\mu}}{\bar{V}_d}}\left(1 + \sqrt{\frac{\bar{V}_d}{\bar{\mu}^3}}\frac{h}{2}\right),\qquad(9.122)$$

while the partition function is

$$Z_{\mathrm{gr}} = \exp\left[\Omega\left(\frac{\bar{\mu}^2}{2\bar{v}_d} + 2h\sqrt{\frac{\bar{\mu}}{\bar{V}_d}}\right)\right].\qquad(9.123)$$

If follows that

$$\langle L_{\mathrm{m}}\rangle \approx \psi_0^2 \Omega \qquad(9.124)$$

and

$$\langle P\rangle \approx \psi_0 h\Omega,\qquad(9.125)$$

so that a typical length is

$$\ell_{\mathrm{av}} \approx \psi_0/h.\qquad(9.126)$$

Since

$$\frac{\langle P\rangle}{\Omega} = n_{\mathrm{H}} = n_{\mathrm{T}} \approx \frac{1}{a_0^2 \ell_{\mathrm{p}}},\qquad(9.127)$$

we find that

$$h \approx \sqrt{\frac{\bar{V}_d}{\bar{\mu}}}\frac{1}{a_0^2 \ell_{\mathrm{p}}}\qquad(9.128)$$

for $\bar{\mu} > 0$.

If $\bar{\mu} < 0$, the strings are short and dilute, so we are in the disordered phase of the model. One can then neglect $\bar{V}_d$ and find that the boson order parameter is

$$\psi_0 \approx h/|\bar{\mu}|,\qquad(9.129)$$

while the grand partition function is

$$Z_{\mathrm{gr}} \approx \exp\left[\Omega\left(\frac{h^2}{|\bar{\mu}|}\right)\right].\qquad(9.130)$$

The total length of the string is now

$$\langle L_{\mathrm{m}} \rangle = \frac{h^2}{\bar{\mu}^2} \Omega,$$ (9.131)

while the average number is

$$\langle P \rangle = \frac{h^2}{|\bar{\mu}|} \Omega.$$ (9.132)

A typical string size is thus

$$\ell_{\mathrm{av}} \approx \frac{1}{|\bar{\mu}|},$$ (9.133)

which increases as the transition is approached from negative $\bar{\mu}$. We obtain h in this regime from Eqs. (9.127) and (9.132),

$$h = \sqrt{|\bar{\mu}|}\, \frac{1}{a_0^2 \ell_{\mathrm{p}}}.$$ (9.134)

The boson order parameter $\psi_0 = \langle \psi_d \rangle = \langle \psi_d^* \rangle$ is thus

$$\psi_0 \approx \sqrt{\frac{\bar{\mu}}{\bar{V}_{\mathrm{d}}}} \left(1 + \frac{\bar{V}_{\mathrm{d}}}{\bar{\mu}^2} \frac{1}{a_0^2 \ell_{\mathrm{p}}} \right), \quad \bar{\mu} > 0,$$ (9.135)

$$\psi_0 \approx \sqrt{\frac{1}{\bar{\mu}} \frac{1}{a_0^2 \ell_{\mathrm{p}}}}, \quad \bar{\mu} < 0.$$ (9.136)

Precisely at the transition, we find $h \sim \bar{V}_{\mathrm{d}}^{1/4} (a_0^2 \ell_{\mathrm{p}})^{-3/4}$ and

$$\psi_0 \sim \frac{1}{\bar{V}_{\mathrm{d}}^{1/4}} \left(\frac{1}{a_0^2 \ell_{\mathrm{p}}} \right)^{1/4}, \quad \bar{\mu} = 0.$$ (9.137)

These expectations are summarized in Fig. 9.17. Note that the transition from a low to a high density of vacancy strings is rounded for any $\ell_{\mathrm{p}} < \infty$. A sharp second-order phase transition is obtained only in the limit of infinitely long polymers, $\ell_{\mathrm{p}} \to \infty$. If the interaction between strings is sufficiently attractive, the defect-proliferation transition will be of first order [55].

Acknowledgements

It is a pleasure to acknowledge stimulating collaborations with P. Le Doussal, R. Kamien, E. Frey and D. Fisher on work described in this chapter. We also benefited from helpful conversations with S. Fraden, R. B. Meyer, A. Parsegian and R. Podgornik. This chapter is adapted with permission from *Observation, Prediction, and Simulation of Phase*

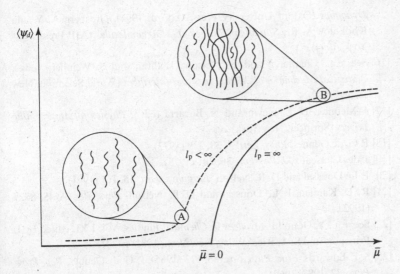

Fig. 9.17. A schematic diagram of the defect order parameter $\langle \psi_d \rangle$ as a function of the chemical potential. Insets depict strings connecting heads and tails. The directed polymers which make up the host crystal for the strings are not shown.

Transitions in Complex Fluids, edited by M. Baus, L. F. Rull and J.-P. Ryckaert (Kluwer, Dordrecht, 1995) pp. 293–335.

References

[1] See the articles in *Statistical Physics and Spatial Statistics,* eds. K. R. Mecke and D. Stoyan (Springer, Berlin, 2000).

[2] D. R. Nelson, *Physica* A **177**, 220 (1991).

[3] R. D. Kamien, P. Le Doussal and D. R. Nelson, *Phys. Rev.* A **45**, 8727 (1992).

[4] B. Maier and J. O. Radler, *Phys. Rev. Lett.* **82**, 1911 (1999); *Macromolecules* **33**, 7185 (2000).

[5] M. C. Marchetti and D. R. Nelson, *Phys. Rev.* B **41**, 1910 (1990).

[6] E. Frey, D. R. Nelson and D. S. Fisher, *Phys. Rev.* B **49**, 9723 (1994).

[7] S. Jain and D. R. Nelson, *Phys. Rev.* E **61**, 1599 (2000).

[8] A. Pertsinidis and X. S. Ling, preprint (http://arXiv.org/abs/cond-mat/0012306).

[9] R. D. Kamien and D. R. Nelson, *Phys. Rev.* E **53**, 650 (1996).

[10] R. D. Kamien, *Phys. Rev.* B **58**, 8218 (1998); see also D. R. Nelson, *Nature* **385**, 675 (1997).

[11] R. D. Kamien, *J. Physique* II **6**, 461 (1996); R. D. Kamien and A. J. Levian, *Phys. Rev. Lett.* **84**, 3109 (2000).

[12] H. H. Strey, J. Wang, R. Podgornik, A. Rupprecht, L. Yu, V. A. Parsegian and E. B. Sirota, *Phys. Rev. Lett.* **84**, 3105 (2000).

[13] See, e.g., J. P. Hansen and I. R. McDonald, *Theory of Simple Liquids* (Academic, New York, 1976).

[14] L. S. Ornstein and F. Zernike, *Proc. Acad. Sci. Amst.* **17**, 793 (1914).

[15] P. G. de Gennes, *Scaling Concepts in Polymer Physics* (Cornell University Press, Ithaca NY, 1979); M. Doi and S. F. Edwards, *The Theory of Polymer*

Dynamics (Oxford University Press, Oxford, 1994); Grosberg A. Y. and Khokhlov A. R., *Statistical Physics of Macromolecules* (AIP Press, New York, 1994).

[16] See, e.g., Chapter 5 and D. Nelson, T. Piran and S. Weinberg (eds.), *Statistical Mechanics of Membranes and Interfaces* (World Scientific, New Jersey, 1989).

[17] J. Meunier, D. Langevin and N. Bocarra (eds.), *Physics of Amphiphillic Layers* (Springer, New York, 1987).

[18] P. G. de Gennes, *Phys. Lett.* A **38**, 339 (1972).

[19] J. des Cloiseaux, *J. Physique* **36**, 281 (1975).

[20] P. Le Doussal and D. R. Nelson, *Europhys. Lett.* **15**, 161 (1991).

[21] R. D. Kamien, P. Le Doussal and D. R. Nelson, *Phys. Rev.* A **45**, 8727 (1992).

[22] See, e.g., Y. Oono in *Advances in Chemical Physics*, Vol. LXI, edited by I. Prigogine and S. A. Rice (Wiley, New York, 1981) p. 301.

[23] S. F. Edwards, *Proc. Phys. Soc.* **85**, 613 (1965); P. G. de Gennes, *Rep. Prog. Phys.* **32**, 1987 (1969).

[24] P. G. de Gennes, in *Polymer Liquid Crystals*, eds. A. Ciferri, W. R. Kringbaum and R. B. Meyer (Academic, New York, 1982) Chapter 5.

[25] R. B. Meyer, in *Polymer Liquid Crystals*, eds. A. Ciferri, W. R. Kringbaum and R. B. Meyer (Academic, New York, 1982) Chapter 6.

[26] T. Odijk, *Macromolecules* **17**, 2313 (1986).

[27] R. Podgornik, D. C. Rau and V. A. Parsegian, *Macromolecules* **22**, 1780 (1989); R. Podgornik and V. A. Parsegian, *Macromolecules* **23**, 2263 (1990).

[28] S. Bhaltacharjee, M. J. Glucksman and L. Makowski, *Biophys. J.* **61**, 725 (1992); J. Tang and S. Fraden, *Phys. Rev. Lett.* **71**, 3509 (1993).

[29] H. Block, *Poly(γ-Benzyl-L-Glutamate) and Other Glutamic Acid Containing Polymers* (Gordon and Breach, London, 1983).

[30] S. Chandrasekhar, B. K. Sadashiva and K. A. Suresh, *Pramana* **9**, 471 (1977).

[31] N. H. Tinh, H. Gasparoux and C. Destrade, *Mol. Cryst. Liq. Cryst.* **68**, 101 (1981); T. K. Attwood, J. E. Lyndon and F. Jones, *Liq. Cryst.* **1**, 499 (1986).

[32] M. E. Cates, and S. J. Candau, *J. Phys. Cond. Matt.* **2**, 6869 (1990).

[33] A. Blumstein, G. Maret and S. Villasagar, *Macromolecules* **14**, 1543 (1981).

[34] P. G. de Gennes, *C. R. Acad. Sci.* B **281**, 101 (1975).

[35] P. G. de Gennes, *Mol. Cryst. Liq. Cryst. (Lett.)* **102**, 95 (1984).

[36] A. Blumstein, in *Liquid Crystalline Order in Polymers*, edited by A. Blumstein (Academic, New York, 1978); E. M. Barrall II and J. F. Johnson, *J. Macromol. Sci. Rev. Macromol. Chem.* **17** 137 (1979).

[37] A. Stroobants, N. W. Lekkerkerker and T. Odijk, *Macromolecules* **19**, 2232 (1986).

[38] S. Fraden, G. Maret, D. L. D. Caspar and R. B. Meyer, *Phys. Rev. Lett.* **63**, 2068 (1989).

[39] R. E. Rosenweig, *Ferrohydrodynamics* (Cambridge University Press, New York, 1989).

[40] T. C. Halsey and W. Toor, *Phys. Rev. Lett.* **65**, 2820 (1990) and references therein.

[41] K. Bedell *et al.*, *Proceedings of the Los Alamos Symposium, 1991: Phenomenology and Applications of High Temperature Superconductors* (Wiley, New York, 1992).

[42] D. R. Nelson, in *Phase Transitions and Relaxation in Systems with Competing Energy Scales*, edited by T. Riste and D. Sherrington (Kluwer, Amsterdam, 1993).

[43] G. Blatter *et al.*, *Rev. Mod. Phys.* (1994).

[44] J. W. Negele and H. Orland, *Quantum Many-Particle Systems* (Addison-Wesley, New York, 1988); V. N. Popov, *Functional Integrals and Collective Excitations* (Cambridge University Press, New York, 1987).

[45] P. G. de Gennes and J. Prost, *The Physics of Liquid Crystals* (Clarendon, Oxford, 1993).

[46] See, e.g., F. R. N. Nabarro and A. T. Quintanilha, *Dislocations in Solids*, edited by F. R. N. Nabarro (North-Holland, Amsterdam, 1980) Ch. 5.

[47] P. Predecki and W. O. Statton, *J. Appl. Phys.* **37**, 4053 (1966).

[48] J. Prost, *Liq. Cryst.* **8**, 123 (1990).

[49] F. Livolant and Y. Bouligand, *J. Phys.* **47**, 1813 (1986); F. Livolant, *J. Mol. Biol.* **218**, 165 (1991).

[50] R. D. Kamien and T. C. Lubensky, *J. Physique* I **3**, 2131 (1993).

[51] A. B. Harris, T. C. Lubensky and R. D. Kamien, *Rev. Mod. Phys.* **71**, 1745 (1999).

[52] P. G. de Gennes, *J. Physique Lett.* **36**, 55 (1975).

[53] J. Selinger and R. Bruinsma, *Phys. Rev. A* **43**, 2910 (1991).

[54] See also D. R. Nelson, *Physica A* **177**, 220 (1991).

[55] E. Frey, D. R. Nelson and D. S. Fisher, *Phys. Rev.* **49**, 9723 (1994).

[56] See, e.g., Chapter 2.

[57] G. J. Vroege and H. N. Lekkerkerker, *Rep. Prog. Phys.* **55**, 1241 (1992).

[58] R. P. Feynman and A. R. Hibbs, *Quantum Mechanics and Path Integrals* (McGraw-Hill, New York, 1965); R. P. Feynman, *Statistical Mechanics* (Benjamin, Reading, MA, 1972).

[59] D. R. Nelson and S. Seung, *Phys. Rev. B* **39**, 9153 (1989).

[60] X. Ao, X. Wen and R. B. Meyer, *Physica A* **176**, 63 (1991).

[61] D. R. Nelson and P. Le Doussal, *Phys. Rev. B* **42**, 10113 (1990); M. C. Marchetti and D. R. Nelson, *Physica C* **174**, 40 (1991).

[62] The equivalent "hydrodynamic" deviation of the Feynman formula (9.55) for quantum bosons is well known. See Chapter 1 of A. Abrikosov, L. Gorkov and I. Dzyaloshinski, *Methods of Quantum Field Theory in Statistical Physics* (Prentice-Hall, New York, 1963).

[63] D. S. Fisher, *Phys. Rev. B* **22**, 1190 (1980), and references therein.

[64] E. H. Brandt and U. Essmann, *Phys. Stat. Sol. B* **144**, 13 (1987).

[65] A more precise theory would substitute a wavevector-dependent *function* for $\mu + \lambda$, e.g., $\mu + \lambda \to B(q_\perp) = [4\pi u_0/(\kappa_D^2 a_0^4)]/(1 + q_\perp^2/\kappa_D^2)$. See, e.g., [64].

[66] I am indebted to discussions with Randall Kamien on this point.

[67] F. Lindemann, *Phys. Z. (Leipzig)* **11**, 69 (1910).

[68] J. V. Selinger and R. F. Bruinsma, *Phys. Rev.* A **43**, 2922 (1991). Because of the uncontrolled approximations in this density-functional theory, freezing occurs when the hard-core diameter of the polymer exceeds the lattice constant!

[69] F. R. N. Nabarro and A. T. Quintanilha, in *Dislocations in Solids*, edited by F. R. N. Nabarro (North-Holland, Amsterdam, 1980).

[70] N. W. Ashcroft and N. D. Mermin, in *Solid State Physics* (Saunders College, Philadelphia, 1976) Ch. 30.

[71] For more on head and tail defects, particularly in the nematic phase of directed polymers, see [7, 21, 25] and V. G. Taratura and R. B. Meyer, *Liq. Cryst.* **2**, 373 (1987).

Index